Marsupial Nutrition

Marsupial Nutrition describes the food resources used by marsupials as diverse as small insectivores and large folivores. It discusses the ways in which their digestive systems and metabolism are designed to cope with foods as different as nectar and fungus, tree sap and tough perennial grasses, and insects and eucalypt foliage. Although the subject species are marsupials the general principles of nutritional ecology and digestive strategies that are introduced at the beginning of the chapters are applicable to all mammals. Advanced undergraduates and graduate students at all levels in the areas of vertebrate zoology, nutrition, ecology and digestive physiology will find *Marsupial Nutrition* particularly instructive. Wildlife biologists, veterinarians and nutritionists will also find much of interest.

IAN D. HUME is Challis Professor of Biology at the University of Sydney. He has carried out research in the field of comparative nutrition in the USA, Germany, Sudan and Japan, as well as in Australia. His book *Digestive Physiology and Nutrition of Marsupials* (1982) won the Whitley Award for Best Text Book from the Royal Zoological Society of New South Wales. He has also co-authored *Comparative Physiology of the Vertebrate Digestive System* (1995), and co-edited *Possums and Gliders* (1984), and *Kangaroos, Wallabies and Rat-kangaroos* (1989). Professor Hume is currently a managing editor of *Journal of Comparative Physiology B* and his research has been widely published in many international journals.

Marsupial Nutrition

Ian D. Hume
Challis Professor of Biology
University of Sydney
Australia

CAMBRIDGE
UNIVERSITY PRESS

PUBLISHED BY THE PRESS SYNDICATE OF THE UNIVERSITY OF CAMBRIDGE
The Pitt Building, Trumpington Street, Cambridge CB2 1RP, United Kingdom

CAMBRIDGE UNIVERSITY PRESS
The Edinburgh Building, Cambridge CB2 2RU, UK   http://www.cup.cam.ac.uk
40 West 20th Street, New York, NY 10011-4211, USA   http://www.cup.org
10 Stamford Road, Oakleigh, Melbourne 3166, Australia

First published 1999

Printed in the United Kingdom at the University Press, Cambridge

Typeset in Monotype Plantin 10.5/13 pt. [VN]

A catalogue record for this book is available from the British Library

Library of Congress Cataloguing in Publication data

Hume, Ian D.
    Marsupial nutrition / Ian D. Hume.
      p.   cm.
    Includes index
    ISBN 0 521 59406 5 (hb). – ISBN 0 521 59555 x (pbk.)
    1. Marsupialia – Nutrition.   2. Marsupialia – Food.   I. Title.
  QL737.M3H86   1999
  573.3'192–dc21   98–24765   CIP

ISBN 0 521 59406 5 hardback
ISBN 0 521 59555 X paperback

Contents

Contents

Preface

Marsupial Nutrition is about the nutritional ecology and digestive physiology of marsupials. The nutritional ecology of a species describes the interface between animals of that species and their food resources. Their digestive physiology determines which resources in the environment are food. The roots of *Marsupial Nutrition* lie in its predecessor, *Digestive Physiology and Nutrition of Marsupials*, published in 1982, but the great increase in knowledge and understanding of the nutritional niches filled by marsupials, and their digestive strategies, over the last 16 years has led to a much more comprehensive treatment of marsupial groups, their nutritional ecology and their digestive physiology in the present book.

Nevertheless, the reasons for writing *Marsupial Nutrition* remain the same: to inform physiologists and nutritionists about how the digestive systems of marsupials work, and to help zoologists and wildlife managers understand how food resources in the environment are utilised by various groups of marsupials, and which elements of the landscape are likely to be critical to the well-being of populations of marsupials in terms of their nutrition.

The ecological niches filled by marsupials are many and varied. Marsupials include many more species and forms than just kangaroos and koalas. In fact there are approximately 180 species of marsupials in Australia and New Guinea, 78 in South America and 1 in North America. They range in body size from 3 g shrew-like planigales (*Planigale* spp.) to 70 kg red kangaroos (*Macropus rufus*). They are found in habitats as diverse as freshwater streams (*Chironectes minimus*, the water opossum of South America), alpine areas (*Burramys parvus*, the mountain pygmy-possum of south-eastern Australia), hot deserts (*Notoryctes*, the marsupial moles of central Australia) and tropical rainforests (dorcopsis wallabies of New Guinea). Their diets range from purely insects to vertebrates, fungi, underground plant roots, bulbs, rhizomes and tubers, plant exudates such as saps and gums, seeds, pollen, terrestrial grasses, herbs and shrubs and tree foliage. Adaptive features of marsupials often have analogues in eutherian (placental) mammals. These are examples of convergent evolution. Some of the convergences are obvious: the marsupial gliders resemble the flying squirrels and lemurs, the Tasmanian tiger or thylacine was dog-like, and marsupial moles are reminiscent of eutherian moles. Other convergences are at the physiological level. For instance, ringtail possums produce two types of faeces and consume only the soft faeces that are higher in nutrient content; eutherian rabbits do the same. The wombats process grasses and sedges in a greatly enlarged colon, as do horses. Numbats feed on termites in much the same way that some eutherian anteaters do.

Marsupial Nutrition opens with a general chapter on metabolic rates of marsupials and their requirements for energy and nutrients such as water and protein. I deal principally with nutrition of the adult animal, and generally include the young only in so far as it affects the energy and nutrient requirements of adults (usually the lactating female). The next seven chapters deal with various groups of marsupials on the basis of their dietary and thus gastrointestinal tract specialisations. They begin with the relatively simple, the carnivores/insectivores, then progress through omnivorous groups to the most complex (several groups of herbivores). Most chapters open by introducing one or more general nutritional concepts pertinent to that part of the book. These concepts hopefully provide the framework for the rest of the chapter that follows. Most chapters conclude with a section on the nutritional ecology of each dietary group. The kangaroos and wallabies are an exception, for a separate chapter is devoted to their dietary niches and nutritional ecology. Chapter 9 provides an opportunity to review and compare the foraging and digestive strategies of the various groups of marsupials in the context of current thinking about marsupial evolution and the possible evolution of digestive strategies in mammals. The final chapter contains suggestions about where research in marsupial nutrition should head in the twenty-first century.

Readers familiar with its predecessor will notice that the material covered in *Marsupial Nutrition* is much more balanced in its treatment of carnivorous, omnivorous and herbivorous marsupials. This is because of the expanded research effort that has gone into marsupials other than kangaroos over the last 16 years. It has also been pleasing to be able to describe the work of, and refer to, many more South American authors than previously. Hopefully this book will stimulate further comparative studies on the digestive physiology, nutritional ecology and metabolism of a still wider range of species from South America in the future.

Within Australia, greater research effort recently has gone into marsupials from more mesic environments. Former research had been concerned more with arid-zone marsupials, not surprisingly so as two-thirds of the Australian land surface is classified as semi-arid or arid. Hopefully the greater coverage of more mesic forms in *Marsupial Nutrition* will stimulate even more studies on a wider range of species from across the broad spectrum of environments from rainforest to desert. Research on the nutrition and metabolism of New Guinean marsupials still lags, but again, recent work brought together in *Marsupial Nutrition* may also stimulate greater research effort on these fascinating animals.

The classification of marsupials to family level adopted in this book is based on that of Woodburne & Case (1996), and to species level on Eisenberg (1989), Redford & Eisenberg (1992), Strahan (1995) and Flannery (1995). In three cases of Australian species I have deviated from Strahan (1995) on the basis of new information. Two species of marsupial moles are now recognised: *Notoryctes typhlops* (southern marsupial mole) and *N. caurinus* (northern marsupial mole) (Maxwell, Burbidge & Morris 1996). The kowari (formerly *Dasyuroides byrnei*) is now subsumed in the genus *Dasycercus* as *D. byrnei* (Maxwell *et al.* 1996). *Antechinus agilis* (agile antechinus) is recognised as a species separate from *A. stuartii* in southern Victoria and south-eastern New South Wales (Dickman *et al.* 1988). The common names used for American species are based on Eisenberg (1989) and Redford & Eisenberg (1992). The common names used for Australian species are based on Strahan (1995), and those for New Guinean species on Flannery (1995). A list of the marsupial species mentioned in the text, with scientific and common names, will be found in the Appendix.

Because of the very recent nature of some of the information contained in this book I have depended heavily on the work of several current graduate students and on that of other colleagues who have generously supplied me with unpublished manuscripts or manuscripts in the process of publication. For allowing me access to their unpublished results, I thank Chris Allen, Bruce Bowden, Don and Felicity Bradshaw, Terry Dawson, Chris Dickman, Bart Eschler, Tim Flannery, Bill Foley, Lesley Gibson, Ross Goldingay, Perdita Hope, Menna Jones, Chris Johnson, Jonathan Kingdon, Steve Lapidge, Ivan Lawler, Geoff Lundie-Jenkins, Diego Moraes, Kylie McClelland, Diane Moyle, David Pass, Georgina Pass, Ken Richardson, Myfanwy Runcie, Felix Schlager, Andrew Smith, Ian van Tets and Mike Wolin.

I am also appreciative of colleagues who have commented on sections of the book: Perry Barboza, John Calaby, Terry Dawson, Chris Dickman, Tim Flannery, Graham Faichney, Bill Foley, Ed Stevens, and Pat Woolley.

Frank Knight provided the silhouettes of each species appearing in the figures, and most of the figures. Paulette Ripikoi and Sylvia Warren assembled most of the tables and references. My wife Desley provided the moral support needed to bring the book to fruition. To all these people my heartfelt thanks.

Ian Hume
Sydney

(1) Metabolic rates and nutrient requirements

1.1 CONCEPTS

This chapter deals with energy and nutrient requirements of marsupials, and how these are related to and can often be predicted from basal metabolic rates. The rest of the book deals with the dietary and foraging habits of the various groups of marsupials, and how food is processed by the animal. Food processing involves prehension and cutting, tearing, crushing or grinding by the teeth, digestion and absorption by the gut, and metabolism of absorbed nutrients in the liver and other body tissues. Available information on all of these aspects of the nutrition and nutritional ecology of marsupials is discussed. The chapters are organised so that the relatively simple digestive systems of carnivorous marsupials are covered first, followed by the more complex systems of omnivores and finally the most complex digestive systems which are found in the herbivores. Problems of defining carnivory, omnivory and herbivory are dealt with in Chapter 2; suffice to say here that for an appropriate sequence of chapters this 'division' of feeding types is convenient and widely understood among biologists.

1.1.1 Nutritional niche

Central to this book is the concept of the nutritional niche of an animal. Hutchinson (1957) introduced the concept of niche width of an organism. Kinnear et al. (1979) applied the concept to herbivores, and demonstrated how symbiotic gut microorganisms effectively expanded the host animal's niche width.

Fig 1.1, adapted from Kinnear et al. (1979), shows the fundamental and realised nutritional niches of a herbivore. The *fundamental nutritional niche* of an animal is described by the range of nutrient concentrations between the minimum required and the maximum tolerated by the species. It is defined in this example by two dimensions, each linearly ordered on the X and Y axes. The lower limits of the dimensions denote the minimum concentrations of each nutrient (for example, an essential amino acid on X and an essential fatty acid on Y) required by the animal. The upper limits denote the maximum levels that can be tolerated without toxicity symptoms appearing. The area, or 2-space (Kinnear et al. 1979), so defined, describes the limits within which the species can survive and persist. A third axis, representing another nutrient, could be added to define a volume or

1

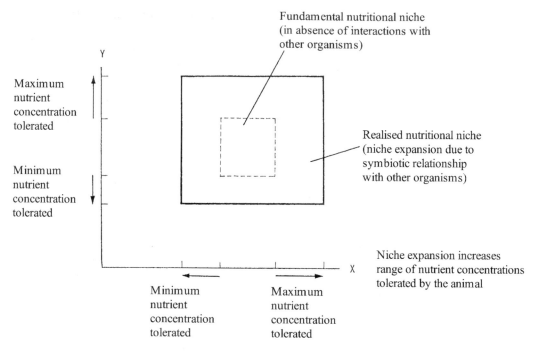

Figure 1.1 The fundamental and realised nutritional niche of an animal, showing the concept of niche expansion due to symbiotic relationship with other organisms. Adapted from Kinnear *et al.* (1979).

3-space, and n axes would define a hypervolume (Hutchinson 1957), and thus a species' fundamental nutritional niche.

The *realised nutritional niche* of an organism is a modified range of nutrient concentrations that can be used by the species because of interactions with other organisms. In the case of the herbivore this means the symbiotic microorganisms resident in its gut. These interactions include biosynthesis of essential nutrients and catabolism of many potentially toxic compounds, and render the host animal more tolerant of both deficiencies and excesses of several nutrients. This is an example of nutritional niche expansion. It expands the range of resources within the environment that the host animal can use as food.

The converse, nutritional niche contraction, results from competition. For instance, a herbivore may be effectively prevented from using highly nutritious food by more efficient competitors, in this case non-herbivores. By harbouring symbiotic microbiota in their gut, herbivores have introduced an additional link into the food chain, which inevitably leads to material and energy losses. Animals without this additional link have an advantage if the food can be digested by the animal's own enzymes (catalytic digestion), but herbivores have an advantage if the food can be digested only autocatalytically (that is, by microbial fermentation). So we find that herbivores usually are associated with poorer quality food resources than non-herbivores even though they are capable of using both good and poor quality foods.

To place the concept of nutritional niche within the framework of this book, a useful working statement is that an animal's nutritional niche can be defined principally by: (a) what it needs in the way of energy and specific nutrients; and (b) how it harvests and extracts those needed nutrients from the food resources available in its nutritional environment. In general, specialist feeders such as carnivores and folivores have narrower nutritional niches than have omnivores and generalist herbivores.

1.2 METABOLIC RATES

Energy and thus food requirements are related to metabolic rate. Three measures of rate of metabolism are relevant here: basal metabolic rate (BMR), field metabolic rate (FMR) and maximum sustained metabolic rate.

The basal metabolic rate of an endotherm is the minimum rate of metabolism compatible with endothermy (McNab 1988b). It can be measured as the rate of oxygen consumption (or heat production) of a non-reproductive, post-absorptive adult animal at rest (but not asleep) in its thermoneutral zone and not experiencing any physical or psychological stress. Some of these conditions are easier to satisfy than others; in herbivores a truly post-absorptive state is never reached without starving the animal because of the continuous nature of digestive function in these animals. Standard metabolic rate (SMR) is the equivalent minimal metabolic rate in an ectotherm at a particular temperature (Withers 1992a).

Field metabolic rate is the energy cost of free existence. It includes basal metabolism along with the costs of maintenance, thermoregulation and activity (Nagy 1994). However, measurements of FMR often include other costs associated with tissue growth, fat storage and reproduction. The latter may include additional activity costs involved with defence of breeding territories, courtship and foraging on behalf of the young. For these reasons FMRs are much more variable for a species than are BMRs. Thus, although FMRs relate directly to the real world, BMRs are widely used for comparisons across species and higher taxons.

The maximum sustained metabolic rate is the highest rate of energy expenditure that an animal can sustain from food intake, without using body energy stores. It has been measured experimentally in small mammals by using combinations of physical activity, cold stress and lactation (Hammond et al. 1994).

1.3 BASAL METABOLIC RATES

Body mass is the major determinant of energy use in endotherms (Nagy 1987, 1994). In eutherian mammals basal energy metabolism has been shown to vary with body mass according to the equation $BMR = a \, Mass^b$. The most appropriate value for the power function 'b' is the subject of much continuing debate. Withers (1992a) summarised the allometric relationships between metabolic rate and body mass for various groups of animals from unicells to vertebrates, and the various arguments for predicting what the power function should be, including that based on geometric similarity. In this theory, as most metabolic activities occur at surfaces, metabolic rate should increase as the square power, whereas body mass increases as the cube power of body size. Thus to compare the BMR of animals of different body size the discrepancy between surface area increase and volume increase should be accommodated by raising body mass to the two-thirds power (i.e. $Mass^{0.67}$), assuming the animal to be a perfect sphere. However, this is rarely the case, and empirically the power function that best fits available data from the smallest to the largest animals has been found to be generally between 0.5 and 1.0, averaging close to 0.75 for interspecific relationships and 0.72 for intraspecific relationships (Withers 1992a).

A more recent model of scaling in energy metabolism is based on the idea that living systems are sustained by the transport of essential materials through space-filling fractal networks of branching tubes. In this model, the terminal tubes do not vary with body size and the energy required to distribute resources through this network is minimised (West, Brown & Enquist 1997). This model suggests that most metabolic functions of animals should be related to body mass by some multiple of the one-quarter power, and that for whole-body parameters the power function should be close to three-quarters because most animals are three-dimensional.

Early work using whole-animal calorimetry yielded interspecific relationships to the power of 0.73 when Brody (1945) used 14 eutherian and 6 avian species, or 0.75 when Kleiber (1961) used 12 eutherians. In each case the work can be criticised for insufficient numbers of animals, inadequate representation of mammalian and avian taxons, and incorrect statistical analysis. A more recent analysis of a much broader data set (248 eutherian species and 42 marsupials) by Hayssen and Lacy (1985) yielded interspecific power functions of 0.70 for eutherians and 0.75 for marsupials. Importantly, interspecific relationships within orders or families often deviated significantly from these power functions. For example, 'b' for 16 heteromyid rodents was 0.91, but for 27 sciurids it was 0.61.

These criticisms notwithstanding, the power function of 0.75 is widely used for interspecific comparisons of metabolic rates and other physiological variables among subsets of eutherian taxons, and for statistical analysis data are often tested for significant deviation from the 'Kleiber line'. When body mass is expressed in kg, the Kleiber line yields a value for 'a' (the intercept) of 70 if the BMR is expressed in kcal $kg^{-0.75}$ d^{-1}, 293 if the BMR is in kJ $kg^{-0.75}$ d^{-1}, or 3.34 if the BMR is in the SI (Système International d'Unités) units of Watts $kg^{-0.75}$.

The traditional view, from Dawson and Hulbert's (1970) comparison of eight Australian marsupial species from five families, has been that the BMR of marsupials is about 30% below that of eutherian mammals. We now know that this is an oversimplification, yet, despite the small number of species represented, Dawson and Hulbert's (1970) 'marsupial line' is still often used as a standard against which other marsupials are compared as the data become available. Their line yields a BMR for the 'average marsupial' of 49 kcal or 204 kJ $kg^{-0.75}$ d^{-1} or 2.33 W $kg^{-0.75}$.

The concept of an average marsupial BMR and a strong taxonomic difference in BMRs between marsupials and eutherians has been challenged by McNab (1978; 1986; 1988b), who concluded that variations in BMR among both marsupials and eutherians are strongly correlated with food habits, activity level and the precision of temperature regulation. In both groups of mammals, feeding on fruit, tree foliage or invertebrates is associated with low BMRs, especially at large body size. This is because these food resources are seasonally unavailable (fruit, invertebrates), poorly digested (tree leaves) or have to be detoxified (tree leaves, some invertebrates). In addition, frugivory and folivory are associated with sedentary, arboreal habits in both mammalian groups. Correlations have also been demonstrated between low mammalian BMRs and fossoriality (burrowing), nocturnal habits and reduced muscle mass (as seen in many arboreal species (McNab 1992)). The question of whether phylogeny or food habits and activity is more important in determining BMR is unresolved, and is likely to remain so until many more marsupial and eutherian species from a wider range of nutritional habitats are examined. At present, the balance of opinion seems to be that there is a basic underlying difference in BMR between eutherians and marsupials (and monotremes), but that the influence of other factors such as food habits and activity is sometimes strong enough to mask phylogeny.

Table 1.1 summarises available data on BMRs of marsupials. Marsupial BMRs tend to form a tight cluster, with about half of the values falling between 65 and 74% of the value expected from an equivalent body mass in eutherians. There are only a few high values, the highest being those of very small species such as the 7 g *Planigale ingrami* (106%) and the 10 g honey possum (*Tarsipes rostratus*) (158%). McNab (1978) also reported a high BMR for the didelphid *Chironectes minimus* (98% of the Kleiber mean), which he attributed to the high rates of heat loss in a semi-aquatic environment. Similarly, among the monotremes, the BMR of the semi-aquatic platypus (*Ornithorhyncus anatinus*) (67% of the Kleiber mean) is higher than that of the terrestrial echidnas *Tachyglossus* (31%) and *Zaglossus* (27%) (Dawson, Grant & Fanning 1979). However, Thompson (1988) re-evaluated the BMR of *Chironectes* and found it to be 64%, not 98%, of the Kleiber mean, and concluded that *Chironectes* was not an exception to the pattern of low BMRs within the Marsupialia. Similarly, Elgar & Harvey (1987) felt

Table 1.1. *Basal metabolic rates (BMR) of marsupials*

Species	Body mass (g)	BMR				Ref.
		mLO_2 $g^{-1} h^{-1}$ [a]	kJ $kg^{-0.75} d^{-1}$ [b]	W $kg^{-0.75c}$	%[d]	
Family Didelphidae						
Marmosa microtarsus	13	1.436	244	2.78	83	1
Monodelphis brevicaudata	76	0.800	211	2.41	72	2
Monodelphis domestica	104	0.608	161	1.83	55	3
Marmosa robinsoni	122	0.800	238	2.71	81	2
Caluromys derbianus	331	0.685	262	2.99	89	2
Metachirus nudicaudatus	336	0.610	234	2.67	80	2
Philander opossum	751	0.450	211	2.41	72	2
Lutreolina crassicaudata	812	0.500	239	2.72	82	2
Chironectes minimus	946	0.580	288	3.28	98	2
Didelphis marsupialis	1329	0.460	249	2.84	85	2
Didelphis virginiana	2403	0.380	238	2.71	81	2
Family Dasyuridae						
Planigale ingrami	7	2.130	310	3.53	106	4
Planigale gilesi	10	1.357	214	2.44	73	4, 5
Planigale maculata	11	1.135	184	2.10	63	6, 7
Sminthopsis crassicaudata	14	1.330	231	2.63	79	6, 8, 9
Antechinomys laniger	24	0.980	195	2.22	67	6
Antechinus stuartii	28	1.278	263	3.00	90	6, 8, 10
Pseudantechinus macdonnellensis	43	0.630	145	1.65	49	6
Dasycercus cristicauda	93	0.505	140	1.60	48	6, 9
Dasycercus byrnei	102	0.760	216	2.46	74	5, 6
Phascogale tapoatafa	157	0.810	257	2.93	88	6
Dasyurus hallucatus	584	0.510	225	2.57	77	6
Dasyurus viverrinus	910	0.450	222	2.53	76	6
Dasyurus geoffroii	1100	0.405	209	2.38	71	11
Dasyurus maculatus	1782	0.330	192	2.19	66	6
Sarcophilus harrisii	5050	0.280	212	2.42	72	6
Family Myrmecobiidae						
Myrmecobius fasciatus	400	0.356	143	1.63	49	12
Family Peramelidae						
Isoodon auratus	428	0.346	138	1.57	47	13
Perameles nasuta	667	0.479	209	2.38	71	8, 14
Isoodon macrourus	1185	0.414	201	2.29	69	8, 14
Macrotis lagotis	1266	0.353	169	1.93	58	14, 15

Table 1.1. *cont.*

Species	Body mass (g)	BMR				Ref.
		mLO$_2$ g^{-1} h^{-1} [a]	kJ kg$^{-0.75}$ d^{-1} [b]	W kg$^{-0.75c}$	%[d]	
Family Peroryctidae						
Echymipera kalubu	695	0.495	218	2.49	74	14
Echymipera rufescens	836	0.470	210	2.39	72	14
Family Phascolarctidae						
Phascolarctos cinereus	4700	0.217	161	1.84	55	16
Family Vombatidae						
Lasiorhinus latifrons	29 920	0.110	130	1.48	44	17
Family Burramyidae						
Cercatetus nanus	70	0.860	223	2.54	76	18
Family Petauridae						
Petaurus breviceps	128	0.692	209	2.38	71	8
Gymnobelideus leadbeateri	166	0.620	199	2.27	68	19
Family Pseudocheiridae						
Pseudocheirus peregrinus	890	0.534	266	2.80	91	20
Pseudocheirus occidentalis	917	0.474	234	2.67	80	15
Petauroides volans	1000	0.417	210	2.39	72	21
Family Tarsipedidae						
Tarsipes rostratus	10	2.900	463	5.28	158	22
Family Acrobatidae						
Acrobates pygmaeus	14	1.067	185	2.11	63	23
Family Phalangeridae						
Trichosurus vulpecula	1982	0.315	188	2.14	64	8
Spilocuscus maculatus	4250	0.240	174	1.98	59	24
Family Potoroidae						
Potorous tridactylus	1035	0.455	231	2.63	79	25
Bettongia penicillata	1070	0.460	236	2.69	81	25
Aepyprymnus rufescens	2870	0.401	263	3.00	90	25
Family Macropodidae						
Lagorchestes conspicillatus	2260	0.320	206	2.35	70	26
Setonix brachyurus	2940	0.304	201	2.29	69	15
Macropus parma	3750	0.367	257	2.93	88	27
Thylogale thetis	4400	0.318	232	2.64	79	27

Table 1.1. (*cont.*)

Species	Body mass (g)	BMR				Ref.
		mLO_2 $g^{-1} h^{-1}$ [a]	kJ $kg^{-0.75} d^{-1}$ [b]	W $kg^{-0.75}$ [c]	%[d]	
Family Macropodidae (*cont.*)						
Macropus eugenii	4878	0.283	212	2.42	72	7, 27
Dendrolagus matschiei	6960	0.205	168	1.92	57	28
Macropus robustus erubescens	30 000	0.178	210	2.40	72	29
Macropus rufus	28 745	0.184	209	2.38	71	7, 29

Note: [a] Mass-specific rate or metabolic intensity
[b] Energetic equivalence of $O_2 = 21$ kJ L^{-1} (Withers 1992b)
[c] $W = 87.72$ kJ d^{-1}
[d] Percentage of predicted value from Kleiber's (1961) equation for eutherians. The 'marsupial mean' is 70% of the eutherian (Dawson & Hulbert 1970).
Source: 1. Morrison & McNab 1962; 2. McNab 1978; 3. Dawson & Olson 1988; 4. Dawson & Wolfers 1978; 5. Dawson & Dawson 1982; 6. MacMillen & Nelson 1969; 7. Morton & Lee 1978; 8. Dawson & Hulbert 1970; 9. Kennedy & Macfarlane 1971; 10. Wallis 1976; 11. Arnold & Shield 1970; 12. McNab 1984; 13. Withers 1992b; 14. Hulbert & Dawson 1974a; 15. Kinnear & Shield 1975; 16. Degabriele & Dawson 1979; 17. Wells 1978a; 18. Bartholomew & Hudson 1962; 19. Smith *et al.* 1982; 20. Munks 1990; 21. Foley 1987; 22. Withers, Richardson & Wooller 1990; 23. Fleming 1985; 24. Dawson & Degabride 1973; 25. Wallis & Farrell 1992; 26. Dawson & Bennett 1978; 27. White, Hume & Nolan 1988; 28. McNab 1988a; 29. Dawson 1973.

that many associations between BMR and dietary category among mammals could equally be described by taxonomic affinities.

Among the lowest marsupial BMRs reported are those of several desert-dwelling species such as the dasyurids *Pseudantechinus macdonnellensis* (only 49% of the rate expected from mass in eutherians) and mulgara (*Dasycercus cristicauda*) (48%), the bilby (*Macrotis lagotis*) (Fig. 1.2) (58%), hairy-nosed wombat (*Lasiorhinus latifrons*) (42%) and golden bandicoot (*Isoodon auratus*) (47%). Also low are several arboreal folivores such as the koala (*Phascolarctos cinereus*) (52%), common spotted cuscus (*Spilocuscus maculatus*) (Fig. 1.3) (59%) and the tree kangaroo *Dendrolagus matschiei* (57%).

1.4 CONSEQUENCES OF A LOW METABOLIC RATE

One consequence of a low BMR is generally a low body temperature (Withers 1992a). A low metabolic rate also has several important consequences for animals in terms of nutrient requirements and thus the width of

Figure 1.2 The bilby (*Macrotis lagotis*), an arid-zone omnivorous marsupial with a basal metabolic rate substantially below that of most other marsupials. (Pavel German)

Figure 1.3 The common spotted cuscus (*Spilocuscus maculatus*), one of several arboreal folivorous marsupials with unusually low basal metabolic rates. (Pavel German)

their nutritional niche. Other consequences, in environmental tolerance and reproductive rate, are related not only to an animal's BMR but also to its metabolic scope, which is the extent to which it can increase metabolic rate above basal to accommodate high rates of heat loss in cold environments and the energetic costs of a high reproductive potential (McNab 1986; Dawson & Olson 1988). Nevertheless, we can confidently predict that a low BMR will mean lower food requirements for maintenance, and that energy reserves will last longer under adverse conditions.

1.5 MAINTENANCE ENERGY REQUIREMENTS OF CAPTIVE MARSUPIALS

In captive wild animals and housed domestic stock, energy additional to basal requirements is needed for feeding, drinking, digestion, absorption and metabolism of absorbed nutrients, and for postural changes, but little is needed for thermoregulation or other activities. Under these conditions, maintenance energy requirements are often approximately double the BMR for the species. Estimated maintenance energy requirements of captive marsupials are listed in Table 1.2. These estimates are from two sources. The first is from feeding experiments in which it is assumed that non-reproductive adult animals at or close to body mass balance eat enough energy to maintain their energy status but no more when offered food *ad libitum*. Total collection of faeces allows calculation of the intake of digestible energy. The second source is from indirect calorimetry measurements of rates of oxygen consumption, and assuming that these are equivalent to metabolisable energy. Metabolisable energy is then converted to digestible energy using appropriate factors. With few exceptions, maintenance requirements are in the range of 150–250% of BMR. There also appears to be a trend for maintenance requirement as a multiple of BMR to decrease with increasing body mass of the species. This may reflect both a greater activity increment and greater requirements for thermoregulation in the smaller species, even under captive conditions.

Comparisons with equivalent eutherians are hampered by a relative lack of data on maintenance energy requirements of captive eutherians. The study by Hume (1974) included sheep with euros and red kangaroos. The estimated maintenance requirement for digestible energy by the sheep was 569 kJ $kg^{-0.75}$ d^{-1}, which is 137% and 125% respectively of those of the euro and red kangaroo. Thus the approximately 30% difference in BMRs between macropodids and their eutherian counterparts, the ruminants, is maintained in maintenance energy requirements. Similarly, the maintenance energy requirements of eutherian carnivores such as the mink (*Mustela vison*) (Farrell & Wood 1968) and the red fox (*Vulpes vulpes*) (Vogtsberger & Barrett 1973) in captivity are significantly higher than those of the

Table 1.2. *Maintenance energy requirements of captive marsupials. Values given as digestible energy (DE)*

Species	Body mass (kg)	Maintenance requirement		Ref.
		kJ kg$^{-0.75}$ d^{-1}	% of BMR	
(a) Estimates from feeding (balance) experiments				
Rufous hare-wallaby (*Lagorchestes hirsutus*)	1.2	326	—	1
Eastern quoll (*Dasyurus viverrinus*)	1.3	545	246	2
Tasmanian devil (*Sarcophilus harrisii*)	3.8	545	257	2
Parma wallaby (*Macropus parma*)	3.8	504	196	3
Tammar wallaby (*Macropus eugenii*)	4.8	320	151	4
Red-necked pademelon (*Thylogale thetis*)	5.0	530	228	4
Koala (*Phascolarctos cinereus*)	5.8	388	241	5
	6.6	330	205	6
Eastern grey kangaroo (*Macropus giganteus*)	20.8	570	—	4
Hairy-nosed wombat (*Lasiorhinus latifrons*)	23.1	140	108	7
Euro (*Macropus robustus erubescens*)	27.0	414	130	8
Common wombat (*Vombatus ursinus*)	27.9	140	—	7
Red kangaroo (*Macropus rufus*)	30.0	456	192	8
(b) Estimates from calorimetry measurements[a]				
Greater glider (*Petauroides volans*)	1.0	580	276	9
Long-nosed potoroo (*Potorous tridactylus*)	1.0	529	229	10
Brush-tailed bettong (*Bettongia penicillata*)	1.1	540	229	10
Common brushtail possum (*Trichosurus vulpecula*)	2.3	370	197	11
Rufous rat-kangaroo (*Aepyprymnus rufescens*)	3.1	386	147	10
Parma wallaby (*Macropus parma*)	4.2	368	143	12
Tammar wallaby (*Macropus eugenii*)	4.5	309	146	12
Red-necked pademelon (*Thylogale thetis*)	4.9	389	168	12

Note: [a]Assumed to be equivalent to metabolisable energy (ME)
Corrected to digestible energy (DE) by the value for ME/DE of 0.95 derived by Wallis & Farrell (1992) for potoroine marsupials, except for the greater glider, which was corrected to DE by 0.55 by Foley (1987).
Source: 1. Bridie, Hume & Hill 1994; 2. Green & Eberhard 1979; 3. Hume 1986; 4. Dellow & Hume 1982a; 5. Ullrey, Robinson & Whetter 1981b; 6. Cork, Hume & Dawson 1983; 7. Barboza, Hume & Nolan 1993; 8. Hume 1974; 9. Foley 1987; 10. Wallis & Farrell 1992; 11. Harris, Dellow & Broadhurst 1985; 12. White, Hume & Nolan 1988.

two marsupial carnivores studied by Green & Eberhard (1979).

Higher energy expenditures have been reported by Cowan, O'Riordan & Cowan (1974) in the alpine dasyurid *Antechinus swainsonii* maintained in cages for eight weeks. At body mass maintenance their digestible energy intake was close to four times their calculated BMR. The authors interpreted this high maintenance estimate as representing the energy cost of maintenance plus activity, as the animals were extremely active in their cages, and they concluded that this total energy expenditure was probably close to the normal energy demand of this species under free-living conditions.

1.6 VOLUNTARY FOOD INTAKE

The lower maintenance energy requirements of captive macropodid marsupials compared with housed domestic eutherian grazer/browsers are often reflected in lower voluntary food intakes of adult animals at or near body mass maintenance. This is illustrated in Table 1.3, which includes data from eight studies in which captive macropodids and ruminants were fed common diets of either chopped lucerne (alfalfa) hay or a chopped barley straw diet. Digestibility of dry matter is often higher in ruminants than in kangaroos (for reasons given in Chapter 6), which means that the difference between the two groups would be even greater if values were expressed as intake of digestible dry matter (equivalent to digestible energy).

1.7 FIELD METABOLIC RATES OF MARSUPIALS

Field metabolic rate, or the energy cost of free existence, is routinely measured by the use of doubly labelled water (Nagy 1980). Water labelled with the stable isotope of oxygen (^{18}O) and either the stable isotope of hydrogen (deuterium) or its radioactive isotope (tritium) is injected into the body water pool. After equilibration with the total body water pool, the rate of washout of the hydrogen isotope is a measure of water flux. The oxygen isotope traces both the water and carbon dioxide in the body, so the difference between washout rates of oxygen and hydrogen is a measure of CO_2 production (metabolic rate). Potential sources of error in the technique are discussed by Nagy (1980).

FMR has now been measured in 28 species of marsupials (Table 1.4). As mentioned earlier, FMR is much more variable within a species than is BMR. The main sources of variation can be readily identified from this table as being sex, season and reproductive state. Nagy (1987) analysed FMRs of 23 species of eutherians and 13 species of marsupials allometrically, and found that the slope of the regression equation relating FMR to body mass was 0.81 for eutherians but only 0.58 for marsupials. The latter exponent is used in Table 1.4. When plotted together the two regression lines cross each other. This means that, unlike BMR, a common scaling factor cannot be used to compare FMRs between the two therian groups. Nagy (1987) concluded that in the body size range of 240–550 g, FMRs of marsupials and eutherians are similar, while at lower body sizes FMRs of eutherians are lower. The only dietary comparison Nagy (1987) was able to make was within the herbivores, for which marsupials and eutherians both

Table 1.3. *Voluntary food intakes of macropodid marsupials and sheep fed chopped lucerne (alfalfa) hay and goats, euros and wallaroos fed a chopped barley straw diet. All values given as g dry matter kg $^{-0.75}$ d^{-1}*

Ruminant		Macropod		Ref.
71.7	(Sheep)	58.1	Red kangaroo	Foot & Romberg
			(*Macropus rufus*)	(1965)
64.1	(Sheep)	38.7	Red kangaroo	McIntosh (1966)
66.5	(Sheep)	48.1	Eastern grey kangaroo	Forbes & Tribe
			(*Macropus giganteus*)	(1970)
79.0	(Sheep)	53.0	Eastern grey kangaroo	Kempton (1972)
91.6	(Sheep)	53.4	Red kangaroo	Hume (1974)
		52.7	Euro (*M. robustus erubescens*)	
62.0	(Sheep)	54.6	Red-necked wallaby	Hume (1977a)
			(*Macropus rufogriseus*)	
		69.2	Red-necked pademelon	
			(*Thylogale thetis*)	
60.3	(Sheep)	56.7	Eastern grey kangaroo	Dellow & Hume
		52.7	Red-necked pademelon	(1982a)
		29.4	Tammar wallaby	
			(*Macropus eugenii*)	
63.0	(Goat)	49.9	Euro	Freudenberger & Hume
		54.8	Wallaroo (*M. robustus robustus*)	(1992)

scaled to 0.64. Herbivorous eutherians generally had higher FMRs than herbivorous marsupials, regardless of body size.

A more recent analysis by Nagy (1994) confirmed a common slope for marsupials of 0.58. Although Green (1997) subsequently proposed different slopes for macropodoid (0.69) and non-macropodoid marsupials (0.52), there is no clear biological basis for lumping all non-macropodoids together, and for this reason the exponent 0.58 is used throughout this book for the purpose of comparing FMRs among marsupials. However, should a more generally acceptable exponent come to light, all tables include sufficient data to enable the reader to recalculate the values on the basis of any other power function.

More useful than FMR for comparative purposes is the ratio of FMR to BMR (calculated by dividing mass-specific FMR by mass-specific BMR) (Koteja 1991). In Nagy's (1987) analysis this ratio decreased with increasing body mass in marsupials, but in eutherians it increased with increasing body mass. The high ratio of FMR to BMR in small marsupials is consistent with their relatively high maintenance energy requirements in captivity (Table 1.2). The high ratio in large eutherians may be partly because most of the large species in Nagy's (1987) analysis were marine mammals, which have higher costs of thermoregulation in water. When Degen & Kam (1995) analysed data from nine marsupial and 24 eutherian species, none of which was aquatic, FMR:BMR ratios were similar at large body size (5–8 kg). However, at small body size (10–20 g) the ratio in marsupials was still twice that of eutherians.

In Table 1.4 the highest FMR:BMR ratios include those of two small dasyurid species (5.0–6.6) and Leadbeater's possum (6.2). Smith *et al.* (1982) calculated that 73% of the Leadbeater's FMR was attributable to activity and specific dynamic action, consistent with the dispersed nature of its food supply and its well-developed territorial social system. The lowest ratios are from an arboreal folivore, the koala (1.7), and two small wallabies measured during the annual summer drought that is characteristic of their Mediterranean-type environment (1.8–1.9). Similarly, Bradshaw *et al.* (1994) recorded an extremely low FMR:BMR ratio in the golden bandicoot on Barrow Island during an extended drought (1.4), but FMRs trebled after cyclonic rains broke the drought a year later.

1.8 METABOLIC SCOPE

The high FMR:BMR ratio of some small marsupial species raises the question of what is the highest rate of metabolism that can be sustained in the long term? Peterson, Nagy & Diamond (1990) defined *sustained metabolic rates* as time-averaged rates of metabolism in free-ranging animals maintaining body mass over periods that are long enough so that metabolism is fuelled by food intake rather than by transient depletion of energy reserves. Sustained metabolic rate is therefore equivalent to the FMR of animals that are in energy balance. They are less than peak, or burst metabolic rates, which are short term and fuelled largely by anaerobic ATP production from energy stores (mainly glycogen). Peak metabolic rates are limited to no more than one or two minutes because of the toxic effects of lactic acid accumulation, but during that time they may be as much as 100-fold the animal's BMR. In contrast, aerobically fuelled sustained metabolic rates are mostly between 2- and 5-fold BMR (Peterson *et al.* 1990), but can be as low as 1.3 (Karasov 1992) and as high as 7.2 in lactating ground squirrels (Kenagy *et al.* 1990). These multiples of BMR are termed the animal's *sustained metabolic scope*.

Metabolic rates higher than the maximum sustained metabolic rate of a species can be maintained over shorter periods (but for at least several hours) in response to severe cold stress. These rates are fuelled aerobically, but the animal may not be in energy balance, although it must be maintaining a stable body temperature. Such rates have been called *summit metabolic rates* (Gelineo 1964), and the difference between summit metabolic rate and the species' BMR is its *metabolic scope*.

Does a low BMR mean a limited metabolic scope? Dawson & Dawson (1982) compared the metabolic scopes of two small dasyurid marsupials with those of two rodents of similar size when exposed to ambient temperatures as low as −13°C. Summit metabolic rates were similar for the four species, but because BMRs were 30% lower for the marsupials, metabolic scopes for the two marsupials were eight to nine times BMR compared with four to six times BMR for the two eutherians. Dawson & Olson (1988) found that summit metabolic rate in the South American didelphid *Monodelphis domestica* was also eight to nine times BMR. In other words, as Hinds & MacMillen (1986) concluded, marsupials have lower metabolic rates than eutherians within their thermoneutral zone but the same metabolic rates as eutherians below thermoneutrality. Garland, Geiser & Baudinette (1988) then reported that marsupials and eutherians did not differ in maximal running speeds. These two lines of evidence indicate that the numerous consequences of a low BMR do not include restricted thermoregulatory or locomotory responses, and that marsupials have greater metabolic scopes than equivalent eutherians.

1.9 TORPOR AND HIBERNATION IN MARSUPIALS

The very high rates of metabolism required for maintenance of endothermy in small mammals at low ambient temperatures are not sustainable unless food supply is constant in quality and quantity. In the absence of food the internal energy stores deplete in a relatively short time while normothermic, and these small endotherms can save large amounts of energy by abandoning regulation of body temperature at their normal high levels. Heterothermy is particularly common in insectivores, both marsupial and eutherian, because a constant supply of insects is unlikely in the wild, and they cannot ameliorate fluctuations in food availability by caching food as granivores (seed eaters) can.

Heterothermy is manifested in two related but distinct ways; shallow daily torpor and hibernation (deep and prolonged torpor) (Geiser & Ruf 1995). The two states are distinct in terms of average maximum torpor bout duration (11 h in daily torpor versus 355 h in hibernation), mean minimum body temperature (17.4°C versus 5.8°C), minimum metabolic rate (0.54 versus 0.04 mL O_2 g^{-1} body mass h^{-1}), and minimum metabolic rate expressed as a percentage of BMR (30% in daily torpor versus 5% in hibernation).

Among marsupials, daily torpor has been observed in South American didelphid opossums, and Australian dasyurids and small possums from the families Petauridae (sugar glider and Leadbeater's possum) and Tarsipedidae (honey possum) (Table 1.5). Hibernation has been recorded in the South American microbiotheriid *Dromiciops australis* and Australian small possums from the families Burramyidae and Acrobatidae (feathertail

Table 1.4. *Field metabolic rates (FMR) of adult marsupials*

	Cohort	Season	Body mass (g)	Field metabolic rate				
				$mLCO_2$ $g^{-1} h^{-1}$	kJ d^{-1}	kJ $kg^{-0.58} d^{-1}$	$\dfrac{FMR}{BMR}$	Ref.
Family Didelphidae								
Marmosa robinsoni	A	Sp	28	3.069	53	422	4.7	1
Family Dasyuridae								
Sminthopsis crassicaudata	A	Sp	17	6.720	69	730	6.6	2
Antechinus stuartii	F	W	19	4.290	60	600	4.9	3
	M	W	29	4.150	75	585	4.0	3
	F	W	24	5.188	77	668	5.0	4
	M	W	54	3.525	117	1021	3.4	4
	L	S	29	5.240	94	730	5.0	4
	NL	S	27	4.730	79	640	4.6	4
Antechinus swainsonii	F	B	53	5.292	173	951	—	5
	M	B	73	3.931	177	808	—	5
	A	S	43	2.870	74	460	—	6
	A	W	40	6.740	162	1035	—	6
Phascogale calura	A	W	34	5.350	112	797	—	7
	F	Sp	35	3.360	73	507	—	7
Dasyurus viverrinus	A	S	1029	1.249	793	780	3.2	1
	A	W	1102	1.720	1169	1105	4.4	1
Sarcophilus harrisii	A	S	7900	0.532	2591	781	2.3	1
	A	W	7100	0.660	2890	927	2.9	1
Family Peramelidae								
Isoodon auratus	A	A (Dry)	307	0.447	72	143	1.4	8
	A	A (Wet)	333	1.395	243	460	4.3	8
Isoodon obesulus	A	A	1230	0.908	644	571	2.7	9
Macrotis lagotis	A	S	928	1.225	617	655	3.7	10
	A	W	848	1.033	480	534	3.0	10
	A	S	1132	0.768	455	423	2.5	11
	A	W	1208	0.991	626	561	3.3	11
Family Phascolarctidae								
Phascolarctos cinereus	NL	W	7800	0.503	2050	623	2.4	12
	M	W	10 800	0.358	2030	511	1.7	12
	F	S	5930	0.480	1442	518	1.7	13
	F	W	6078	0.485	1495	532	1.8	13
	NL	Sp	6030	0.570	1748	621	2.4	13
	L	Sp	6730	0.550	1855	624	2.4	13

Table 1.4. (*cont.*)

	Cohort	Season	Body mass (g)	Field metabolic rate			FMR/BMR	Ref.
				mLCO$_2$ g^{-1} h^{-1}	kJ d^{-1}	kJ kg$^{-0.58}$ d^{-1}		
	M	S	7400	0.383	1470	462	1.5	14
	M	W	7800	0.440	1659	501	1.7	14
Family Petauridae								
Petaurus breviceps	F	Sp	112	2.563	153	545	3.9	15
	M	Sp	135	2.671	192	613	4.1	15
Gymnobelideus leadbeateri	F	Sp	117	3.100	219	760	6.2	16
	M	Sp	133	2.890	232	748	6.1	16
Family Pseudocheiridae								
Pseudocheirus peregrinus	NL		968	1.142	561	572	2.2	17
	L		993	1.515	759	762	2.9	17
	M		994	1.244	643	645	2.5	17
Petauroides volans	F	W	940	1.029	492	512	2.5	18
	M	W	1050	1.024	547	531	2.5	18
Hemibelideus lemuroides	A	Sp	1026	1.293	675	665	—	19
Pseudochirulus herbertensis	A	Sp	1103	0.795	446	421	—	19
Family Tarsipedidae								
Tarsipes rostratus	A	W	10	6.682	34	491	2.7	20
Family Potoroidae								
Potorous tridactylus	F	S	852	1.032	473	519	2.1	21
	M	S	824	0.997	453	507	2.0	21
	F	Sp	757	1.214	512	602	2.6	21
	M	Sp	868	1.337	629	683	2.8	21
Bettongia penicillata	A	S	1100	0.936	524	496	2.4	22
	A	A	1100	1.054	590	558	2.5	22
	A	W	1100	1.242	695	658	3.2	22
Bettongia gaimardi	A	S	1700	1.011	874	642	—	1
Aepyprymnus rufescens	A	S	2860	0.994	1363	741	3.3	23
	A	W	2890	1.011	1495	808	3.4	23
Family Macropodidae								
Lagorchestes hirsutus	A	S	1351	0.753	531	446	—	24
	A	W	1453	0.870	661	532	—	24
Setonix brachyurus	A	S	1900	0.574	548	378	1.8	25
Macropus eugenii	A	S	4380	0.518	1150	488	1.9	25

Table 1.4. (*cont.*)

	Cohort	Season	Body mass (g)	Field metabolic rate				Ref.
				mLCO$_2$ g^{-1} h^{-1}	kJ d^{-1}	kJ kg$^{-0.58}$ d^{-1}	$\dfrac{FMR}{BMR}$	
Thylogale billardierii	A	S	5980	0.532	1630	578	2.2	26
Petrogale xanthopus	A	S	8900	0.488	2209	622	2.2	22
Macropus giganteus	M	S	43 900	0.369	8170	911	2.5	26

Note: Cohort: A, adult; F, female; M, male; L, lactating; NL, non-lactating
Season: A, autumn; B, breeding; S, summer; Sp, spring; W, winter
Rate of CO_2 production converted to kJ using the equivalents of 25.7kJ per LCO_2 for carnivores (Nagy *et al.* 1988), 23.8 for omnivores (Nagy, Bradshaw & Clay 1991) and 21.2 for herbivores (Munks & Green 1995).
FMR/BMR calculated as mass-specific FMR divided by mass-specific BMR (Degen & Kam 1995).
Daily Energy Expenditure (DEE) in kJ d^{-1} converted to a metabolic body mass basis using kg$^{0.58}$ (Nagy 1987, 1994).
References: 1. Green 1997; 2. Nagy *et al.* 1988; 3. Nagy *et al.* 1978; 4. Green *et al.* 1991; 5. Nagy 1987; 6. Green & Crowley 1989; 7. Green, King & Bradley 1989; 8. Bradshaw *et al.* 1994; 9. Nagy, Bradshaw & Clay 1991; 10. Gibson 1999; 11. Southgate, cited by Green 1997; 12. Nagy & Martin 1985; 13. Krockenberger, 1993; 14. Ellis *et al.* 1995; 15. Nagy & Suckling 1985; 16. Smith *et al.* 1982; 17. Munks & Green 1995; 18. Foley *et al.* 1990; 19. Goudberg 1990; 20. Nagy *et al.* 1995; 21. Wallis, Green & Newgrain 1997; 22. Green 1989; 23. Wallis & Green 1992; 24. Lundie-Jenkins, cited by Green 1997; 25. Nagy, Bradley & Morris 1990; 26. Nagy, Sanson & Jacobsen 1990.

glider) (Table 1.5) (Geiser 1994). All but one of these species are either insectivorous or omnivorous, feeding on a mixture of plant exudates and arthropods. The one exception is the honey possum, which feeds only on nectar and pollen (Withers *et al.* 1990). Torpor and hibernation in these various small marsupials is discussed in relation to their nutritional ecology in Chapter 2 (carnivores, including insectivores) and Chapter 3 (omnivores).

1.10 WATER TURNOVER

An animal's requirement for water can be determined by measuring its rate of water turnover (WTR). If most marsupials have lower BMRs than equivalent eutherians, then it might be expected that water turnover rates in marsupials would also be low, at least when both are measured under standard conditions. For such measurements Nicol (1978) suggested that the ambient temperature should be at the lower end of the animal's thermoneutral zone, since a higher temperature may result in increased water loss for evaporative cooling, while a lower temperature will increase

Table 1.5. *Torpor and hibernation in marsupials. A more complete list of carnivorous marsupials that enter torpor is given in Table 2.5 (Chapter 2)*

Family and species	Body mass (g)	Minimum body temperature (°C)	Torpor duration (h)	Torpor pattern
Didelphidae				
4 species (see Table 2.5)	13–111	16–27	8	Torpor
Microbiotheridae				
Dromiciops australis	30	—	120	Hibernation
Dasyuridae				
18 species (see Table 2.5)	7–1000	11.0–28.2	2–20	Torpor
Myrmecobiidae				
Myrmecobius fasciatus	500	—	—	Torpor
Notoryctidae				
Notoryctes typhlops	60	—	—	Torpor
Petauridae				
Petaurus breviceps	130	15.6	15.5	Torpor
Gymnobelideus leadbeateri	130	—	—	Torpor
Burramyidae				
Cercartetus nanus	21	1.3	552	Hibernation
Cercartetus concinnus	18	4.7	264	Hibernation
Cercartetus lepidus	12	5.9	144	Hibernation
Cercartetus caudatus	30	—	<24	Hibernation(?)
Burramys parvus	63	2.4	336	Hibernation
Acrobatidae				
Acrobates pygmaeus	12	2	120	Hibernation
Tarsipedidae				
Tarsipes rostratus	10	5	10	Torpor

Source: After Geiser (1994).

metabolic rate and thus increase water turnover. Water must be available *ad libitum*, for water deprivation lowers rates of metabolism and water turnover (Hulbert & Dawson 1974b). Alternatively, food containing a high proportion of water will supply adequate amounts. Water turnover measured under these conditions might then be described as the standard water turnover rate of the species (Nicol 1978).

Water turnover rate can be estimated from the dilution rate of a single dose of tritiated or deuterated water in blood, urine or evaporative water (Rübsamen, Nolda & Engelhardt 1979). If evaporative water is used there will be a small but significant error introduced by the differential movements of hydrogen, tritium and deuterium across membranes (Rübsamen *et al.* 1979). Water turnover rates in free-living animals are routinely measured during measurement of FMR using doubly labelled water.

Using tritiated water, Richmond, Langham & Trujillo (1962) found that the standard water turnover rate in seven species of captive eutherians ranging in size from a 21 g house mouse to a 399 kg horse was 134 ± 32 mL $kg^{-0.80}$ d^{-1}. Denny & Dawson (1975a) subsequently showed that the mean water turnover rate in five macropodid marsupial species under similar conditions was 98 ± 21 mL $kg^{-0.80}$ d^{-1}. Although variation around the mean is considerable in both studies, the macropodid mean is 27% lower than the eutherian mean. These and other values for water turnover rate in captive marsupials are listed in Table 1.6. With few exceptions, the data support the concept of a generally low standard water turnover rate in marsupials. However, Nicol (1978) examined 27 eutherian and 13 marsupial species and concluded that habitat had a far greater effect on standard water turnover rate than did phylogeny. Although the ecological significance of water turnover rates measured under standard conditions is likely to be limited, when compared over a wide range of species it does seem that standard water turnover rates can be useful in separating desert-adapted species from others.

Nagy & Peterson (1988) examined scaling relationships between water turnover rate and body mass across a wide range of animal taxons, both in captivity and in the field. The slope of 0.95 for humans and 96 other eutherians, either captive or domestic, was higher than the slope for 16 captive marsupials (0.77), while the intercept value was lower, making direct comparisons between marsupials and eutherians difficult. Nevertheless, their conclusion was similar to that of Nicol (1978) that habitat and dietary category play major roles in setting a species' standard water turnover rate.

Water turnover rates measured in the field are much more meaningful ecologically. Here, Nagy & Peterson (1988) concluded that mass-corrected water turnover rates were determined much more by dietary habits than by phylogeny. They found a lower slope for free-living marsupials (0.60) than for captive marsupials (0.77), mainly because of the much higher water influx rates of small dasyurids and bandicoots in the field than in captivity, so

Table 1.6. *Water turnover rates (WTR) of captive marsupials with water available* ad libitum

Species	Habitat	Body mass (g)	Water Turnover Rate		Ref.
			mL d^{-1}	mL kg$^{-0.80}$ d^{-1}	
Family Dasyuridae					
Sminthopsis crassicaudata	T-A	15	7.5	216	1
Antechinomys laniger	A	18	6.7	167	2
Dasycercus cristicauda	A	86	11.5	82	3
Dasycercus byrnei	A	127	16.8	88	4
Dasyurus viverrinus	T	1340	162.0	128	5
Sarcophilus harrisii	T	3840	383.0	131	5
	T	5250	393.0	104	6
Family Peramelidae					
Perameles nasuta	T	972	68.3	70	7
Macrotis lagotis	A	1080	48.1	45	7
Isoodon macrourus	T	1470	131.0	96	7
Family Vombatidae					
Lasiorhinus latifrons	A	25 000	433.4	33	8
Family Potoroidae					
Potorous tridactylus	T	1400	137.0	105	9
Family Macropodidae					
Thylogale thetis	T	3520	522.7	191	10
Macropus eugenii	MA	5420	525.7	136	10
	MA	6500	291.0	65	9
Macropus giganteus	T-A	22 100	937.0	79	9
Macropus rufus	A	23 400	1 430.0	115	9
M. robustus erubescens	A	24 100	1 224.0	96	11
M. robustus robustus	T	31 000	1 850.0	119	9

Note: Habitat: A, arid; T, temperate; MA, maritime arid.
References: 1. Morton 1980; 2. Macfarlane 1975; 3. Kennedy & Macfarlane 1971; 4. Haines *et al.* 1974; 5. Green & Eberhard 1979; 6. Nicol 1978; 7. Hulbert & Dawson 1974b; 8. Wells 1973; 9. Denny & Dawson 1975a; 10. Dellow & Hume 1982b; 11. Denny & Dawson 1973.

it is debatable as to whether this exponent should be applied over the whole range of marsupial body masses. When the same data were divided on the basis of diet, analysis of covariance yielded a common slope of 0.71 for 28 herbivores and 23 carnivores, but the latter group had a higher intercept, no doubt because it contained mainly small species. Therefore, with the emphasis in this book on ease of comparison, the data in Table 1.7 on water turnover rates in free-living marsupials are based on an exponent of 0.71.

21

Table 1.7. *Water turnover rates (WTR) in free-living marsupials*

Species	Cohort	Season	Body mass (g)	WTR mL d⁻¹	WTR mL kg⁻⁰·⁷¹ d⁻¹	Ref.
Family Didelphidae						
Marmosa robinsoni	A	Sp	28	9.4	119	1
Family Dasyuridae						
Sminthopsis crassicaudata	A	S	20	22.4	360	2
	A	W	13	19.8	432	2
	A	Sp	17	13.3	244	3
Antechinus stuartii	A	W	26	13.9	186	4
	NL	S	27	13.2	173	5
	L	S	29	24.9	308	5
Phascogale calura	A	W	34	11.2	123	6
	F	Sp	35	14.0	153	6
Antechinus swainsonii	NL	Sp	47	23.1	202	7
	L	Sp	54	72.5	576	7
	A	S	43	23.4	218	8
	A	W	40	24.5	240	8
Dasyurus viverrinus	A	S	1120	202	186	9
	A	W	920	261	277	9
	L	Sp	984	332	336	9
Sarcophilus harrisii	A	S	7900	724	167	1
	A	W	7100	743	184	1
Family Peramelidae						
Isoodon auratus	A	Au (Dry)	307	27.9	65	10
	A	Au (Wet)	333	47.0	103	10
Isoodon obesulus	F	Au	1060	102	98	11
	M	Au	1370	104	83	11
Isoodon macrourus	A	Au	1410	354	277	12
Macrotis lagotis	A	S	928	73.2	77	13
	A	W	848	66.6	74	13
	A	S	1132	68.0	62	14
	A	W	1208	56.0	49	14
Family Phascolarctidae						
Phascolarctos cinereus	F	Sp	7800	358	83	15
	M	Sp	10 800	475	88	15
	F	S	5930	323	91	16
	F	W	6193	321	88	16
	NL	Sp	5900	248	71	16
	L	Sp	6140	298	82	16

Table 1.7. (*cont.*)

Species	Cohort	Season	Body mass (g)	WTR mL d^{-1}	WTR mL kg$^{-0.71}$ d^{-1}	Ref.
Family Vombatidae						
Lasiorhinus latifrons	A	S	22 200	555	61	17
	A	W	22 200	777	86	17
Family Petauridae						
Petaurus breviceps	F	Sp	112	21.7	103	18
	M	Sp	135	40.6	168	18
Gymnobelideus leadbeateri	F	W	95	44.5	237	19
	M	W	133	42.8	179	19
Family Pseudocheiridae	NL	Sp	968	117	119	20
Pseudocheirus peregrinus	L	Sp	993	158	159	20
	M	S	1046	101	98	20
	M	W	951	124	129	20
Pseudochirulus herbertensis	A	Sp	1103	140	131	21
Petauroides volans	F	W	934	80.4	84	22
	M	W	1042	98.8	92	22
Hemibelideus lemuroides	A	Sp	1026	155	152	21
Family Tarsipedidae						
Tarsipes rostratus	A	W	10	9.1	241	23
Family Phalangeridae						
Trichosurus vulpecula	NL	–	1520	134	100	24
	L	–	1590	163	117	24
Family Potoroidae						
Potorous tridactylus	A	S	816	131	151	25
	A	Sp	784	174	207	25
Bettongia penicillata	A	S	1100	67.1	63	26
Aepyprymnus rufescens	A	S	2850	398	189	26
	A	W	2900	373	175	26
Family Macropodidae						
Setonix brachyurus	A	S	1900	90.5	57	27
Lagorchestes conspicillatus	A	Sp	2230	90.8	51	26
Petrogale inornata	NL	—	3200	251	110	12
Petrogale rothschildi	A	Sp (Dry)	3350	149	63	26
	A	S (Wet)	2740	365	178	26
Macropus eugenii	A	S	4380	270	95	27
Thylogale billardierii	A	S	5980	585	164	28

Table 1.7. (*cont.*)

Species	Cohort	Season	Body mass (g)	WTR mL d^{-1}	WTR mL kg$^{-0.71}$ d^{-1}	Ref.
Petrogale xanthopus	A	S	8900	475	101	26
	A	W (Dry)	8900	497	105	26
	A	W (Wet)	8900	1304	276	26
Macropus rufus	A	S	21 800	861	97	29
M. robustus erubescens	A	S	28 100	1107	104	29
Macropus giganteus	A	S	43 900	2600	177	28

Abbreviations: A, Adult; F, female; M, male; L, lactating; NL, non-lactating
Au, autumn; S, summer; Sp, spring; W, winter.

References: 1. Green 1997; 2. Morton 1980; 3. Nagy *et al.* 1988; 4. Nagy *et al.* 1978; 5. Green *et al.* 1991; 6. Green, King & Bradley 1989; 7. Nagy & Peterson 1988; 8. Green & Crowley 1989; 9. Green & Eberhard 1983; 10. Bradshaw *et al* 1994; 11. Nagy, Bradshaw & Clay 1991; 12. Hulbert & Gordon 1972; 13. Gibson 1999; 14. Southgate, cited by Green 1997; 15. Nagy & Martin 1985; 16. Krockenberger 1993; 17. Wells 1973; 18. Nagy & Suckling 1985; 19. Smith *et al* 1982; 20. Munks & Green 1995; 21. Goudberg 1990; 22. Foley *et al.* 1990; 23. Nagy *et al.* 1995; 24. Kennedy & Heinsohn 1974; 25. Wallis, Green & Newgrain 1997; 26. Green 1989; 27. Nagy, Bradley & Morris 1990; 28. Nagy, Sanson & Jacobsen 1990; 29. Dawson *et al* 1975.

Highest water turnover rates are found in the small carnivorous/insectivorous dasyurids. The range of some of these species extends into the arid zone (e.g. *Sminthopsis crassicaudata*), but their food of animal tissue contains enough water that special measures for water conservation beyond fossoriality and nocturnality are not necessary for survival. Lowest water turnover rates are found in desert-adapted omnivores such as the golden bandicoot and herbivores, including the spectacled hare-wallaby and Rothschild's rock-wallaby. Low water turnover rates are also seen in Mediterranean-zone species at the end of the long annual summer drought; examples in Table 1.7 include the omnivorous southern brown bandicoot and the herbivorous small wallabies *Setonix brachyurus* (quokka) on Rottnest Island and *Macropus eugenii* (tammar wallaby) on Garden Island near Fremantle, Western Australia. Low water turnover rates are also characteristic of the strictly arboreal folivores *Phascolarctos cinereus* (koala) and *Petauroides volans* (greater glider).

Water use increases when more water becomes available, either directly or indirectly, because metabolic rates increase as more food also becomes available and is processed by the animal; three examples (one bandicoot and two rock-wallabies) in Table 1.7 illustrate this. Other factors that increase metabolic rate also increase WTR. Thus WTRs tend to be higher in males than in non-lactating females (males generally have higher metabolic rates), and higher in winter (increased thermoregulatory demands) than in summer. The largest increases in WTRs are generally seen in females during lactation, when there are significant increases in both metabolic rate and excretion of water in milk.

1.11 PROTEIN TURNOVER AND NITROGEN REQUIREMENTS

One of the major components of BMR is protein synthesis and degradation in the body. The energy costs directly involved in whole-body protein synthesis, together with other processes associated with protein synthesis, such as RNA turnover, amino acid activation and intermediary metabolism, account for a significant fraction of an animal's BMR (Reeds, Fuller & Nicholson 1985). Thus there should be a close relationship between BMR and rates of whole-body protein synthesis. This indeed appears to be so. White, Hume & Nolan (1988) found that protein turnover rates in three wallabies, the tammar, parma wallaby (*M. parma*) and red-necked pademelon (*Thylogale thetis*) were 23-47% lower than those reported in six eutherian species by Waterlow (1984), in line with the generally lower BMRs of macropodid marsupials. Wombats (Chapter 4) have even lower BMRs and protein turnover rates that are 57–74% lower than Waterlow's (1984) eutherian values (Barboza, Hume & Nolan 1993).

Among the wallabies, whole-body protein synthesis rates were significantly lower in the tammar than in the other two species (White *et al.* 1988), in line with a lower fed (but not basal) metabolic rate and a lower maintenance energy requirement (Table 1.2).

Differences in protein turnover rates are manifested in the whole animal in differences in rates of inevitable loss of nitrogen, mainly in the faeces (as metabolic faecal nitrogen) and urine (as endogenous urinary nitrogen). Metabolic faecal nitrogen and endogenous utinary nitrogen account for the bulk of an animal's maintenance nitrogen requirement. Maintenance nitrogen requirements are measured in captive animals fed a range of diets that ideally vary only in their content of protein, or nitrogen. The maintenance nitrogen requirement of the species is then the nitrogen intake that leads to zero nitrogen balance, where nitrogen balance is the difference between nitrogen intake and nitrogen excretion (urine plus faeces). Other avenues of nitrogen loss from the body, such as shed hair and sloughed skin, are negligible and usually ignored.

Importantly, compared to energy, the total requirement for nitrogen is much less affected by additional requirements for free existence such as activity and thermoregulation. Thus estimates of the maintenance requirement of captive animals for nitrogen are likely to be a realistic reflection of the needs of adult animals in the wild. Only growth and reproduction impose significant increments on the total protein requirement (Fig 1.4),

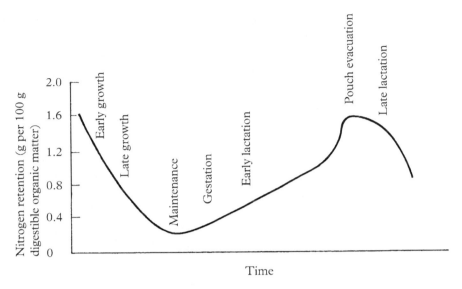

Figure 1.4 Total protein requirement of a female marsupial, measured as the ratio of nitrogen retention to digestible organic matter intake. Although the curve is adapted from a ruminant model, it is probably applicable to all therian mammals, even though the relative lengths of gestation and lactation differ between marsupials and eutherians; in marsupials gestation is short and its protein requirement is low, but lactation is extended and expensive in terms of both protein and energy.

but protein requirements for these physiological functions in marsupials have not been investigated to any extent. Wallis & Hume (1992) found that total nitrogen requirements of breeding female *Aepyprymnus rufescens* (rufous rat-kangaroos or rufous bettongs) at peak lactation were four times those of adult male animals.

Maintenance nitrogen requirements that have been established for marsupials are listed in Table 1.8. Values are given for both dietary and truly digestible nitrogen. Dietary requirements depend on several factors, among them differences in digestibility of the nitrogen in the diet. Truly digestible requirements at least partially allow for such differences. Most estimates of the maintenance requirement for truly digestible nitrogen of terrestrial herbivorous marsupials seem to be within the range 150–250 mg kg$^{-0.75}$ d^{-1}. In contrast, many equivalent eutherians have requirements calculated on the same basis that are twice these levels (Robbins 1983). Thus it seems that the basic phylogenetic difference seen in BMR between marsupial and eutherian grazer/browsers is reflected also in maintenance nitrogen requirements.

Moreover, just as with BMRs, diet and habitat also seem to play a role in shaping maintenance nitrogen requirements. Habitat effects are clearly seen among the Macropodidae, with the highest requirements associated with moist forest habitats in the case of the red-necked pademelon and parma wallaby and the lowest with the arid-zone euro. Diet effects are demonstrated by the data for the arboreal folivores. Common brushtail possums (*Trichosurus vulpecula*) maintained on *Eucalyptus melliodora* (yellow box) foliage required more than twice the amount of nitrogen as on a semi-purified diet containing no plant secondary metabolites (PSMs). Plant secondary metabolites are compounds synthesised for purposes other than general metabolism, such as for defence against invasion by pathogens or ingestion by herbivores. The main plant secondary metabolites found in eucalypt foliage are phenolics and terpenes (essential oils). Their anti-herbivore effects are discussed in Chapter 5. The dry matter of *E. melliodora* foliage contains about 26% of total phenolics, which resulted in large losses of non-dietary faecal nitrogen (largely microbial cells) (Foley & Hume 1987b). The reason for the high maintenance requirement of greater gliders fed *E. radiata* (narrow-leafed peppermint) foliage is probably the high content (11%) of terpenes in this species. In this case there were unusually high losses of nitrogen in the urine, mainly as ammonium ions because of the large amount of glucuronides excreted as conjugates of terpenes, and the need to maintain the acid–base balance of the animal (Foley 1992). The mechanism involved is treated in greater detail in Chapter 5.

The only non-herbivores in Table 1.8 are the didelphid *Caluromys philander*, the eastern pygmy-possum (*Cercartetus nanus*) and the sugar glider (*Petaurus breviceps*). *C. philander* feeds mainly on fruit but also nectar, gums and invertebrates (see Chapter 3), and its maintenance nitrogen requirements are relatively low. The sugar glider feeds on a mixture of *Acacia* gum, eucalypt sap and insects. Its low maintenance dietary nitrogen requirement of 87 mg kg$^{-0.75}$ d^{-1} is partly explained by an unusually low loss of metabolic faecal nitrogen (0.7 mg g^{-1} dry matter intake) compared with an average value in herbivorous marsupials of 2.8 (Smith & Green 1987). Metabolic faecal nitrogen is a mixture of undigested bacteria containing nitrogen of endogenous origin, abraided gut mucosal cells and unabsorbed digestive enzymes. The absence of indigestible fibre in the diets of the three non-herbivores would minimise bacterial growth and gut abrasion, and thus metabolic faecal nitrogen.

Endogenous urinary nitrogen (EUN) loss was also low in the sugar gliders (25 mg kg$^{-0.75}$ d^{-1}) compared with an average value in macropod marsupials of 54, although euros average 32 mg kg$^{-0.75}$ d^{-1} (see Table 1.9). Common wombats also have low EUN losses (31 mg kg$^{-0.75}$ d^{-1}). Endogenous urinary nitrogen is related more closely to the animal's metabolic rate than to any aspect of its protein metabolism. In eutherians, Smuts (1935) has shown that about 2 mg of EUN is excreted for each kcal of basal heat production. Assuming that a similar relationship holds in marsupials, EUN should be lower in macropod marsupials than in equivalent eutherians. The data in Table 1.9 demonstrate that this is so; the average value for

Table 1.8. *Maintenance nitrogen requirements of marsupials. All data in mg N kg $^{-0.75}$ d $^{-1}$*

| | | | Maintenance requirement | | |
			Dietary	Truly digestible	Ref.
Species	Body mass (kg)	Diet			
Family Didelphidae					
Caluromys philander	0.43	Fruit, casein	176	146	1
Family Phascolarctidae					
Phascolarctos cinereus	5.1–8.3	Eucalypt foliage	283	271	2
Family Vombatidae					
Vombatus ursinus	27.9	Chopped oat straw, grain	158	71	3
Lasiorhinus latifrons	23.1	Chopped oat straw, grain	201	116	3
	29.0–31.0	Chopped oat straw, grain	205	—	4
Family Burramyidae					
Cercartetus nanus	24	Pollen	46	43	5
	24	Mealworms	147	127	5
Family Petauridae					
Petaurus breviceps	0.15	Honey, pollen	87	—	6
Family Pseudocheiridae					
Pseudocheirus peregrinus	0.6–0.7	Eucalypt foliage	380	290	7
Petauroides volans	1.1	Eucalypt foliage	700	560	8

Family Phalangeridae					
Trichosurus vulpecula	1.3–2.5	Semi-purified	203	189	9
	2.5	Eucalypt foliage	560	420	8
Family Potoroidae					
Potorous tridactylus	0.9	Grain-based	—	199	10
Bettongia penicillata	1.1	Grain-based	—	199	10
Aepyprymnus rufescens	2.8–3.1	Grain-based	—	199	10
Family Macropodidae					
Thylogale thetis	3.8–4.8	Chopped alfalfa, sucrose	600	530	11
Macropus parma	3.7–3.9	Chopped alfalfa, sucrose	566	477	12
Macropus eugenii	4.2–5.4	Chopped oat hay, casein	290	250	13
	3.8–5.1	Chopped alfalfa, sucrose	240	230	11
M. robustus robustus	11.8–17.7	Chopped oat hay, casein	300	240	14
M. r. erubescens	10.5–15.6	Chopped oat hay, casein	290	160	15
M. giganteus	18.6–30.3	Chopped oat hay, casein	350	270	14

References: 1. Foley, Charles-Dominique & Julien-Laferriere 1999; 2. Cork 1986; 3. Barboza, Hume & Nolan 1993; 4. Wells, 1968; 5. van Tets 1996; 6. Smith & Green 1987; 7. Chilcott & Hume 1984b; 8. Foley & Hume 1987b; 9. Wellard & Hume 1981b; 10. Wallis & Hume 1992; 11. Hume, 1977b; 12. Hume 1986; 13. Barker 1968; 14. Foley, Hume & Taylor 1980; 15. Brown & Main 1967.

Table 1.9. *Endogenous urinary nitrogen (EUN) in seven marsupial and five eutherian herbivores*

Species	Body mass kg	EUN mg kg $^{-0.75}$ d^{-1}	Ref.
Marsupials			
Quokka (*Setonix brachyurus*)	2.0–3.6	43	Brown (1968)
Black-footed rock-wallaby (*Petrogale lateralis*)	3.9	49	Brown (1968)
Tammar wallaby (*Macropus eugenii*)	4.2–5.3	58	Barker (1968)
Euro (*M. robustus erubescens*)	8.5–19.7	32	Brown (1968)
Red kangaroo (*M. rufus*)	14.4	87	Brown (1968)
Hairy-nosed wombat (*Lasiorhinus latifrons*)	23.1	42	Barboza *et al.* (1993)
Common wombat (*Vombatus ursinus*)	27.9	31	Barboza *et al.* (1993)
Eutherians			
Guinea pig (*Cavia porcellus*)	0.3–0.5	145	Brody (1945)
Rabbit (*Oryctolagus cuniculus*)	1.2–2.8	148	Brody (1945)
Goat (*Capra hircus*)	24.2–62.0	115	Hutchinson & Morris (1936)
Sheep (*Ovis aries*)	31.8–42.0	87	Smuts & Marais (1938)
Camel (*Camelus dromedarius*)	250	60	Schmidt-Nielsen *et al.* (1967)

five macropods is less than half that for five herbivorous eutherians. Note-worthy, however, is the camel which overlaps the marsupial range; its BMR is also substantially below the Kleiber (1961) prediction for eutherians (Schmidt-Nielsen *et al.* 1967).

A similar argument cannot be used to explain the unusually low EUN of sugar gliders, because their BMR is within the macropod range (Table 1.1). An alternative explanation is that part of their endogenous nitrogen is retained by being recycled to the digestive tract. Herbivores excrete less urea, one component of EUN, than non-herbivores, because of recycling of endogenous urea to their gut, where resident microorganisms rapidly de-grade it and use a portion of the ammonia released for the synthesis of their own protein. The amount of ammonia trapped in this way depends on the energy available to the microbes. Although the sugar glider is an omnivore, the gums on which it feeds may be largely fermented in the hindgut. The caecum of the sugar glider is surprisingly large for an omnivore (see Chapter 3), which supports this view. Their high-energy diet may be expected to result in efficient trapping of recycled nitrogen, resulting in lower urea excretion rates than those of herbivores feeding on lower-energy plant material. This would lower their EUN.

Another major component of EUN is creatinine. Creatinine is a metabolic end-product of creatine, a precursor of the high-energy compound phosphocreatine found in muscle. Creatine is synthesised in the liver. Surplus creatine from the reversible interconversion of creatine and phosphocreatine in muscle is converted to creatinine and excreted by the kidneys. The rate of creatinine excretion by a healthy animal fed diets free of creatine and creatinine (as those of herbivores are) appears to be equivalent to the rate of creatine synthesis in the liver, which proceeds at a rate proportional to BMR. Unlike urea, an end-product of nitrogen metabolism, creatinine does not appear to enter the digestive tract to be degraded by microorganisms. Nor does its rate of excretion appear to be affected by muscular activity or other factors that increase metabolic rate, although it is disturbed by anything that raises deep body temperature (Mitchell 1962).

Thus in healthy herbivores any differences in the rate of creatinine excretion should reflect differences in BMR. Table 1.10 presents data on creatinine excretion from seven marsupial and five eutherian herbivores. Although variation around the mean values for some species is high, it does appear that, in general, marsupial levels are below those of the eutherians listed, consistent with their lower BMR. Among the marsupials it is of interest that the euro is the lowest and the red-necked pademelon is among the highest in terms of both creatinine excretion rate and maintenance nitrogen requirement. Thus there are several links between nitrogen and energy metabolism.

1.12 OTHER NUTRIENTS

Relative to energy, water and protein, there is only limited information on the requirements of marsupials for the micronutrients (vitamins, minerals and essential fatty acids). There is no evidence of unusually high requirements for any micronutrient among the marsupials, but there are suggestions that several micronutrients are required by some marsupials in extremely small amounts. An example is provided by quokkas on Rottnest Island. Early attempts to graze sheep on the island failed when the sheep suffered from deficiencies of several trace elements, including copper and cobalt. In contrast, studies by Barker (1960, 1961a, b) found no evidence of deficiencies of either mineral in quokkas on Rottnest, despite the poor mineral status of the sandy, limestone-derived soils. Requirements for the two minerals by the quokkas were calculated to be less than 50% of the requirements of sheep (see Chapter 7), resulting in a much wider nutritional niche for the native herbivore.

Another example is provided by the low ash content of *Eucalyptus* foliage (see Fig. 5.1), suggestive of a low mineral status. The absence of any reports of mineral deficiencies in koalas and greater gliders, two arboreal marsupials that feed almost exclusively on eucalypt leaves (Chapter 5), provides indirect evidence that mineral requirements of these two folivores are also

Table 1.10. *Creatinine excretion in seven marsupial and five eutherian species*

Species	Body mass (kg)	Creatinine excretion (mg kg$^{-0.75}$ d^{-1})	Ref.
Marsupials			
Long-nosed potoroo (*Potorous tridactylus*)	1.2	41	Nicol (1976)
Quokka (*Setonix brachyurus*)	3.3	34	Ramsay (1966); Kinnear & Main (1969)
Tammar wallaby (*Macropus eugenii*)	2.4–5.9	29	
	3.8	30	Fraser & Kinnear (1969)
	4.3	24	Wilkinson (1979)
Red-necked pademelon (*Thylogale thetis*)	4.3	39	Wilkinson (1979)
Euro (*M. robustus erubescens*)	14.6	24	Fraser & Kinnear (1969)
Hairy-nosed wombat (*Lasiorhinus latifrons*)	23	36	Barboza *et al.* (1993)
Common wombat (*Vombatus ursinus*)	28	25	Barboza *et al.* (1993)
Eutherians			
Rabbit (*Oryctolagus cuniculus*)	2.0	55	Brody (1945)
Pig (*Sus scrofa*)	24–79	56	Smuts (1935)
Sheep (*Ovis aries*)	37–50	62	Fraser & Kinnear (1969)
Cow (*Bos taurus*)	322	110	Brody (1945)
Camel (*Camelus dromedarius*)	515	85	Brody (1945)

low. Similarly, Ullrey, Robinson & Whetter (1981a) reported concentrations of phosphorus, sodium, selenium, zinc and copper in eucalypt foliage consumed by koalas at San Diego Zoo which would have been inadequate for sheep or horses.

A report by Barboza & Vanselow (1990) on copper toxicity in a southern hairy-nosed wombat (*Lasiorhinus latifrons*) maintained in captivity on formulated diets containing a commercial supplement designed for growing pigs also suggests that the mineral requirements of wombats may be much lower than those of domestic animals.

Although largely anecdotal, these examples indicate that the marsupial–eutherian differences in energy, water and protein requirements discussed earlier in this chapter may well apply also to the micronutrients. Not much more can be said though until more direct comparisons between marsupials and equivalent eutherians are conducted.

1.13 SUMMARY AND CONCLUSIONS

The concept of the nutritional niche of an organism was introduced in this chapter to set the framework for the rest of the book. The nutritional niche of a species can be defined principally by what it needs in terms of energy and specific nutrients, and how it harvests and extracts those needed nutrients from the food resources available. In general, more generalist feeders have wider nutritional niches than specialists; for example, species that feed on fruit and leaves have wider nutritional niches than specialist folivores. The amount of any particular nutrient required has two components: the amount needed for maintenance of the adult animal, and additional amounts needed for growth, reproduction and free existence. Maintenance requirements are often closely related to the species' basal metabolic rate, but the extent to which requirements are increased above maintenance in different physiological states and by environmental factors is dependent on many factors. Sound knowledge of the basic biology and ecology of the species is necessary before the likely relative importance of these various physiological and environmental factors can be appreciated. This applies particularly to the total energy and thus total food requirements of free-living animals. Information from captive animals studied under controlled conditions is vital for describing and understanding mechanisms. Information from free-living animals in different seasons, different physiological states and different environments is equally vital for interpreting captive results and for testing extrapolations from captivity to the wild state.

There is now enough information from captive and field studies to suggest that basal metabolic rates of marsupials have a phylogenetic base that is modified in many cases by food habits and activity levels. Generally though, marsupials have lower BMRs than their eutherian counterparts.

This trend is often reflected in lower maintenance requirements for energy, protein and water, but at the level of FMRs marsupial–eutherian comparisons are limited by insufficient data. Nevertheless, it appears that summit metabolic rates of small marsupials are similar to those of small eutherians, and thus small marsupials have greater metabolic scopes. Greater metabolic scopes in marsupials mean that a low basal metabolic rate does not translate into limited capacity for thermoregulation or locomotory responses. However, in inadequate environments a low BMR serves to maximise the life of energy stores.

A generally low standard water turnover rate in captive marsupials cannot be translated easily into relative field requirements because of the different scaling relationships associated with marsupials and eutherians. However, it is clear that phylogeny plays a subordinate role to environment in determining total water requirements in free-living marsupials and eutherians.

Less information is available on nitrogen requirements, but in this case maintenance requirements determined in captivity are more directly applicable to the free-living animal. What is needed is more information on the addition costs of growth and reproduction of a range of marsupial species. The next seven chapters deal with the ways in which the various groups of marsupials satisfy their energy and other nutrient requirements from different foods. The groups are separated by their food habits and mode of digestion, beginning with the carnivores.

2 Carnivorous marsupials

2.1 CONCEPTS

Chapter 1 dealt with one aspect that helps to define an animal's nutritional niche, its requirements for energy and specific nutrients. Chapter 2 introduces the other main aspect that defines an animal's nutritional niche, viz. how it harvests and extracts those needed nutrients from the food resources available in its nutritional environment; that is, its foraging and digestive strategies.

2.1.1 Optimal digestion theory

Foraging and digestive strategies of animals are closely linked. Optimal foraging theory predicts that the foraging behaviour that endows an animal with the greatest fitness is that which maximises the net rate of energy (or nutrient) intake (Townsend & Hughes 1981). In an analogous way, optimal digestion theory predicts that the digestive strategy that endows an animal with the greatest fitness is that which maximises the net rate of energy (or nutrient) release and absorption from ingested food. This concept is modelled for a continuous-flow digestive system in Fig 2.1.

In this model, the net energy released is initially negative until the food's defences, such as the chitinous exoskeletons of arthropods or the lignified cell walls of plants, are overcome (e.g. by mastication). This is followed by a period of rapid digestion (of haemolymph and soft tissues of arthropods, and the contents of plant cells), but thereafter digestion rate falls as digestion is progressively confined to less tractable dietary components such as structural proteins of animal tissues and structural carbohydrates of plant cell walls. Optimal digestion time is given by the straight line from the origin tangential to the curve (Sibly 1981).

Several predictions arise from the model. First, optimal digestion time will vary among foods, being longer for poor quality foods (e.g. adult insects) than those of higher quality (soft-bodied insect larvae). Second, because of longer digestion times, animals routinely eating poorer quality foods should have larger digestive tracts (e.g. herbivores versus carnivores). Third, if gut capacity is limiting (as in young animals), the optimal strategy is to maximise digestion rate by selecting only high-quality food items. Finally, at any given level of intake, an animal should maximise retention of food in order to maximise the amount of energy obtained (Sibly 1981), although there is a holding time beyond which net loss occurs. As will be

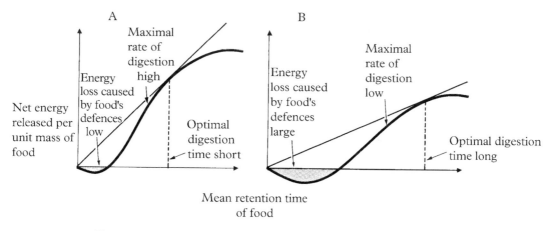

Figure 2.1 Model of digestion in a continuous-flow system for a high-quality (A) and a low-quality (B) food. Modified from Sibly (1981) by Hume (1989).

seen, the digestive strategies found in the range of marsupials described in this book are broadly consistent with these predictions.

2.1.2 Modelling the digestive tract

That there is a close relationship between optimal foraging and optimal digestive strategies was recognised by Penry & Jumars (1987) in their work with marine deposit feeders. They used principles of chemical reactor theory to formulate optimisation constraints that operate in digestive reaction rate equations. From these they developed models of digestion for a variety of marine deposit feeders. These and similar models have since been applied to digestion in mammalian herbivores (Hume 1989; Alexander 1991), nectar- and fruit-eating birds (Martinez del Rio & Karasov 1990) and fish (Horn & Messer 1992).

Penry & Jumars (1987) considered three basic types of chemical reactors: batch reactors, plug-flow reactors (PFR) and continuous-flow, stirred-tank reactors (CSTR) (Fig 2.2). Chemical reactors are classified on the basis of: (a) whether input is discontinuous or continuous; and (b) whether reactants are brought together with or without mixing (Levenspiel 1972). Batch reactors process reactants (substrates) in discrete batches; all reactants are added instantaneously (i.e. input is discontinuous) and are well mixed. The reaction is allowed to proceed for a set period, after which reaction products and unreacted materials are all removed. The reactor may then remain empty for a period or be refilled. Extent of reaction can be high, depending on the time materials are left in the reactor, but material flow is interrupted and low overall, which results in low production rate capabilities, unless reactor volume is very high. In ideal batch reactors (those that can be described accurately by simple equations), reactants are instantaneously

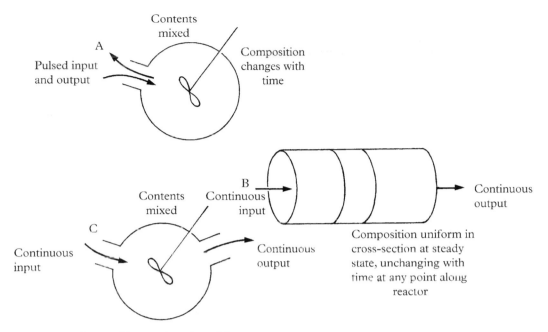

Figure 2.2 Models of three types of chemical reactors that have analogues in the marsupial digestive tract. A. Batch reactor (BR), with pulsed input and pulsed output, which describes digestive systems that process food in batches, such as the stomach of some carnivores; B. Plug-flow reactor (PFR), with continuous flow through a tubular reaction vessel, which describes the small intestine; and C. Continuous-flow, stirred-tank reactor (CSTR), with continuous flow through a well-mixed sacciform reaction vessel, which describes diverticula of the gut such as the sacciform forestomach region of kangaroos, wallabies and rat-kangaroos, and the hindgut caecum. From Stevens & Hume (1995).

and continuously mixed, so that concentrations are spatially uniform and vary only with time. Although batch reactors usually have only one opening (Fig. 2.2), batch processing may be found in digestive systems with two openings such as the stomach of some carnivores. Batch-reactor guts may be comparatively flexible under varying food supply, and can be emptied and refilled quickly when better quality food becomes available. For instance, Cochran (1987) has applied batch-reactor theory to the problem of optimal food retention times for carnivores that partially consume individual prey. More applicable to parts of the herbivore digestive system, particularly the caecum, are semi-batch reactors, which feature pulsed inputs but continuous output (Penry 1993).

Plug-flow reactors feature continuous, orderly flow of material through a usually tubular reaction vessel. In ideal PFRs, material does not mix along the flow axis, but there is perfect radial mixing, so that at steady state there is a continuous decline in reactant concentrations from inlet along the reactor to outlet, and a continuous increase in concentration of products. The small intestine is the part of the digestive tract that can be best modelled as a PFR, even though we know that flow through the small intestine is not perfectly continuous, that there is some, though limited, axial mixing, and that absorption across the wall reduces volume flow and changes its composition. Plug flow provides the greatest rate of digestive product formation in the minimum of time and volume under most conditions (Penry & Jumars 1987), although extent of digestion may be low. For these reasons PFRs are best suited to food of high quality. It is therefore not surprising that the carnivore digestive tract is usually dominated by the small intestine.

Continuous-flow, stirred-tank reactors are characterised by continuous flow through a usually spherical reaction vessel of minimal volume. Mixing is continuous, so that, at steady state, reactant concentrations are uniform throughout the vessel and with time. Reactant concentration is diluted immediately upon entry into the vessel by materials circulating in the reactor. This reduces reaction rate, but extent of reaction can be high if material flow is low enough. Diverticula of the gut of herbivores can sometimes be modelled as CSTRs; they retain food and symbiotic microbes and thereby provide the conditions and time required for the slow process of plant cell wall fermentation. Extent of fermentation in such diverticula is maximised by large size of the fermentation chamber and by mechanisms to increase retention times of food and microbes. However, CSTRs are not applicable to the carnivore digestive system.

2.2 DIETS OF CARNIVORES

As mentioned in Chapter 1, division of dietary categories into only three types, viz. carnivory, omnivory and herbivory, is in many instances an oversimplification because of the problem of defining boundaries between carnivory and omnivory, and between omnivory and herbivory. For instance, many carnivores are seen to eat plant material, either regularly or seasonally when prey species are scarce or unavailable. Although the brown antechinus (*Antechinus stuartii*) is regarded as a carnivore, Statham (1982) found that in a sclerophyll forest in north-eastern New South Wales, flowers of epacrid species constituted up to 41% of the diet of *A. stuartii* during winter when insects, their principal food item, were present in only small numbers. Thus it is not always easy to defend marginal examples of carnivory or herbivory.

Terminology and definition also differ among authorities. For example, Langer & Chivers (1994) prefer faunivory over carnivory, and refer to omnivores as either fruit and animal eaters or fruit and foliage eaters. Eisenberg (1981) defined 53 macroniches available to mammals according to their different dietary and habitat specialisations. Only 18 of these macroniches are clearly filled by marsupials (Lee & Cockburn 1985), based on present knowledge (Table 2.1). The lack of fully aquatic and flying marsu-

Table 2.1. *Marsupial genera classified by Lee & Cockburn (1985) according to macroniches defined by dietary and habitat specialisations by Eisenberg (1981)*

Macroniche	Marsupial examples
Terrestrial herbivore/grazer	*Lagorchestes, Macropus, Peradorcas, Petrogale, Onychogalea*
Terrestrial herbivore/browser	*Dorcopsis, Dorcopsulus, Lagostrophus, Setonix, Thylogale, Wallabia, Wyulda*
Terrestrial frugivore/omnivore	*Hypsiprymnodon*
Terrestrial carnivore	*Thylacinus, Sarcophilus*
Terrestrial myrmecophage	*Myrmecobius*
Semi-fossorial herbivore/grazer	*Lasiorhinus, Vombatus*
Scansorial herbivore/browser	*Dendrolagus*
Arboreal herbivore/browser	*Petauroides, Phalanger, Phascolarctos, Pseudocheirus, Trichosurus*
Arboreal frugivore/omnivore	*Caluromys, Caluromysiops, Glironia*
Terrestrial frugivore/granivore	*Burramys*
Terrestrial fungivore/omnivore	*Aepyprymnus, Bettongia, Potorous*
Semi-aquatic insectivore	*Chironectes*
Arboreal gumivore	*Gymnobelideus, Petaurus*
Terrestrial insectivore/omnivore	*Antechinus, Caenolestes, Chaeropus, Dasycercus, Dasykaluta, Dasyuroides, Dasyurus, Echymipera, Isoodon, Lestodelphis, Lestoros, Lutreolina, Macrotis, Metachirus, Microperoryctes, Monodelphis, Murexia, Myoictis, Neophascogale, Ningaui, Parantechinus, Perameles, Peroryctes, Phascogale, Phascolosorex, Planigale, Pseudantechinus, Rhyncholestes, Rhynchomeles, Satanellus, Sminthopsis*
Scansorial insectivore/omnivore	*Didelphis, Dromiciops, Philander*
Arboreal insectivore/omnivore	*Dactylopsila, Distoechurus, Marmosa*
Arboreal nectarivore	*Acrobates, Cercartetus, Tarsipes*
Fossorial insectivore/omnivore	*Notoryctes*

Source: Lee & Cockburn 1985.

pials, and the paucity of fossorial, semi-aquatic, granivorous and diurnal marsupials accounts for much of this reduction in the number of macroniches utilised.

For the purposes of discussion in this book a carnivorous species is one which eats mainly, though not necessarily only, animal material. This animal material may consist of either vertebrate or invertebrate prey, or both. Thus insectivory is not distinguished from carnivory.

Carnivore diets are characterised by a high content of protein, water, vitamins and minerals, a variable amount of fat and a low level of carbohydrate. The high ratio of protein to carbohydrate in the diet means that little hexose is absorbed from the small intestine, and, instead, the animal's glucose and energy requirements are met largely from amino acids. Thus, compared with omnivores, the activity of the gluconeogenic enzymes pyruvate kinase and phospho-enol-pyruvate carboxykinase in the liver of carnivores is always high (Rogers, Morris & Freedland 1977). Transaminase activity is also consistently high and as a result the maintenance protein requirements of strict carnivores are usually higher than those of omnivores, at least among the Eutheria. In the absence of information to the contrary the same is assumed to hold among marsupials. The other characteristic of carnivore diets is the high digestibility of muscle and viscera, but low digestibility of the exoskeletons of insects and other invertebrates, the teeth and some bones of vertebrates, and reptile dermal scales, bird feathers and mammalian hair.

Redford & Dorea (1984) compiled data on the nutritional composition of some terrestrial invertebrates (Table 2.2), and concluded that although most invertebrates were of similar food value, there were two exceptions in which the nutritional quality was substantially higher. The first are the winged ants and termites, which are high in fat; these reproductive castes must fly from the nest and found new colonies, living off their own fat reserves for extended periods while they do so. They are the ant and termite castes eaten by the greatest number of animals, not only because of their higher energy content but because they are more vulnerable to predation away from their nest and soldier defence. The second exception are the larval and pupal forms of insects, which also tend to have higher fat contents, and also they have less chitin and are less well defended than adults of the same species.

Ash contents are generally low, except for earthworms and geophagous termite workers; also, the high ash content of some termite soldiers indicates that inorganic matter ingested by workers may also be passed to soldiers (Redford & Dorea 1984). Crude protein values for invertebrates are problematic, because they assume that all nitrogen present is in the form of amino acids. This is not true, because of the presence of nitrogen in the chitinous exoskeleton. Chitin, which makes up a large part of the exoskeleton of most arthropods, is a structural polysaccharide composed of N-acetyl-D-glucosamine residues linked by β-1,4-glycosidic bonds, which are difficult to hydrolyse. Several eutherian carnivores secrete a chitinase from their gastric mucosa and/or pancreas (Stevens & Hume 1995), but whether this holds for any marsupials is unknown. The amount of nitrogen thus made available from the arthropod exoskeletons also remains to be determined.

The low digestibility of invertebrate exoskeletons and the fibrous proteins of scales, feathers and hair, principally keratins, means that examination of scats (faeces) of carnivores is a good way of working out their dietary habits. The scale arrangement and other features of mammalian

Table 2.2. *Nutritional composition of some terrestrial invertebrates*

Species	Water (% wet weight)	Ash	% of dry matter		
			Total nitrogen	Crude protein[a]	Fat
Annelida					
Oligochaeta					
Earthworms	82–85	9–23	9–11	54–69	4–9
Arthopoda					
Orthoptera	—				
Locusts	57–71	4–9	7–12	43–76	5–50
Crickets	71–76	8–9	9–11	57–67	17–22
Coleoptera					
Mealworm larvae	66	7	9	55	33
Weevil adult	52	—	10	63	5
Lepidoptera					
Waxmoth larvae	56	2	5	31	62
Silkworm larvae	80–82	—	9	58–59	28–34
Diptera					
Housefly pupae	—	5–12	10	61–63	9–16
Isoptera					
Termites – workers	66–80	9–61	3–8	16–49	2–7
– soldiers	70–78	7–38	4–12	25–74	2–14
– winged	34	11	6	39	42
Hymenoptera					
Ants – winged females	60	—	1–3	8–19	24–60
– winged males	60	—	4–10	25–63	3–8
Bee – larvae	77	3	11	67	16
– pupae	70	2	15	92	8

Note: [a] Crude protein assumed to approximate total nitrogen × 6.25.
Source: After Redford & Dorea 1984.

hair are excellent diagnostic tools, and keys for the identification of mammalian hair are available (e.g. Brunner & Coman 1974). The task is complicated though among insectivores by the habit of many small species of eating only the softer parts of insects. For example, planigales (*Planigale* spp.) dextrously use their forepaws to manipulate the hard exoskeleton of the prey so that it can be discarded. The soft parts do not appear in the faeces, leading to underestimation of their dietary contribution. Conversely, any plant structural material eaten is poorly digested in the carnivore gut and its dietary contribution can be easily overestimated from its proportion in the scats. A more accurate way of analysing dietary habits is to examine stomach contents, but this usually means killing the animal.

Read (1987a) acknowledged the shortcomings of diet assessment on the basis of scat rather than stomach content analysis, but defended its use on the basis of its non-destructive nature: First, it allows continual assessment of diet without altering population densities. Second, it has no direct influence on other concurrent studies on small populations. Third, it allows investigation of ontogenetic changes in the feeding preferences of individual animals in the population. Also, it has the advantage that invertebrate fragments are more compact in faeces than in the stomach, which reduces the time the investigator spends searching for recognisable parts (Dickman & Huang 1988).

From their evaluation of a shrew and three insectivorous dasyurid marsupials (*Antechinus stuartii, A. swainsonii* and *Sminthopsis griseoventer*), Dickman & Huang (1988) also concluded that faecal analysis is a relatively reliable method for determining the diet of generalist insectivores that eat hard-bodied prey if the results are expressed as the number of animals in which a prey item is found rather than as the minimum number of prey eaten per animal (i.e. as the per cent frequency of occurrence over all samples rather than the minimum frequency of occurrence). However, the problem of overlooking soft-bodied prey remains. Nevertheless, given the several advantages of scat over stomach analysis, most studies of the dietary habits of mammals are based on analysis of scats rather than of stomach contents.

2.3 DIETS OF CARNIVOROUS MARSUPIALS

The marsupial carnivores belong to one of two groups. The first group consists of the three American families, the Caenolestidae (shrew-opossums, seven species), the monospecific Microbiotheriidae (*Dromiciops australis*) and some members of the Didelphidae (opossums, 70 species). Other didelphids are omnivorous rather than carnivorous, but the distinction in many cases is blurred.

The second group of carnivores consists of the Australasian family Dasyuridae (51 species in Australia; 14 species in New Guinea), the Australian family Notoryctidae, with two species (the northern and southern marsupial moles), and two monospecific Australian families, the Myrmecobiidae (*Myrmecobius fasciatus*, the numbat) and the Thylacinidae (*Thylacinus cynocephalus*, the thylacine or Tasmanian tiger).

2.3.1 South American species

The Caenolestidae are usually described as shrew-like insectivores limited in their distribution to the Andean zone of western South America (Hunsaker 1977). All five species of *Caenolestes* are terrestrial inhabitants of forests. Osgood (1921) found mainly insect and arachnid remains in the stomachs of three *Caenolestes obscurus* that he examined, and Hunsaker (1977) listed caterpillars, beetles, ants, centipedes and also spiders as staple food items of the caenolestids. However, Kirsch & Waller (1979) considered from their observations of the feeding behaviour of a captive male *C. obscurus* when given live rats that this caenolestid may also prey on small vertebrates: 'The animal would move toward a rat, sniffing vigorously, seize and lift the rat with its forepaws or pin it to the substrate, and bite it several times quickly with its incisors. The caenolestid would then commence eating the rat by biting off a section of the head with its cheek teeth and take successive bites posteriorly.' Unfortunately the few caenolestid stomachs they opened in the field were empty, and thus they were not able to confirm their observations on captive *C. obscurus* in the wild. In Peru, Barkley & Wittaker (1984) found that lepidopteran larvae, centipedes and arachnids made up 76% by volume of the diet of *C. fuliginosus*.

Lestoros inca, from southern Peru, feeds on insects and other small invertebrates, as does *Rhyncolestes raphanurus* (Chilean shrew opossum) (26 g body mass) in southern Chile. The tail of *Rhyncolestes* may be seasonally incrassated, suggesting that it stores fat there as insurance against a shortage of invertebrate prey in the cold winters. However, they may consume more non-animal material as well; in one study, the stomachs of 31 *Rhyncolestes* contained 55% by volume of invertebrates (including 8% earthworms), but 40% plant and fungal material as well (Redford & Eisenberg 1992). Both *Lestoros* and *Rhyncolestes* are terrestrial.

The only extant microbiotheriidid, *Dromiciops australis* (Monito del Monte) (22–29 g) is arboreal in its habitat of dense, humid forests in southern Chile. Its diet consists mainly of insects, especially Coleoptera, and other arthropods. In one study, the stomachs of 38 *Dromiciops* contained 59% by volume of adult arthropods (Redford & Eisenberg 1992).

The Caenolestidae possess peculiar lip flaps (Osgood 1921) which have been variously suggested to function to hold, convey inwards or eject food items. Kirsch & Waller (1979) suggested from their observations on a captive *C. obscurus* that they may help the sensory vibrissae and fur at the side of the mouth from becoming clogged with blood and dirt, as might result from this caenolestid's method of feeding, and also prevent dirt from entering the mouth.

The didelphids which have been described as being largely carnivorous include *Philander opossum* (gray four-eyed opossum), *Chironectes minimus* (water opossum or yapock), *Lutreolina crassicaudata* (little water opossum or thick-tailed opossum), *Monodelphis domestica* (gray short-tailed opossum) and *Lestodelphis halli* (Patagonian opossum). Although the *Marmosa* group (the murine or mouse opossums) are described by Streilein (1982) as omnivorous, species inhabiting colder regions are thought to be more carnivorous (Hunsaker 1977). For example, an examination of the stomachs of three *M. cinerea* (76 g body mass), which occurs throughout a broad altitudinal range, showed only insects (Redford & Eisenberg 1992). Meserve (1981) found from analysis of the stomach contents of *Thylamys elegans* (Chilean mouse-opossum) (29 g) that 90% of its annual diet was insects. *Philander opossum* (330–440 g body mass) feeds on small mammals, frogs, birds, bird eggs, carrion and some fruit (Fleming 1972); it is primarily terrestrial, but is often found near water. In French Guinea the stomach contents of four individuals contained 85% animal matter and 15% fruit and seeds (Eisenberg 1989). Fruit is probably an important source of water for *P. frenata* in dry months in Brazil (Santori *et al.* 1996). *Chironectes minimus* (390– 90 g) (Fig. 2.3) is almost entirely aquatic,has webbed hind feet, and feeds on insects, crustaceans, fish and frogs, but may also eat some aquatic vegetation, and fruit (Hunsaker 1977). Prey are captured with either the front feet or the mouth. *Lutreolina crassicaudata* (430 – 640 g) forages terrestrially on small mammals, reptiles, fish, molluscs and insects (Redford & Eisenberg 1992). *Monodelphis domestica* (70 g) is an accomplished predator, feeding primarily on invertebrates, but it also takes small rodents, carrion and even fruits and seeds. In captivity it avidly consumed

Figure 2.3 *Chironectes minimus*, the water opossum or yapock, from South America, with webbed feet is the most aquatic of all marsupials. (Diego Astúa de Moraes)

living and dead rodents, frogs, snakes and lizards, a variety of insects and other invertebrates, and most fruits, leading Streilein (1982) to suggest that some degree of opportunistic feeding (omnivory) may be found to be a general trait among the Didelphidae. This suggestion serves to highlight the problem of defining limits to carnivory, particularly when so little is known about the food habits and general ecology of the animals, as is the case with many South American marsupial species. Busch & Kravetz (1991) analysed the stomach contents of 23 *Monodelphis dimidiata* (eastern short-tailed opossum) from suburban Buenos Aires, and found insects to be the most frequent item, with smaller amounts of arachnids, mammals and plants. When five captive *M. dimidiata* were offered a variety of foods (insects, snails and slugs, isopods, earthworms, fruit, meat and live and dead rodents), the animals ate insects, snails and slugs, isopods and earthworms, with the larger animals tending to take the larger prey. The authors concluded that *M. dimidiata* was a food generalist, but primarily insectivorous; when insects are scarce in autumn and winter they probably switch to eating small rodents that are at their highest abundance at this time of the year.

In *Lestodelphis halli* (Patagonian opossum) (76 g) the canines are exceptionally long, suggesting a carnivorous habit. Captive animals kill mice at lightning speed, eating everything (bones, teeth and fur). Redford & Eisenberg (1992) suggested that it may forage under the snow in winter for small rodents and other animals.

2.3.2 Australian species

MARSUPIAL MOLES, NUMBAT AND THYLACINE

The diets of the two species of marsupial moles (*Notoryctes*) and the thylacine have been little studied. Stirling (1891) remarked that ants were conspicuous in the stomach contents of four of the earliest recorded specimens of *Notoryctes*, consistent with their typically insectivorous dentition. The contents of the digestive tracts of ten museum specimens of *Notoryctes* were examined by Winkel & Humphery-Smith (1988). One was empty. In the other nine, predatory and seed-eating ants were both well represented, as were termites. Grass seed material was present in eight specimens. The finding of large numbers of ant eggs in several specimens suggests that the moles were feeding within ant nests. The high energetic cost of burrowing in desert sands means that feeding activity should be concentrated in areas where food is relatively abundant, such as at ant and termite nests. The presence of seed material in the digestive tracts can also be explained by feeding associated with nests of seed-harvesting ants. Weevil larvae, beetles and spiders were also present in several specimens. The weevil larvae could well have been associated with *Acacia* roots or with the seeds. Parasitic nematodes were recorded in three specimens. One of them was *Nicollina peregrina*. Apart from marsupial moles, *Nicollina* is restricted to the short-beaked echidna (Beveridge & Spratt 1996).

The ecology and feeding habits of the numbat have been reported by Calaby (1960). Its preferred habitat in south-west Western Australia is eucalypt woodland in which the heartwood of most trees has been eaten out by termites and the woodland floor is strewn with fallen hollow limbs and logs. The numbat is one of only two diurnal marsupials (the other being the musky rat-kangaroo (*Hypsiprymnodon moschatus*) in the tropical rain forests of north Queensland). It is also generally solitary. From scat analysis Calaby (1960) determined that the numbat's diet consisted mainly of termites scratched from the upper 5 cm of soil. All species of termites were eaten, roughly in proportion to their abundance. Ants made up as much as 15% of the diet, but the bulk of the ants eaten were the small predatory species which are probably ingested incidentally when they swarm in to prey on the exposed termites; numbats apparently did not deliberately search for ants. Other invertebrates were almost completely absent from Numbat scats.

Friend & Whitford (1993) calculated that workers of the termite *Nasutitermes exitiosus* contained 74% water. The dry matter contained 53% total crude protein, 34% carbohydrate, 8% fat and 5% ash. Not all the crude protein is available however, because a significant amount of nitrogen is bound up in indigestible chitin in the exoskeleton; usable crude protein was estimated by enzymatic assay to be 37% of dry weight. The milk produced by free-living numbats late in lactation contains 2% hexose, 14% protein and 12% fat, which is quite similar to late lactation milk produced by tammar wallabies (*Macropus eugenii*). In numbat milk nearly all the fat is triglyceride. Most notable is the high level (60%) of oleic acid (C18:1) in the triglyceride (Griffiths *et al.* 1988).

The limited dietary breadth of the numbat may explain in part the low diversity of the helminth fauna (nematodes, cestodes, digeneans and acan-thocephalans) found in this myrmecophage (ant eater). In contrast, the Dasyuridae harbour the widest range of helminth families of any group of marsupials (Beveridge & Spratt 1996).

The natural food of the thylacine consisted of wallabies and other smaller marsupials, rats, birds, and possibly lizards and echidnas (Troughton 1965). Its alleged attacks on domestic sheep and poultry resulted in what appears to have been a successful extermination of the species. The thylacine had the longest snout of all mammalian carnivores, dasyuroid or canid alike (Werdelin 1986). A long snout means a weak bite at the canines. From the relatively low fracture rate of its canines, Jones & Stoddart (1998) concluded that the thylacine was probably a slow-running predator that took small (1–5 kg) prey relative to its body size (15–30 kg). It would have killed its prey with a crushing bite. Its canines were larger than those of similarly sized eutherian carnivores, which would have increased the efficiency of the thylacine as a predator to some extent by partially compensating for its weak bite.

The large canines of the thylacine resulted from overeruption (Jones 1995). Unlike eutherians, marsupial carnivores have only one set of teeth, which erupts into a tiny juvenile jaw but which must also serve the individual as an adult. Teeth cannot grow with body size once the roots have closed, but the phenomenon of overeruption (when enamel encases only the top of the tooth, leaving the underlying dentine near the gumline exposed and able to expand) results in an increase in canine height and diameter with increasing age and body size even though the root has closed. The canines of all carnivores, both eutherian and marsupial, over-erupt to some extent, but this occurred to the greatest extent in the thylacine.

SMALL DASYURIDS

Dasyurids range in body size from the tiny shrew-sized planigales (5–12 g) to the 7–9 kg Tasmanian devil (*Sarcophilus harrisii*). There is a strong positive correlation between the body size of dasyurids and that of their invertebrate prey (Fisher & Dickman 1993). For instance, Woolnough & Carthew (1996) found that the 7 g *Ningaui yvonneae* (southern ningaui) consistently selected small prey items over large, and that handling time increased exponentially with prey size. Thus smaller members of the Dasyuridae are almost wholly insectivorous, while the larger members take vertebrate prey as well.

Several studies of sympatric dasyurid species have been published. The diets of two planigales, *Planigale gilesi* (Giles' planigale) at 7–12 g body mass, and *P. tenuirostris* (narrow-nosed planigale) at 5–7 g, were compared by Read (1987a) at Fowler's Gap, western New South Wales. Scat analysis over a two-year period showed that in all seasons both species were generalists with respect to both prey type and size; prey items in the diet occurred in the same proportions as those available. However, body size was a factor in separating the diets of the two species. All animals took prey in the small size classes, but only the larger animals of both species took the larger prey. Because *P. gilesi* tended to be larger, most of the largest prey were taken by this species. Coleopterans (beetles), arachnids (spiders) and isopods (slaters) were the most important prey taxons in most seasons. Vertebrates were only a small component of the diet of *P. gilesi* and especially of the smaller species *P. tenuirostris*. Plant matter was not found in any sample.

Morton, Denny & Read (1983) compared the diets of two desert-dwelling members of the genus *Sminthopsis*, *S. crassicaudata* (fat-tailed dunnart) (Fig. 2.4) at 12 g and *S. macroura* (stripe-faced dunnart) at 16 g body mass living sympatrically at two sites, one in western Queensland and one in north-western New South Wales. At both sites both species subsisted almost entirely on invertebrates, but *S. crassicaudata* took more orthopterans (grass-hoppers) and *S. macroura* took more isopterans (termites).

Figure 2.4 *Sminthopsis crassicaudata*, the fat-tailed dunnart, a small carnivorous/insectivorous marsupial which relies on caudal fat stores as one means of maintaining energy balance in winter. (David Walsh)

There were no differences in the size of prey taken by the two species, probably because the difference in body size between the two species is so slight.

Hall (1980) used both stomach content and scat analysis in his study of the diets of two sympatric species of *Antechinus* in forest habitat in southern Victoria. *A. agilis* (agile antechinus) (formerly *A. stuartii*) at 20–35 g is smaller than *A. swainsonii* (dusky antechinus) at 40–65 g. *A. agilis* took more, and a wider size range of, weevils (Curculionidae) than did *A. swainsonii*, but the larger *A. swainsonii* took prey that were on average 23% longer and 75% heavier than did *A. agilis*. Overall, though, it seemed that both species were generalists and opportunistic feeders. In alpine environments, the diet of *A. swainsonii* includes approximately 68% adult invertebrates, 10–21% insect larvae and worms, and 5–13% plant material (Dickman *et al*. 1983; Mansergh *et al*. 1990). *Phascogale tapoatafa* (brush-tailed phascogales) at 110–270 g body mass were found to be primarily arboreal insectivores by Traill & Coates (1993) in open sclerophyll forest in northern Victoria. Beetles, spiders and ants made up over 75% of all items in four stomach contents and 12 scat samples, and animals were observed foraging on the trunks and major branches of eucalypt trees. Grass seeds and anthers were presumably taken incidentally while foraging in the herb layer. Only one vertebrate prey item, a house mouse, was detected in 200 scats examined, but the authors thought that nectar may be seasonally important for phascogales. Eucalypt nectar was also an important energy source for the sympatric dasyurid *Antechinus flavipes* (yellow-footed antechinus) (Traill & Coates 1993).

Although the number of species studied is a small proportion of the total number of small dasyurids, it can be predicted that arthropods will be the major diet component throughout the group. At the upper end of the body size range more vertebrate material may be taken. Thus mulgaras (*Dasycercus cristicauda*) at 100–170 g body mass have been found to take rodents at over one-third of the diet by frequency at certain times of the year, and will eat other dasyurids, reptiles, birds and frogs as well (Chen, Dickman & Thompson 1998).

LARGE DASYURIDS

Larger dasyurids include even more vertebrate material in their diet. The five extant species of large dasyurids (four *Dasyurus* spp. (quolls) and the Tasmanian devil (*Sarcophilus harrisii*)), together with the thylacine, are defined as a guild on the basis of a similar diet of primarily vertebrate prey (Jones 1997). Based on the ratio of prey to predator size, Jones & Barmuta (1998) grouped Tasmanian species into two dietary categories. The first group consists of devils at 7–9 kg and male spotted-tailed quolls (*Dasyurus maculatus*) at up to 7 kg; they take relatively large prey species. At Cradle Mountain National Park, devils fed primarily on large mammals such as wallabies and wombats, and secondarily on medium-sized mammals (Jones & Barmuta 1998). Male spotted-tailed quolls took mainly medium-sized mammals, followed by large mammals.

The second group consists of eastern quolls (*D. viverrinus*) at 1–2 kg and female spotted-tailed quolls at up to 4 kg; they take relatively small prey species. At Cradle Mountain, female spotted-tailed quolls took mainly birds, followed by small mammals. Eastern quolls fed on a wide range of prey; medium-sized and small mammals, birds, reptiles and invertebrates were eaten in roughly equal proportions by mass.

Thus spotted-tailed quolls experience the greatest dietary overlap; males overlap with adult devils in winter and with subadult devils in summer; more limited data suggest a similar pattern of overlap between female spotted-tailed quolls and eastern quolls. Devils experience the least dietary overlap. The high levels of competition experienced by spotted-tailed quolls may explain their low density at Cradle Mountain; devils were two and a half times more abundant than eastern quolls, and five times more abundant than spotted-tailed quolls (Jones & Barmuta 1998).

Jones and Barmuta's (1998) findings from Cradle Mountain in Tasmania on the prey sizes taken by the large dasyurids are in general accord with those of other workers on devils and quolls. For instance, Guiler (1970) found that devils in a natural area depended almost exclusively on wallabies. Two other populations in farming areas made use of a wider food spectrum, including introduced as well as native species, depending on availability. Native species found in stomach contents included birds, mainly in younger animals; devils, especially young animals, can climb trees. Introduced species eaten included rabbits and sheep. Indeed, devils have a poor reputation among farmers for killing domestic stock. However, the presence of blowfly larvae (maggots) in some samples indicates that they were eaten as carrion (Green 1967), and Buchmann & Guiler (1977) remarked on 'a general ineptitude among devils for killing'. Injured or moribund sheep may well be taken, and there is no doubt that devils are able to take quick and opportunistic advantage of natural mortality, and of weak lambs and offal.

The diet of the spotted-tailed quoll (or tiger quoll) was investigated by Belcher (1995) in East Gippsland, Victoria. Scats were collected from two latrines over a two-year period. From the analysis of these scats, *D. maculatus* was found to be a predator of vertebrates, largely medium-sized mammals (500 g to 5 kg). The most important prey species were the European rabbit, the common brushtail possum (*Trichosurus vulpecula*) and the common ringtail possum (*Pseudocheirus peregrinus*). Other prey included *Antechinus* species, bush rats, echidnas, macropod marsupials, wombats, birds, reptiles and invertebrates. Subadult quolls consumed significantly more small mammals, reptiles and invertebrates and fewer rabbits than did adult quolls, consistent with the notion that larger predators take larger prey.

Eastern quolls feed mainly on insects, lizards and small birds and mammals, including mice, rats and young rabbits (Troughton 1965). The importance of insects in the diet of *D. viverrinus* was confirmed by Blackhall (1980) in southern Tasmania, but plant material, mainly grasses and herbs, consistently occupied the greatest proportion of the faecal samples examined (28–41%). Other components included fruit, mainly introduced blackberries (*Rubus fruticosus*), and hair from house mice and eastern quolls. Blackhall (1980) thought that the quoll hair probably came mainly from loose hair shed or rubbed off in the traps used for capturing the animals. Little avian material was detected in the scats, indicating that *D. viverrinus* feeds almost exclusively on the ground.

The western quoll *(D. geoffroii)* also forages primarily on the ground and at night, but also occasionally climbs small trees. In desert habitats its diet includes mammals up to the size of rabbits, as well as lizards, frogs and invertebrates. In forested habitats the diet is broader, and includes insects such as winged termites, cockroaches and beetles, freshwater crustaceans, terrestrial reptiles, mammals up to the size of bandicoots and birds up to the size of parrots (Soderquist & Serena 1994). The shortcomings of scat analysis are again highlighted by the observation of Archer (1974) that captive *D. geoffroii* and *Phascogale calura* (red-tailed phascogale) regurgitated food, particularly after rapid ingestion of a large meal. Soft material was reswallowed, but *P. calura* removed hard pieces of invertebrate cuticle from regurgitated material in its mouth with its forepaws, in the same manner frequently observed during feeding by planigales.

MARSUPIAL LIONS

One of the largest and most widely distributed groups of marsupial carnivores were the marsupial lions (*Thylacoleo* spp.), which date from the Pliocene (Archer & Dawson 1982). The dietary habits of the Pleistocene species *Thylacoleo carnifex* have attracted much speculation because of its unique dentition (Fig 2.5) and the fact that it has no living relatives. In this case diet must be inferred entirely from the dental and masticatory systems of fossils, on the assumption that there are close relationships among diet, feeding structures and digestive tract morphology. This assumption should be reasonably robust if the link between an animal's foraging and digestive strategies is as close as was suggested in Section 2.1.1. There is no doubt that diet, dentition and gut morphology are usually tightly linked, but there are exceptions. In some cases dietary differences are not reflected in dental patterns, because numerous modifications of the digestive tract compensate, as in the Ruminantia (Hofmann 1973). In other cases dietary specialisation is reflected in neither dentition nor the digestive tract, as in the Gelada baboon (*Theropithecus*) in which a highly skilled and selective feeding behaviour compensates.

There is no evidence from extant species that compensations among dentition, gut morphology and behaviour are predictable (Sanson 1991). Thus when the dental and masticatory systems of fossils are used to speculate on the diet of extinct species, there is always room for debate. A good case in point is *Thylacoleo carnifex*, which is variously described as having been carnivorous, omnivorous and herbivorous. The problem is that the thylacoleonids are of herbivorous phalangeroid stock yet their dentition has few if any features of extant herbivores. The upper and lower incisors meet at their tips. There is only one (upper) canine. The first two upper and lower premolars are vestigial. The third premolars are huge sectorial (cutting) blades. The first lower molar contributes to the sectorial blade while the first upper molar and the second molars are very reduced. The lower incisors are often described as caniniform, the premolars as carnassials, and the reduced molars as having similarities in form to those of felids (Sanson 1991). There is no modern equivalent of this dental pattern.

From a study of tooth wear patterns and a reconstruction of jaw mechanics, Wells, Horton & Rogers (1982) concluded that *Thylacoleo* was a

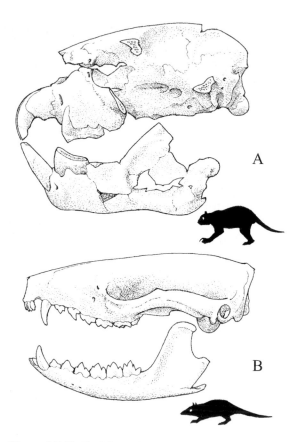

Figure 2.5 Skull of the extinct *Thylacoleo carnifex* (A) showing the greatly modified dentition relative to that of *Dasyurus* (B). From Woods (1956) and Dawson *et al.* (1989).

specialised flesh eater and probably consumed little bone. The caniniform lower incisors may have been used for stabbing prey, and the sectorial premolars for shearing muscle, tendons and skin (Sanson 1991). A lack of blunt cusps anywhere and a lack of well-developed anterior premolars argue against an ability to crack bones. As its limb adaptations were for climbing and grasping rather than running, Wells *et al.* (1982) thought that *Thylacoleo* was probably a leopard-like predator that partly avoided competition with the thylacine and with scavenging forms such as *Sarcophilus* by hauling its prey into trees. Its large body size also allowed it to select larger prey, which also would have minimised competition (Case 1985).

POSSUMS

In addition to the dasyurids, several small extant Australasian possums have been described as having a largely insectivorous diet (e.g. *Dactylopsila trivirgata*, the striped possum of Cape York and Papua New Guinea (Smith 1982a); *Burramys parvus*, the mountain pygmy-possum of the Australian Alps (Mansergh *et al.* 1990); *Cercatetus caudatus*, the long-tailed pygmy-possum of high altitude rainforests of North Queensland and New Guinea (Smith 1986)). However, they are treated in Chapter 3 along with other small possums. Their distinguishing feature could simply be that they use insects rather than pollen as a source of protein.

2.4 **DENTITION**

Despite the assertion that within the smaller dasyurids there is little or no feeding specialisation (Fox 1982), analogies with herbivores suggest that observed subtle differences in dasyurid dentitions may reflect differences in diet selection. This is because of the effort required to breach the barrier imposed by arthropod exoskeletons of different thickness and toughness. Small insectivores tend to chop food finely by many small cutting edges, but such a mechanism may not be appropriate when dealing with larger prey; the cuticle of some larger insects may be too hard for small insectivores to pierce. This is particularly so for dasyurids eating desert beetles; the cuticle of some beetles is too hard for most dasyurids to pierce. Thus at least some dasyurids cannot afford to be dietary opportunists, but have to be selective in their choice of prey (Sanson 1985).

The upper molars of dasyurids are roughly triangular, resulting in a V-shaped embrasure formed between each upper and lower molar. There are three main cusps on the upper molar (the protocone, metacone and paracone), and two triangular juxtaposed basins on the lower molar (the trigonid and talonid basins) (Archer 1976). Occlusion involves two distinct phases. The first is a puncture-crushing phase in which there is tooth-food-tooth contact as food is caught and pulverised between the protocone of the upper molar and the talonid basin. The second phase is a shearing phase, in which the anterior-most crests of the lower molars shear past the posterior crests of the preceding upper molars; food is also ground during this phase (Moore & Sanson 1995).

Comparison of the dental morphology and action of the kowari, *Dasycercus* (*Dasyuroides*) *byrnei*, with that of an omnivore of similar size, the sugar glider (*Petaurus breviceps*), indicated that the more limited shearing of the sugar glider's dental action means that they can only compress insects and not break them down into small pieces as kowaris are able to do. Sugar gliders can extract the haemolymph and soft tissues of the larvae by their compression action, but they discard the exoskeleton, and about 33% of the total larval nitrogen along with it. When mealworm (*Tenebrio*) larvae were cut into pieces of varying sizes, the extent to which trypsin removed nitrogen from the larvae *in vitro* increased from near zero for whole larvae to 70–85% when larvae of the same size were cut into eight pieces, sizes which match the exoskeleton particle size found in kowari faeces (Moore & Sanson 1995). Thus the finer the cuticle can be comminuted, the more nitrogen can be extracted from insect prey.

Although this comparison across feeding strategies is rather extreme, it serves to illustrate the point that subtle differences in dentition among the Dasyuridae may be reflected in dietary differences that have not been identified so far, and may not be in any future studies that are based only on scat analysis.

2.5 DIGESTIVE TRACT MORPHOLOGY

Carnivores are universally distinguished from nearly all omnivores and herbivores not only by their dentition but also by the morphology of their gastrointestinal tract. Correlated with the generally highly digestible nature of the food swallowed, the overall impression of the carnivore digestive system is one of simplicity. The stomach is usually simple, without diverticula or development of a forestomach fermentation region, although it may be quite capacious in those species that consume large prey. Crisp (1855) noted that the stomach of *Thylacinus* was very muscular and capable of considerable distension. The small intestine is short, as is the large intestine. The colon is usually of similar diameter to that of the small intestine, and may or may not be distinguished from the small intestine by the presence of a caecum.

Osgood's (1921) monographic study of *Caenolestes obscurus* includes a detailed description of the alimentary canal. This has been supplemented by a more recent microscopic examination of the caenolestid stomach by Richardson, Bowden & Myers (1987). The stomach (Fig 2.6) is roughly elliptical in shape, and can be divided into oesophageal, cardiac, proper gastric and pyloric regions that occupy 1%, 1%, 87% and 11% of the gastric mucosa, respectively (Richardson *et al.* 1987). In the oesophageal region the mucosa is organised into thin lamellae surrounding the opening of the oesophagus (cardia) and extending along the lesser curvature to a point almost midway between the oesophagus and pylorus. It has a non-glandular, stratified squamous epithelium of about nine cell layers and with a total thickness of about 100 μm. The basal cells are cuboidal, while those at the luminal surface are flattened. The cardiac gland region forms a narrow diffuse ring around most of the oesophageal opening. It consists of mucus-secreting branched tubular glands up to 350 μm deep.

The proper gastric gland region consists of two parts, a simple glandular mucosa about 550 μm thick that occupies most of the region, and a compound cardiogastric gland of similar mucosa drawn into complex folds (Fig. 2.7a). The gland is about 1400 μm thick, is bilobed and surrounds the cardia on the lesser curvature of the stomach. From the exterior it is clearly evident through the thin muscular stomach wall. There are 40–60 slit-like (Osgood 1921) or punctate (Richardson *et al.* 1987) openings on the mucus-secreting columnar epithelium of the cardiogastric gland surface. The unbranched gastric glands opening into the orifices are arranged into

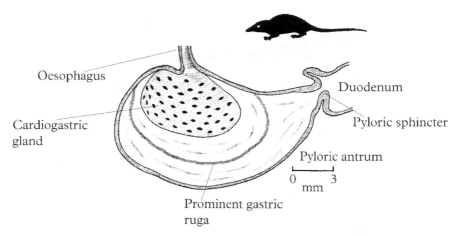

Figure 2.6 Schematic representation of the mucosal surface of the stomach of *Caenolestes obscurus*. From Richardson, Bowden & Myers (1987).

basal, central and apical zones. The basal zone contains mainly pepsinogen-secreting chief cells, the central zone predominantly acid-secreting parietal cells, and the apical zone a mixture of parietal and mucous neck cells.

The pyloric region is about 300 μm thick, with glandular epithelial cells arranged around invaginations which create gastric pits that extend deep into the lamina propria towards the muscularis mucosae (Fig. 2.7b). Both chief and parietal cells are absent, the only secretion being of mucus.

Aggregations of the mucosa of the proper gastric gland region into a cardiogastric gland separates *Caenolestes* from all other carnivorous marsupials that have been examined, both American and Australian, but similar structures are found in wombats and the koala, as well as in several disparate eutherians including the beaver, dugong, pangolins and grasshopper mouse (*Onychomys leucogaster*). Its function is unknown, but Richardson *et al.* (1987) have suggested that at least in *Caenolestes* its role may be to supply large quantities of pepsin, hydrochloric acid and mucus to cope with the more-or-less constant intake of high protein food by this small carnivorous marsupial. Why other small carnivores don't have a similar structure is unexplained.

The small intestine of *Caenolestes*, although short, dominates the digestive tract, being 87% of total gastrointestinal length (Richardson *et al.* 1987). This is consistent with the idea that high-quality diets are best processed in a plug-flow reactor-type organ such as the small intestine. The duodenum appeared to Osgood (1921) to have slightly thicker walls than the distal small intestine. The large intestine is very short, the colon being only 7% of gastrointestinal length. The caecum is small, 'scarcely more than a vermiform appendix' (Osgood 1921), and only 3–5 mm in length. Its mucosal lining resembles that of the ileum in being villous, in contrast to that of the colon which is organised into large longitudinal folds. Hill &

Figure 2.7 Sections through the stomach of *Caenolestes obscurus*. Magnification 80:1.
(a) Photomicrograph of a longitudinal section through the cardiogastric gland and adjacent gastric mucosa. A. cardiogastric gland orifice; B. cardiogastric gland invagination; C. proper gastric gland region. (b) Photomicrograph of an oblique section of the pyloric gland region. A. longitudinal muscle; B. circular muscle; C. muscularis mucosae; D. gastric pits; E. mucus. From Richardson, Bowden & Myers (1987).

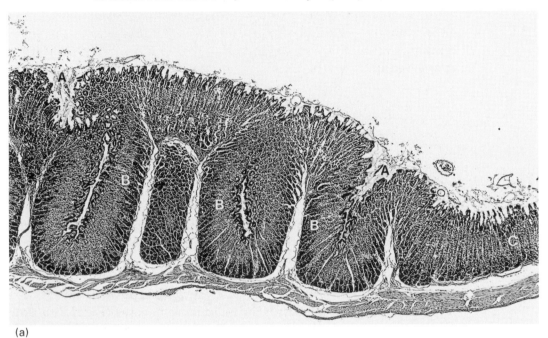

(a)

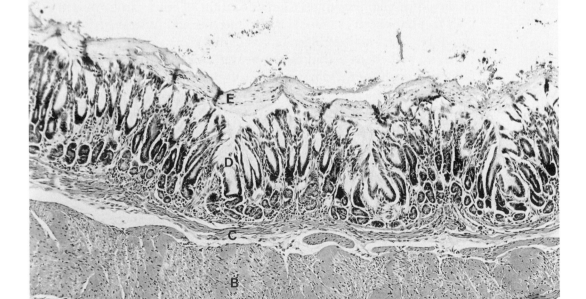

(b)

Rewell (1954) remarked on the cranially directed caecum of *Caenolestes*, when the caecum of all other American marsupials is directed to the right and caudally.

Schultz (1976) has provided drawings of the digestive tracts of three other American carnivores, *Marmosa* sp., *Chironectes minimus* and *Philander* (*Metachirops*) *opossum*. These are shown in Fig 2.8. These three didelphids have in common a simple stomach without any development of a specialised cardiogastric gland, a short small intestine, an even shorter colon, and a small though distinct caecum. Barnes (1977) has described the digestive tract anatomy of *Marmosa robinsoni*. The stomach is simple with glandular mucosa throughout, and there is minimal differentiation of cardiac and fundic areas. The pyloric region secretes only mucus, and the pylorus is clearly defined externally by the presence of a white collar of duodenal or Brunner's glands.

Brunner's glands are peculiar to mammals, but within the Mammalia they have been found in all monotremes, marsupials and eutherians (Krause 1972). They secrete an alkaline fluid containing mucin which protects the proximal duodenal mucosa from the mechanical trauma of digesta moving through the intestinal tract and from the ulcerating effects of acid-pepsin secretions of the stomach. In eutherians, they extend a short distance along the duodenum of carnivores, a longer distance in most omnivores, and the greatest distance in herbivores (Stevens & Hume 1995). (The monotremes differ from the other mammals in that their gastric epithelium is non-glandular throughout (Griffiths 1978) and Brun-

Figure 2.8 Gastrointestinal tracts of (a) *Marmosa* sp. (a mouse opossum), (b) *Philander* (*Metachirops*) *opossum* (gray four-eyed opossum), and (c) *Chironectes minimus* (water opossum or yapock). A small caecum is present in each of these South American carnivorous species. Redrawn from Schultz (1976).

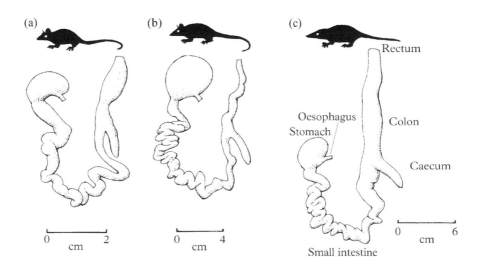

ner's glands are confined to the submucosa of the distal stomach. Thus there is no peptic digestion in the monotreme stomach, nor is there a distinct pylorus, although there is an abrupt change from squamous epithelium to glandular mucosa at the proximal end of the duodenum.) In eutherians, Brunner's glands are generally confined to the submucosa of the duodenum and their ducts empty into intestinal glands (crypts of Lieberkühn) (Krause 1972). In American marsupials, Brunner's glands drain into large mucosal depressions or stomata which the surrounding Brunner's glands encompass. In Australian marsupials the ducts empty directly into the duodenal lumen. The position and size of Brunner's glands in *Dasyurus hallucatus* (northern quoll) are shown in Fig. 2.9.

In *Marmosa robinsoni* the short, uncomplicated small intestine has two histologically distinct regions, the duodenum and the ileum. The duodenum has a large calibre and long, finger-like villi without duodenal crypts. The ileum is of slightly smaller calibre but has a larger lumen owing to its shorter villi. Glandular crypts are short and mucin-secreting goblet cells gradually decrease in number toward the ileocaecal valve. In contrast to *Caenolestes*, the short caecum of *Marmosa* is histologically similar to the colon in that villi are completely absent. Crypts lined with abundant goblet cells form the mucosa. The colon lacks haustra (non-permanent sacculations) or other special features.

Thus *Marmosa*, and presumably *Chironectes* and *Philander* also, differ from *Caenolestes* in both the absence in the stomach of a cardiogastric gland and the presence in the hindgut of a more than vestigial caecum. Osgood (1921) regarded the short colon of *Caenolestes* as a primitive condition, and the tiny caecum a secondary one. The only American marsupial reported to

Figure 2.9 Stomach and duodenum of *Dasyurus hallucatus* (northern quoll) to show the position and size of the Brunner's glands. After Oppel (1896).

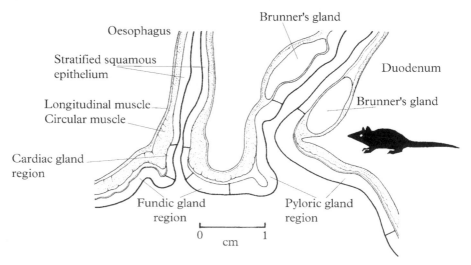

have no caecum is *Dromiciops* (Barbour 1977). This is consistent with the inclusion of *Dromiciops* in the cohort Australidelphia (see Chapter 9), as no Australian marsupial carnivore has a caecum either. The Australian carnivores *Notoryctes*, *Myrmecobius*, *Sminthopsis* and *Antechinomys laniger* are shown in Fig 2.10, and *Dasyurus maculatus*, *Dasycercus* (*Dasyuroides*) *byrnei* and *Phascogale tapoatafa* in Fig. 2.11 in order to illustrate the remarkable uniformity in their gastrointestinal tracts. Except for the absence of a caecum, they are similar to those of the didelphids just described, and to those of most eutherian carnivores (Stevens & Hume 1995).

Owen (1868), one of the early European anatomists interested in comparative aspects, considered the stomach of carnivorous marsupials to be 'relatively more capacious' and 'better adapted for the retention of food' than in the Carnivora, but the detailed information necessary to test this statement is lacking. The Brunner's glands of the Dasyuridae form a thick collar in the duodenum immediately distal to the pylorus (Krause 1972). The glands are composed of numerous lobules aggregated to form large lobes which drain through a complex duct system directly into the duodenal lumen. In the numbat and the marsupial mole the Brunner's glands are less extensive and the lobes are smaller than in the Dasyuridae.

The lack of a caecum in the Dasyuridae, Myrmecobiidae, Thylacinidae and Notoryctidae was considered by Mitchell (1905) to be an example of secondary loss of a primitive organ, in the same way that Osgood (1921) regarded the very small caecum of *Caenolestes* to be a secondary condition. Hume & Warner (1980) argued that the first mammals in the Jurassic, which are thought to have been all small, nocturnal insectivores, most probably had a caecum, derived from reptilian ancestors. Secondary loss of

Figure 2.10 Gastrointestinal tracts of (a) *Notoryctes typhlops* (southern marsupial mole), (b) *Myrmecobius fasciatus* (numbat), (c) *Sminthopsis* sp. (a dunnart), and (d) *Antechinomys laniger* (kultarr). All Australian carnivorous marsupials lack a caecum. Redrawn from Schultz (1976).

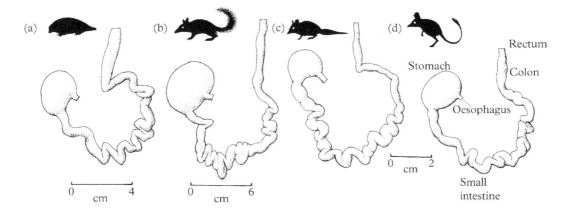

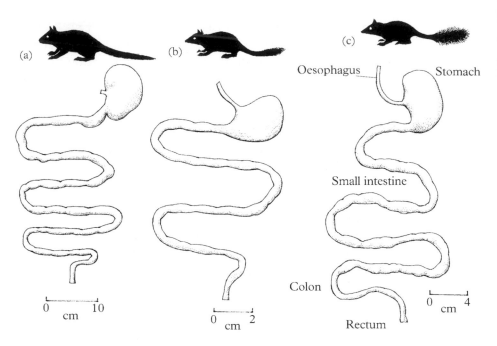

Figure 2.11 Gastrointestinal tracts of (a) *Dasyurus maculatus* (spotted-tailed quoll), (b) *Dasycercus* (*Dasyuroides*) *byrnei* (kowari) and (c) *Phascogale tapoatafa* (brush-tailed phascogale).

this primitive organ appears to have occurred not only in some carnivorous marsupials but also in some eutherian carnivores such as the mink (*Mustela vison*).

2.6 DIGESTIVE FUNCTION

Because of their short, uncomplicated digestive tract, the rate of passage of food residues through eutherian carnivores is rapid. A transit time (i.e. time of first appearance in the faeces) of only 30 min was recorded for various prey items in the shrew *Sorex araneus* by Pernetta (1976). The prey items had been completely excreted within 3 hours (100% excretion time). Read (1987b) used the exoskeletons of insects as indigestible markers in two species of planigales, *Planigale gilesi* (Giles' planigale) and *P. tenuirostris* (narrow-nosed planigale), maintained on minced meat. The insects were eaten within 5–10 min of being offered, and so were a pulse dose. Transit times were only 30–60 min after dosing, and 100% excretion times were 3–4 hours. These are similar to the similarly small eutherian shrews. T. J. Dawson & A. C. Paizs (pers. comm.) measured rate of passage of stained mealworms in 18 g *Sminthopsis crassicaudata* (fat-tailed dunnart) and 140 g *Dasycercus* (*Dasyuroides*) *byrnei* (kowari) maintained on a mealworm diet. Mean retention time (the average time for the marked mealworms to pass through the entire tract) was significantly longer in the larger species (87 min versus 53 min in *S. crassicaudata*). In *D. byrnei* held at a lower ambient temperature (21 °C instead of 32 °C) mean retention time increased to 150 min, despite significantly greater food intakes, suggesting that gut volume increased to accommodate the 29% extra food ingested (Karasov & Hume 1997).

Excretion times generally increase with increasing body size. Thus in the 1 kg hedgehog and the 5 kg raccoon 95% excretion times for solute and particulate markers were close to 12 and 24 hours respectively (Stevens & Hume 1995).

Moyle (1999) used pulse doses of two indigestible markers in a study with eastern quolls (*Dasyurus viverrinus*) fed a basal diet of commercial small carnivore mix with either mealworms or ground lucerne hay added. Cobalt-EDTA marked the solute phase of digesta; this substance is completely soluble in water, associates only with fluid, is non-toxic, and is not absorbed by the gut to any appreciable degree. The particulate phase was marked with plant fibre (cell walls) isolated from ground oaten hay and mordanted with chromium. The mordanting procedure renders the particles indigestible, and so particles of specific size can be tracked. In this case the mordanted particles were passed through a nest of screens of decreasing apertures, and only those that passed through the 1.0 mm screen but were retained on the 0.5 mm screen were used. These marked particles were assumed to resemble fragments of insect cuticle or plant fibre.

The two markers (0.07 g Co-EDTA, 0.12 g Cr-cell walls) were given as a pulse dose at sunset by mixing them into a small amount of the usual food. As soon as this was eaten the animals received their unmarked food. Their cages were checked at regular intervals over the next 90 hours, and any faeces present were collected, dried, acid digested to remove all organic matter, and analysed for Co and Cr content by atomic absorption spectroscopy (Caton *et al.* 1996). Mean retention time was similar for the two markers, suggesting no selective retention of either phase of digesta in the quoll's simple digestive tract. In summer, mean retention times averaged 13.0 hours on the mealworm diet and 17.1 hours on the plant diet. In winter, low ambient temperatures evidently increased gut motility without increasing gut volume, based on shorter mean retention times (7.3 hours and 5.9 hours on the insect and plant diets respectively). As might be expected, rate of digesta passage in *D. viverrinus* is slower than in the smaller planigales.

When fed exclusively on ground house mice, apparent digestibility (apparent assimilation efficiency) of dry matter in *A. swainsonii* averaged 80%, and of energy 87% (Cowan *et al.* 1974), illustrating the ease with which carnivore diets can be digested. Green & Eberhard (1979) fed Tasmanian devils and eastern quolls on whole dead laboratory rats and recorded apparent dry matter digestibilities of 79% and 81% in the two species

Table 2.3. *Intestinal disaccharidase activity in marsupials*

Family and species	Disaccharidase activity (units[a] g^{-1} wet weight of mucosa)					
	Maltase	Isomaltase	Sucrase	Lactase	Trehalase	Cellobiase
Family Dasyuridae						
Antechinus stuartii	43	—	10	0.1	8	0.1
Dasyurus maculatus	69	38	5	1.3	24	0.1
Family Peramelidae						
Perameles nasuta	19	12	5	0.7	11	0.3
Isoodon obeselus	—	—	0	—	4	0.2
Family Phalangeridae						
Trichosurus vulpecula	41	23	7	0.5	7	0.2
Family Pseudocheiridae						
Pseudocheirus peregrinus[b]	0.2	0.1	0.1	11.2	—	0.1
Family Phascolarctidae						
Phascolarctos cinereus	12	5	2	0.7	0	1.4
Family Macropodidae						
Macropus giganteus	0.3	0.1	0	0.1	0.1	0
Macropus giganteus[b]	0.6	0.1	0	5.6	—	0.1

Note: [a] One unit of disaccharidase activity hydrolyses 1 μmol substrate min^{-1}.
[b] Pouch young.
After Kerry 1969.

respectively, and apparent energy digestibilities of 87% and 89%, remarkably similar to the estimates of Cowan *et al.* (1974) in *A. swainsonii*. Apparent digestibility of the dry matter and energy of *Tenebrio* larvae by *A. stuartii* was 84% and 87% respectively (Nagy *et al.* 1978). Apparent digestibility does not account for faecal material of endogenous or metabolic origin. True digestibilities, which do account for this material, would be even higher than these apparent digestibilities.

The reliance of dasyurids on invertebrates as food is reflected in the high levels of activity of the disaccharidase trehalase in their small intestine. Trehalose is a storage disaccharide found only in insects. Kerry (1969) measured the activities of numerous disacharidases in six marsupial families (Table 2.3). Although the relatively high activities of maltase, isomaltase and sucrase found in both dasyurid species examined could be taken as an indication that their natural diet is not exclusively animal in origin, it must be remembered that there is another source of sucrose to dasyurids, and that is in the digestive tracts of insect prey that feed on sap, nectar and fruits.

When Sabat, Bozinovic & Zambrano (1995) studied the response of the strictly insectivorous Chilean mouse-opossum, *Thylamys* (*Marmosa*) *elegans*, to fruit pulp and live insects, they were surprised to find that the activities of sucrase, maltase and trehalase in the small intestine were all modulated by their respective substrates, but if the sucrose in the digestive tracts of insect prey is available to *T. elegans*, the results are not unexpected. Alternatively, the three enzymes form a complex which is under a common substrate-dependent regulatory mechanism (Galand 1989), but this alternative is not supported by the finding of Sabat & Bozinovic (1994) that sucrase activity was higher in summer than in winter, whereas trehalase activity was constant. This finding may also reflect seasonal shifts in the type of insect prey taken by *T. elegans*.

The assumption of Sabat *et al.* (1995) that *T. elegans* is a strict insectivore is supported by their finding that it was not able to obtain sufficient energy to satisfy maintenance energy requirements on an exclusive diet of fruits, but was able to on an exclusive diet of insects. Although apparent digestibility of dry matter was higher (91%) on fruit than on insects (59%), dry matter intake on fruit was only half that on insects, so that digestible energy intake was only 316 kJ $kg^{-0.75}$ d^{-1} on fruit versus 522 on insects. Maintenance energy requirements for this species have not been published, but those of two dasyurid marsupials are 545 kJ $kg^{-0.75}$ d^{-1} (see Table 1.2). Presumably food passage rate on fruit is too slow to allow more food to be processed by *T. elegans*, but this should be measured.

2.7 WATER AND ELECTROLYTE METABOLISM

Some of the highest water turnover rates (WTRs) in free-living marsupials have been reported in small dasyurids (see Table 1.7), even though the range of some of these species extends into the arid zone. One reason is related to small body size. Small mammals, because of their high mass-specific metabolic rates, have high rates of evaporative and respiratory water loss, and have higher WTRs than do large mammals (Beuchat 1990a). Another reason might be that their diet of animal tissue, with its high water content, exerts little selective pressure for water conservation measures. Also, their diet is high in protein content, which means that water is required for excretion of more urea than in non-carnivores. There is interest therefore in the abilities of small dasyurid and didelphid carnivores to conserve water, particularly through modification of kidney function.

The ability of the mammalian kidney to conserve water by concentrating urine is reflected in various measures of kidney morphology, particularly the ratio of medulla to cortex. This is because urine concentration takes place in the countercurrent multiplier system established by the loops of Henle and collecting ducts in the medulla. The magnitude of the osmotic

Table 2.4. *Relative medullary thickness (RMT) of 25 dasyurid marsupials*

Species (number of animals)	RMT	Habitat
Antechinus swainsonii (2)	3.7	Alpine
Dasyurus maculatus (1)	5.3	Moist temperate
Phascogale tapoatafa (1)	5.5	Mediterranean, summer drought
Sminthopsis dolichura ★ (2)	5.6	Mediterranean, summer drought
Antechinus flavipes ★ (3)	5.6	Mediterranean, summer drought
Sminthopsis griseoventer (1)	5.7	Mediterranean, summer drought
Phascogale calura (3)	5.8	Mediterranean, summer drought
Antechinus melanurus (1)	6.0	Moist temperate
Antechinomys laniger (2)	6.1	Arid
Dasyurus hallucatus (3)	6.2	Tropical, winter drought
Sminthopsis murina ★ (2)	6.2	Semi-arid
Sminthopsis virginiae ★ (1)	6.4	Tropical, winter drought
Sminthopsis granulipes ★ (2)	6.6	Mediterranean, summer drought
Dasykaluta rosamondae ★ (1)	7.2	Moist temperate
Pseudantechinus woolleyae ★ (1)	7.7	Arid
Planigale maculata ★ (1)	7.9	Moist sub-tropical
Sminthopsis leucopus ★ (1)	7.9	Moist temperate
Sminthopsis hirtipes ★ (2)	8.0	Arid
Dasycercus (Dasyuroides) cristicauda ★ (2)	8.4	Semi-arid to arid
Sminthopsis crassicaudata ★ (4)	8.9	Semi-arid to arid
Sminthopsis youngsoni ★ (1)	9.0	Semi-arid
Sminthopsis macroura ★ (2)	10.1	Arid
Ningaui timealeyi ★ (1)	10.8	Arid
Ningaui ridei ★ (1)	11.1	Arid
Pseudantechinus macdonnellensis ★ (1)	11.5	Arid

Note: ★ indicates extrarenal papilla is present.
Source: After Brooker & Withers 1994.

gradient maintained by this system depends on the length of the collecting ducts and loops of Henle; increasing loop length generally results in greater urine concentrating capacity. Of the various indices of renal morphology that have been suggested, the oldest, the relative medullary thickness (RMT) is perhaps the most widely used for comparative purposes. RMT is defined by Sperber (1944) as: 'The medullary thickness × 10 divided by kidney size where kidney size equals the cube root of the dimensions of the kidney.' The dimensions of the kidney are length, breadth and width. Arid-habitat species tend to have higher RMTs than mesic-habitat species of similar body size, and to have greater urine concentrating capacities.

Brooker & Withers (1994) measured RMT in 25 species of dasyurid marsupials (Table 2.4), and found that arid-dwelling species such as *Ningaui ridei* (wongai ningaui) and *Pseudantechinus macdonnellensis* (fat-tailed pseudantechinus) had the highest RMTs (11.1 and 11.5), while *Antechinus swainsonii* (dusky antechinus) collected from alpine environments had the lowest (3.7). Thus, despite diets of high water contents in all habitats, dasyurids in more arid environments generally have a greater need to conserve water than do their more mesic relatives. Kidneys with greater RMTs help in this regard. As can be seen in Fig. 2.12, the medulla of the arid-zone species extends as a papilla into the top of the ureter.

When Beuchat (1990a) compiled information from the literature on urine concentration, RMT and body mass for 245 species of mammals, she found that the maximum recorded urine concentration (U_{max}, in mOsm kg^{-1} water) declined with body mass as $U_{max} = 2564 \, M^{-0.097}$. This relationship means that only a few species larger than 10 kg can concentrate urine to 3000–4000 mOsm kg^{-1}, and only species less than 400 g can achieve urine concentrations substantially greater than this level.

Beuchat (1990b) also showed that U_{max} was directly proportional to RMT, but the regression accounted for only 59% of the variability in urine concentrating capacity among species. Thus there are substantial interspecific differences that are not attributable to the length of the loops of Henle. On the basis of examples presented by Beuchat (1990a), diet is a likely candidate to explain some of these differences.

Figure 2.12 Outlines of midsaggital longitudinal sections through the kidneys of (a) *Antechinus swainsonii* (dusky antechinus), (b) *Dasyurus maculatus* (spotted-tailed quoll) (two moist temperate species), (c) *Ningaui ridei* (wongai ningaui) and (d) *Pseudantechinus macdonnellensis* (fat-tailed pseudantechinus) (two arid-zone species). The scale bar is 2 mm in each case. The cortex is shown as white, outer medulla as stippled, and inner medulla as black. The relative area of the medulla is greater in more arid species. From Brooker & Withers (1994).

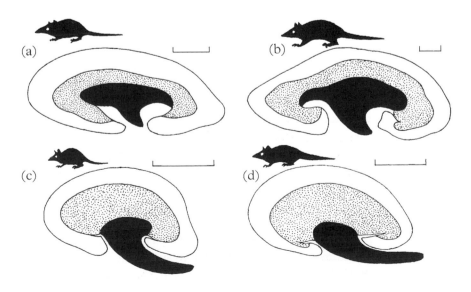

Similarly, when Brooker & Withers (1994) used Beuchat's (1990b) relationship to predict U_{max} for five small dasyurid species, the highest recorded urine concentrations measured under laboratory conditions were substantially below predicted values for four of the species. In this case the difference could be due to the high water content of their natural food, together with their nocturnal habits; these small carnivores do not normally experience water stress, even in desert habitats.

Only two didelphid species have been examined in this regard, the carnivorous *Philander opossum* (gray four-eyed opossum) by Fonseca & Cerqueira (1991) and the omnivorous *Didelphis virginiana* (Virginia opossum) by Plakke & Pfeiffer (1970). U_{max} values were both approximately 3000 $mOsm\,kg^{-1}$. A U_{max} close to this (2800 mOsm kg^{-1}) would be predicted for *P. opossum* on the basis of body mass, and thus the data are consistent with the preferred mesic habitat of this South American species in which there is little need to conserve water. However, for *D. virginiana*, a U_{max} of only 2250 would be predicted on the same basis, suggesting that the Virginia opossum has greater urine concentrating ability, which should enable it to exploit a wider range of habitats than *P. opossum*, as it does (see Chapter 3).

2.8 ENERGY RELATIONSHIPS

Many of the energetic features of didelphid and dasyurid marsupials can be related to the small body size of most of them, and therefore to their relatively high mass-specific metabolic rates. For instance, Hallam, Holland & Dawson (1995) concluded from a study of four dasyurids that haemoglobins with a low affinity for oxygen, an adaptation to an active lifestyle, were a general characteristic of dasyurid marsupials.

Although basal metabolic rates of dasyurids are generally lower than those of eutherian carnivores of similar size (Chapter 1), they have greater metabolic scopes than their eutherian counterparts (Dawson & Dawson 1982). That is, they are capable of increasing their metabolic rate to the maximal levels measured in small eutherians (Dawson & Olson 1988). However, if food is not available to fuel these high metabolic rates, didelphid and dasyurid marsupials reduce their rate of heat loss in several ways, including the use of torpor, and nest-sharing. Caudal fat storage also helps many small dasyurids to meet the energetic demands of low ambient temperature during periods of food shortage, even though caudal fat may only provide a day or so of reserve energy. Brown adipose tissue (BAT), the principal site of non-shivering thermogenesis in eutherian hibernators and neonates, does not appear to play a significant role in thermoregulation in cold-acclimated carnivorous marsupials (May 1997). Nevertheless, May (1997) detected some form of non-shivering thermogenesis in cold-acclimated adult *Dasycercus* (*Dasyuroides*) *byrnei*, possibly in the liver. The only reports of BAT in marsupials appear to be those of Loudon, Rothwell & Stock (1985) in pouch-young red-necked wallabies (*Macropus rufogriseus*), and of Hope et al. (1997) in adult *Sminthopsis crassicaudata*. The amount of BAT measured in *S. crassicaudata* was tiny, only 60–67 mg. It seems that BAT is not usually present in adult marsupials, but could be important in pouch-young animals as they develop thermoregulatory capabilities. This is an area that needs a lot more research.

Geiser (1994) reviewed the occurrence of shallow, daily torpor and deep, prolonged torpor (hibernation) in marsupials; the differences between these two energy-conserving states are discussed in Chapter 1 (Section 1.9). The incidence of torpor and hibernation in carnivorous marsupials is summarised in Table 2.5. The only species in this group that has been described as a hibernator is the microbiotheriid *Dromiciops australis* (Rosenmann & Ampuero 1981; Grant & Temple-Smith 1987). *Dromiciops* fattens in autumn and shows a hibernation season during winter. During hibernation metabolic rates fall to 1% of those of normothermic animals and hibernation bouts last for about five days.

All the other species listed in Table 2.5 use daily torpor (four didelphids, 18 dasyurids, plus the numbat (*Myrmecobius fasciatus*) and southern marsupial mole (*Notoryctes typhlops*)). Torpor bouts last from as little as 2 hours to as much as 20 hours, and metabolic rates fall to 10–60% of BMR.

The energy savings associated with daily torpor in free-living fat-tailed dunnarts (*Sminthopsis crassicaudata*) (Fig. 2.4) during winter (the non-breeding season) were calculated by Frey (1991) and compared with those of other energy-conserving mechanisms in this species, including reduced activity above ground, basking, huddling in groups, use of nests, choice of thermally favourable nesting sites, and lowered resting body temperature. The field site at Werribee in southern Victoria was an area of flat tussock grassland devoid of trees and shrubs, but with numerous scattered boulders that were used by *S. crassicaudata* for shelter. The relative shortage of invertebrate prey at this time of the year (June–July) is illustrated in Fig 2.13. Lower than usual ambient temperatures resulted in minimal numbers of invertebrates on the soil surface at night in the absence of rain, and the frequency and duration of torpor increased. However, at all temperatures, rainfall during the night increased the available biomass of invertebrates (especially slugs and earthworms), and *S. crassicaudata* did not enter torpor. This strongly suggests that torpor in *S. crassicaudata* is induced by food shortage and not by low ambient temperature.

The average energy savings calculated by Frey (1991) are shown in Table 2.6. Savings from huddling (nest sharing) increase with group size, and can be considerable, as can those from entering torpor and by reducing activity on the surface on dry nights. Total energy savings reached 20–25% of daily energy expenditure in the absence of energy-saving mechanisms after rainy

Table 2.5. *Torpor in carnivorous marsupials*

Family Species	Body mass (g)	Minimum body temperature(°C)	Torpor duration (h)	Torpor pattern
Didelphidae				
Marmosa microtarsus	13	16	8	Torpor
Thylamys elegans	30	–	–	Torpor
Marmosa robinsoni	122	23	–	Torpor
Monodelphis brevicaudata	40–111	27	–	Torpor
Microbiotheriidae				
Dromiciops australis	30	7.1	120	Hibernation
Dasyuridae				
Dasyurus geoffroii	1000	23.1	—	Torpor
Dasycercus (Dasyuroides) byrnei	120	20.4	8	Torpor
Dasycercus cristicauda	70–110	14	12	Torpor
Phascogale tapoatafa	100–235	—	—	Torpor
Antechinus flavipes	30–70	24.5	6	Torpor
Antechinus stuartii	20–60	19.9	9	Torpor
Antechinus swainsonii	50–100	28.2	–	Torpor
Sminthopsis murina	18	15.0	8	Torpor
Sminthopsis ooldea	11	—	—	Torpor
Sminthopsis longicaudata	15–20	—	—	Torpor
Sminthopsis crassicaudata	17	13.0	20	Torpor
Sminthopsis macroura	20–28	14.0	18	Torpor
Antechinomys laniger	27	11.0	16	Torpor
Planigale maculata	10–16	19.6	—	Torpor
Planigale ingrami	6–9	—	2–4	Torpor
Planigale tenuirostris	7	—	2–4	Torpor
Planigale gilesi	8	14.3	15	Torpor
Ningaui yvonneae	10–13	15.3	12	Torpor
Myrmecobiidae				
Myrmecobius fasciatus	500	—	—	Torpor
Notoryctidae				
Notoryctes typhlops	60	—	—	Torpor

Note: The data include the minimum body temperature
and the longest duration of torpor bout reported for adults of each species.
Source: After Geiser 1994.

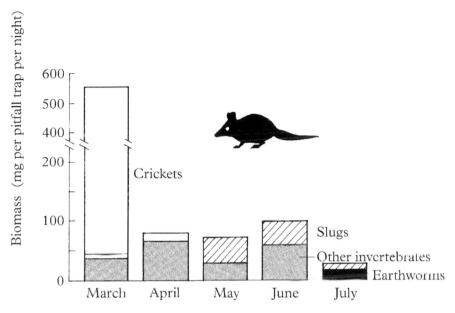

Figure 2.13 Food available to free-living *Sminthopsis crassicaudata* (fat-tailed dunnart) at Werribee, Victoria. From Frey (1991).

or mild (minimum 6 °C) dry nights, the major contributors being the use of a nest and huddling. After cold (minimum 0 °C) nights the savings can reach 40–50%, primarily due to torpor and reduced activity on the surface.

Despite these considerable energy savings, if there is no feeding on one night, the animals must cover their energy expenditure for a period of about 36 hours because they never forage in the daytime. For a 10 g *S. crassicaudata* this would amount to 25–35 kJ, but the energy stored in the tail (4–17 kJ, according to Morton (1978)) plus the rest of the body (approximately 29 kJ) should be enough to cover this expenditure and sustain the animal until the following night. This example, plus the existence of caudal fat storage in many small mammals, including 10 dasyurids (Morton 1980) shows that there are strong selective forces favouring energy storage. The selective pressure seems to be short-term variability in insect abundance. Caudal fat stores, along with torpor and huddling, provide an important buffer for *S. crassicaudata* against unpredictable periods of food shortage, particularly in winter.

The fact that torpor, nest sharing and caudal fat storage were found in *S. crassicaudata* at both Fowler's Gap in the arid zone and Werribee, suggests that the biology of this small dasyurid includes a series of adaptations to counter variability in food supply, and that the variability is a major problem facing a nocturnal insectivore inhabiting open environments, regardless of the degree of aridity. Extraordinary mobility in response to food variability is a further likely adaptation of arid-zone dasyurids. For instance,

Table 2.6. *Average daily energy savings in free-living* Sminthopsis crassicaudata *(fat-tailed dunnarts) at Werribee, Victoria during Winter (the non-breeding season)*

Mechanism	Minimum overnight temperature (°C)	Energy (J g⁻¹ d⁻¹)	Savings (%)
Use of nest (mean group size 4.3)	6	170	5
Lowered body temperature plus selection of a warmer nest site	6	115	3
Basking (1.5 h d⁻¹)	6	120	3
Huddling (nest-sharing):			
Rainy nights (mean group size 4.2)	6	450	13
Rainy nights (mean group size 3.0)	0	350	9
Dry nights (mean group size 2.3)	6	280	8
Dry nights (mean group size 1.2)	0	0	0
Torpor:			
Dry nights (6 h d⁻¹)	0	400	11
Reduced activity:			
Dry nights (8 h d⁻¹)	0	370	10

Source: After Frey 1991.

S. *youngsoni* (lesser hairy-footed dunnart) and *S. dolichura* (little long-tailed dunnart) can move more than 10 km toward localised rainfall (C. R. Dickman, pers. comm.).

Two other characteristics of *S. crassicaudata*, that is, high mobility (the home range is more than 600 m in diameter (Morton 1978), compared with 100–200 m for *Antechinus stuartii* in a rainforest habitat (Wood 1970)), and repeated breeding within an uncertain optimum period (Morton 1978), are also consistent with habitats which are open and relatively unprotected from sudden changes in the weather and therefore from unpredictable short-term shortages in insect availability.

At the other end of the scale, *A. stuartii* lives in well-watered forest habitats with a highly predictable cycle in insect availability. Related to this it has a most unusual life history, with breeding occurring only once per year, in late winter. All males die after mating, while females all survive into their second and some into their third year (Fig. 2.14). There is a high degree of synchrony in mating, births (usually confined to a period of a few days in a particular population, two to three weeks after male die-off (Wood 1970)), and dispersal.

Woollard (1971) followed changes in body mass, energy intake and nitrogen balance of five male and five female captive *A. stuartii* over the period March–October (autumn, winter and spring). The animals were fed minced meat, and urine and faeces were collected on blank newsprint on the cage floor. Results are summarised in Fig. 2.15. The pattern of body mass change was similar to that observed by Wood (1970) in the field. In young animals energy intake (kJ d^{-1}) was slightly greater in males than in females, which can be explained by the higher body mass of the males. From June, energy intake of females stabilised while that of males increased. Similar

Figure 2.14 Population changes in *Antechinus stuartii* (brown antechinus) in a Queensland rainforest. From Wood (1970).

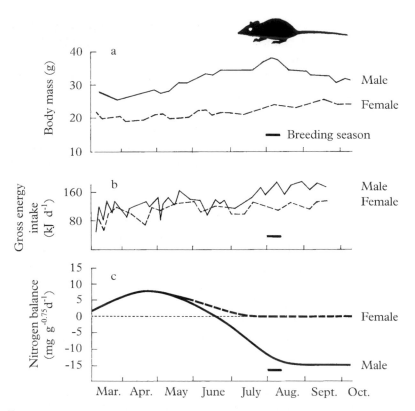

Figure 2.15 Changes in body mass, energy intake and nitrogen balance of a laboratory colony of *Antechinus stuartii* during winter and spring. After Woollard (1971).

levels of energy expenditure were recorded in free-living *A. stuartii* at Jervis Bay in New South Wales by Green *et al.* (1991).

Nitrogen balance was positive and increasing in March and April, but then declined. Females were close to balance from June but, despite increased food consumption, males went into negative nitrogen balance in June, and remained in substantial negative balance until the end of the study in late September. The kidneys of wild-caught male (but not female) *A. stuartii* were smaller in August than predicted, with marked changes in microscopic morphology, including a reduction in the number of glomeruli per mm^2 of cortex and hypertrophy of the proximal tubules (McAllan, Roberts & O'Shea 1996). These changes in renal structure are suggestive of renal failure at the end of mating. That the captive males of Woollard (1971) survived to late September is explained by the fact that they were housed individually and therefore deprived of the opportunity of mating and thus not exposed to the usual stresses of free-living *A. stuartii*. Males brought into the laboratory during the last week of mating die at the expected time (Lee & Cockburn 1985).

The endocrine changes involved in the annual die-off of male *A. stuartii* have been elucidated by Bradley, McDonald & Lee (1980) and McDonald, Lee, Bradley & Than (1981), and are summarised by Lee & Cockburn (1985). Basically the die-off phenomenon in *A. stuartii* results from the failure of the glucocorticoid feedback mechanism. Similar reproductive strategies have been described for seven other species of *Antechinus* and two species of *Phascogale*, all of which are confined to the forests, woodlands and heaths of southern, eastern and northern Australia, where the climates are both highly seasonal and predictable (Lee & Cockburn 1985). However, *Dasykaluta rosamondae* (little red kaluta), a small dasyurid from the seasonally unpredictable Pilbara region of north-western Australia, also exhibits annual male die-off (Woolley 1991).

There are thus fundamental differences between the life histories of the two small dasyurids *A. stuartii* and *S. crassicaudata* which can be related directly to differences between their respective habitats, principally in the variability in availability of invertebrates, their staple food resource.

In the alpine zone of the Snowy Mountains, *A. swainsonii* (dusky antechinus) survives the winter by spending the coldest parts of the night in nests excavated in creek banks or just below the soil surface. Surface activity is mainly diurnal, which allows the animals to bask in the sun (Green & Crowley 1989). This behaviour would save considerable energy. However, they appear to forgo the energetic benefits of nest-sharing, and their FMR increases by a factor of 2.4 between summer and winter (Table 1.4), but the advantage of nesting solitarily may mean that competition for scarce food resources (mainly soil invertebrates) is reduced.

2.9 TOLERANCE TO FLUOROACETATE

The climax vegetation communities in the south-west of Western Australia contain an abundance of species of woody shrubs in the leguminous genera *Gastrolobium* and *Oxylobium* (Kitchener 1981). These plants are heavily defended against herbivory by fluoroacetic acid. Its sodium salt, sodium monofluoroacetate (Compound 1080) is widely used in many countries against vertebrate pests. Most mammals are fatally poisoned by less than 1 mg of 1080 kg^{-1} body mass. Its toxicity is due to its conversion in the body to fluorocitrate which is a competitive inhibitor of the enzyme aconitate hydratase [EC 4.2.1.3]; 1080 blocks the Krebs citric acid cycle at the citrate stage. The two plant genera contain fluoroacetate at up to 2.65 g kg^{-1} fresh leaf (Mead *et al.* 1985), and to more than 6.50 g kg^{-1} seeds (Twigg & King 1991). Although carnivorous marsupials do not eat the leaves of *Gastrolobium* and *Oxylobium*, they feed on insects that do. Estimated values for LD$_{50}$ (the dose required to kill 50% of individuals) were much higher (7.5–17.5 mg kg^{-1}) for three carnivores from south-western Australia (*Dasyurus geoffroii*, *Antechinus flavipes* and *Phascogale calura*) than for *D. viverrinus* and *A. flavipes* from south-eastern Australia (1.5–3.5 mg kg^{-1}), where there is no natural exposure to fluoroacetate (King, Twigg & Gardner 1989). *D. hallucatus* from Nourlangie in the Northern Territory, where toxic *Gastrolobium grandiflorum* is common, also exhibited higher resistance to 1080 (LD_{50} of 7.5 mg kg^{-1}).

Clearly a secondary tolerance to fluoroacetate has evolved in marsupial carnivores from areas where their prey feed on fluoroacetate-bearing vegetation. The highest tolerance among the marsupial carnivores has been recorded for *P. calura*, which is endemic to south-western Australia. Its close association with fluoroacetate-bearing vegetation appears to be the reason for its high tolerance (LD_{50} of 17.5 mg 1080 kg^{-1}). Its persistence in this region is probably due to the protection provided by these plants against introduced predators, all of which are highly susceptible to 1080 (Twigg & King 1991).

The degree to which fluoroacetate tolerance has developed within native animal populations is in the order herbivores > omnivores > carnivores. In the case of the herbivorous insects eaten by marsupial carnivores in south-western Australia, species that specialise in feeding on *Eucalyptus* leaves (which do not produce fluoroacetate) are 40–150 times more sensitive to the toxin than species that include fluoroacetate-bearing vegetation in their diet (Twigg & King 1991). Tolerance to fluoroacetate appears to have evolved on all three continents where some native plants contain significant levels of the toxin (Australia, Africa and South America). The opossum *Didelphis marsupialis*, where it coexists with the fluoroacetate-bearing *Palicourea marcgravii* in South America, has an LD_{50} of 60 mg 1080 kg^{-1} (Atzert 1971, cited by Twigg & King 1991).

2.10 SUMMARY AND CONCLUSIONS

Carnivorous marsupials have relatively narrow nutritional niches. This applies particularly to the smaller species with their high mass-specific requirements for energy and water. The larger species meet their energy and nutrient requirements by feeding mainly on vertebrate prey, while the smaller species take mainly invertebrates. However, invertebrates (and ectothermic vertebrates – amphibians and reptiles) are available on a seasonal basis, often with marked summer peaks and winter troughs in areas of predictable rainfall. In semi-arid and arid areas food availability is much less predictable. Thus even though their invertebrate prey have high water contents, water conservation has been shown to be more important in small dasyurids from the arid zone than in those from more mesic environments. Energy conservation is also of critical importance in small dasyurids. Responses to unpredictable energetic stresses in open habitats of both arid and mesic environments include the use of torpor, nest sharing and caudal fat storage. Their reproductive strategy includes repeated breeding within an uncertain optimum period. This contrasts with the high degree of synchrony in breeding in small dasyurids in well-watered forest habitats with a highly predictable cycle in invertebrate prey availability.

The most dramatic dental modifications among carnivorous marsupials are seen in the extinct *Thylacoleo carnifex*, but the functional significance of subtle differences in the dentition of modern dasyurids has not been adequately explored. The gastrointestinal tract of carnivorous marsupials is simple, and dominated by the small intestine. This is consistent with the high quality of most of the dietary items taken by dasyurids and carnivorous didelphids, and with the performance characteristics of a plug-flow reactor, many of which are exhibited by the small intestine. The high trehalose content of invertebrates is reflected in the high and constant levels of trehalase activity in the small intestine of both dasyurids and carnivorous didelphids. Similar information is needed on chitinase and chitobiase activities in insectivorous marsupials, and on the nutritional value of the chitin consumed. The presence of a cardiogastric gland in the South American *Caenolestes obscurus* is intriguing, but its function is unexplained. Rate of food passage through the marsupial carnivore tract is rapid, with no suggestion of differential passage of fluid or particulate digesta. Mean retention times are shortest in the tiny planigales, and longer in the larger eastern quoll.

The seasonal changes in food availability faced by many carnivores are accommodated more easily by omnivores because of their broader diet spectrum, and their ability to switch between alternative foods as availabilities fluctuate. Omnivorous marsupials are discussed in Chapter 3.

3   Omnivorous marsupials

Omnivory, by definition, includes ingestion of plant and/or fungal as well as animal material. This means that greater amounts of indigestible residues will be consumed by omnivores than by carnivores. This has several important nutritional consequences. One of these is the requirement for greater lubrication to protect the gut lining from physical trauma during passage of plant residues (Hume & Warner 1980). Another is that plant residues provide an additional substrate for bacteria and other microbes resident in the gut, primarily in the hindgut caecum. Thus, compared with carnivores, we usually see in the omnivore digestive tract an increased caecal capacity, together with an increase in small intestinal length and in colon length and diameter.

Although Lee & Cockburn (1985) classified many marsupial genera as some sort of omnivore in Eisenberg's (1981) list of mammalian macro-niches (see Table 2.1), marsupial omnivores fall quite neatly into three main groups. The first consists of most members of the New World family Didelphidae, the opossums, of which *Didelphis* is the best-known member. The second consists of the bandicoots and bilbies (families Peramelidae and Peroryctidae). The third group consists of numerous Australasian species which feed on a mixture of non-foliage plant materials and invertebrates. These include the family Burramyidae (five species of Australian pygmy-possums, of which one, *Cercatetus caudatus* (long-tailed pygmy-possum) is found also in New Guinea); the arboreal family Petauridae (six species in Australia and five in New Guinea, of which *Dactylopsila trivirgata* (striped possum) and *Petaurus breviceps* (sugar glider) are shared with Australia); and the family Acrobatidae, represented by the feathertail glider (*Acrobates pygmaeus*) in Australia and the feathertailed possum (*Distoechurus pennatus*) in New Guinea. One other family, the monospecific Tarsipedidae, containing the honey possum (*Tarsipes rostratus*), is included here, although its diet consists almost entirely of plant material (pollen and nectar).

3.1   DIDELPHID MARSUPIALS

One of the most successful didelphid marsupials is the Virginia opossum, *Didelphis virginiana*. Its present range extends from Nicaragua to as far north as southern Ontario, Canada (Hunsaker 1977). The northern limits to its distribution are explained by its evolution from a tropical lowland ancestor (*D. marsupialis*) and its opportunistic spread northwards to the limit of its thermoregulatory ability. Its thermal conductance is twice the expected value for a mammal of its size, and its thermoneutral zone is high (29–35 °C). Its northern limit is close to the −2 to −7 °C January isothermal line. It does not hibernate, but at ambient temperatures below −7 °C it seldom leaves its den to forage. Although it does not store much fat, over extended periods of severe winter weather *D. virginiana* relies on mobilisation of body stores, and up to one-third of its winter energy requirements are met through body tissue catabolism (Harder & Fleck 1997). The life of body stores is increased by an endogenous winter metabolic depression. Hsu, Harder & Lustick (1988) found that female Virginia opossums held captive under ambient conditions and fed near *ad libitum* lost 27% of their initial body mass of 3.7 kg between autumn and spring. This endogenous winter weight loss accommodates the trough in food availability during the coldest months, and appears to be the main physiological adaptation of *D. virginiana* to low winter temperatures (Harder & Fleck 1997).

The other barrier to the spread of *D. virginiana* in North America is aridity. As Plakke & Pfeiffer (1965, 1970) demonstrated, the Virginia opossum is not able to raise its urine concentration to compensate for reduced water intake any more than would be predicted from its body mass (Beuchat 1990a). This is consistent with a relative medullary thickness (RMT) of 5.7. The deserts of north-western Mexico and south-western USA were barriers to the western expansion of the range of the Virginia opossum. It was reliant on human transportation to cross these barriers into California and then northwards into Oregon and Washington. Within western, central and south-eastern USA, *Didelphis* is found mainly in temperate woodlands. Its success in this large range is due in part to its very broad dietary spectrum, as well as its extremely adaptable behaviour (Hunsaker 1977).

The Virginia opossum is basically a terrestrial species, and uses arboreal habitats primarily when foraging. Literature reports on food preferences differ between regions, suggesting that *D. virginiana* is an opportunistic omnivore. For instance, in eastern Texas, Lay (1942) found the stomach contents of *D. virginiana* to consist of about 60% animal material (insects, worms, mammals, birds, crayfish and snails) and 40% plant (fruit, green leaves, leaf and log litter, acorns and grass seeds). On the other hand, Fitch and Sandidge (1953) found that scats of *D. virginiana* in north-eastern Kansas in autumn and winter contained mainly fruit; other food items were crayfish, insects, corn, rabbit carrion, young snakes, snails, frogs and lizards. However, scat analysis may be expected to bias results toward the less digestible items relative to stomach content analysis, as discussed in Chapter 2. In the case of omnivores this bias will usually favour plant materials. Some rapidly digested items do not appear in the scats at all. Thus Wood (1954) found that of 39 different food items used by *D. virginiana* in eastern Texas, 36 were recorded in the stomach but only 10 appeared in the faeces.

Despite the problems associated with diet analysis, it is clear that the Virginia opossum has an extremely broad food spectrum. The digestive tract is also typically omnivorous in form (Fig. 3.1). The salivary glands are represented by large mandibular glands and smaller parotid and sublingual glands (Flower 1872). The mucosa of the distal oesophagus is raised into transverse rugae (Sonntag 1921). The stomach is simple and globular in form. The gastric mucosa is largely occupied by fundic glands, the remainder being pyloric glands, with a very narrow zone of cardiac glands at the oesophageal opening (Bensley 1902). Krause, Yamada & Cutts (1985) examined the distribution of enteroendocrine cells along the gastrointestinal tract of *D. virginiana*. The enteroendocrine cells, together with the endocrine cells of the pancreas, contain specific regulatory peptides or amines that help to control digestive functions such as gastric acid secretion, pancreatic secretion of electrolytes and enzymes, and contraction of the gall bladder (Stevens and Hume 1995). Common gut hormones include gastrin, gastric-inhibitory polypeptide (GIP), secretin, cholecystokinin and pancreozymin. In the stomach of *D. virginiana*, 90% of the enteroendocrine cells are confined to the pyloric glands. The pyloric gland

Figure 3.1 Gastrointestinal tract of *Didelphis virginiana* (Virginia opossum). Redrawn from Schultz (1976).

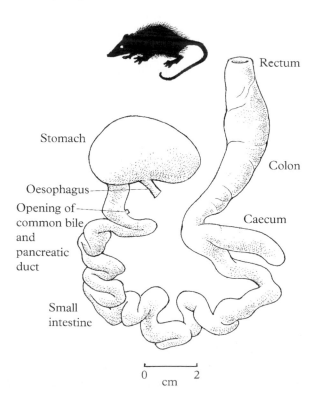

region contained 73% of the gastrin-containing cells and 60% of the somatostatin-containing cells (Krause *et al.* 1985). Similar distributions of these and other enteroendocrine cells have since been shown in the white-bellied opossum (*Didelphis albiventris*), a South American species, by Barbosa *et al.* (1990), and both have been found to be similar to patterns seen in most eutherian mammals.

The small intestine of *D. virginiana* is about 250% of body length (Owen 1868; Flower 1872), and a collar of lobed glandular tissue (Brunner's glands) is found in the submucosa immediately distal to the pylorus. Brunner's glands secrete an alkaline fluid and mucus. In *Didelphis* the glands empty into large mucosal depressions surrounded by the glands themselves (Krause & Leeson 1969). Enteroendocrine cells are found in all sections of the small and large intestines, in nearly equal numbers, although the distribution of each type varies. For instance, gastric and pancreatic secretions are influenced mainly by the endocrine population of the proximal duodenum. Most of the endocrine cell types found in the eutherian intestine are also present in the intestine of *Didelphis* (Krause *et al.* 1985).

The caecum of *D. virginiana* is simple and conical in shape, and is 20–40% of body length. The colon is approximately 150% of body length, has only a loose mesenteric attachment and so is quite mobile (Owen 1868).

Most of the research on *D. virginiana* has been concerned with the functioning of the lower oesophageal sphincter, pyloric sphincter and ileo-caecal junction muscles. This has been motivated by the fact that smooth muscle arrangements in these regions resemble the situation in humans. Little research has been done on digestive function in any didelphid species.

One of the few nutritional experiments with *D. virginiana* was conducted by Maller, Clark & Kare (1965). Opossums were offered a basal diet (ground commercial dog food) of 42% carbohydrate, 23% protein, 9% fat, 21% ash and 5% water, with a gross energy content of 14 kJ g^{-1}. Intakes measured on the basal diet were 38–40 g kg$^{-0.75}$ d^{-1}. The energy concentration was then lowered by the addition of three graded levels of cellulose, and raised by enrichment with three levels of lard. Gross energy thus ranged from 7 to 21 kJ g^{-1}. Intakes of both food and energy were lowest on the diluted diets. That is, the animals were not able to compensate for the lower energy density of the food by eating more. Enrichment of the basal diet also depressed food intake, but not enough to maintain a constant energy intake; as a result, gross energy intakes were 17–28% above basal levels. These results can be used to suggest that, because *D. virginiana* faces unpredictable fluctuations in food quality and availability in the wild, it has not evolved mechanisms for precise short-term control of energy intake.

The three other species of *Didelphis*, *D. marsupialis* (black-eared or common opossum), *D. albiventris* (white-eared or white-bellied opossum), and *D. aurita*, can also be described as omnivores. Although *D. albiventris* (1500 g body mass) will concentrate its foraging efforts on fruit when available

(Streilein 1982), all are probably as opportunistic in their feeding habits as *D. virginiana*. In Argentina, *D. albiventris* stomachs were found to contain earthworms, ants, small birds and eggshells together with plant material, especially fruit (Redford & Eisenberg 1992). Where *D. marsupialis* coexists with fluoroacetate-bearing plants it has evolved a high tolerance to the poison, as have several Australian dasyurid marsupials (see Chapter 2, Section 2.9). *Didelphis aurita* in south-eastern Brazil fed mainly on invertebrates (mostly insects), as well as fruits and vertebrates, based on per cent occurrence in faeces of 100%, 54% and 31% respectively (Santori, Astúa de Moraes & Cerqueira 1995; Freitas *et al.* 1997).

Caluromyids are the most highly arboreal of the new world opossums; they seldom forage on the ground, depending instead on fruits, nectar and insects in the canopy of rainforests (Eisenberg 1989). The subfamily Caluromyinae consists of *Caluromys* (woolly opossums, three species), *Caluromysiops irrupta* (black-shouldered opossum), and *Gironia venusta* (bushy-tailed opossum). *Caluromys philander* (350 g body mass) (Fig. 3.2) feeds primarily on fruit but also on nectar, gums and invertebrates in the canopy of multistratal forests (Charles-Dominique 1983). *Gironia* feeds on insects, fruits and seeds (Hunsaker 1977).

Metachirus nudicaudatus (brown four-eyed opossum) (240–350 g) (Fig. 3.3) has been described as either frugivorous/omnivorous (Hunsaker 1977) or insectivorous/omnivorous (Robinson & Redford 1986). More detailed studies by Santori *et al.* (1995) and Freitas *et al.* (1997) in south-eastern

Figure 3.2 *Caluromys philander* (bare-tailed woolly opossum), an omnivorous South American opossum that includes a lot of fruit in its diet. (Diego Astúa de Moraes)

Figure 3.3 *Metachirus nudicaudatus* (brown four-eyed opossum), another omnivorous South American opossum but one which includes more insects in its diet. (Diego Astúa de Moraes)

Brazil support the latter description; they found that percentage of occurrence in the faeces of *M. nudicaudatus* was highest for invertebrates (84%), mainly ants, termites, cockroaches and beetles, followed by vertebrates (26%) and then fruit at 11%.

Most mouse opossums (*Marmosa, Marmosops, Thylamys*) are also described by Streilein (1982) as being omnivorous, but they probably include more animal than plant material in their diets. For instance, Vieira & Palma (1996) found that mean occurrence in the faeces of *T. velutinus* from a savanna region of central Brazil was 44% arthropods, 31% unidentified animal material and 25% plant matter. Both *M. robinsoni* and *M. fuscata* (dusky mouse opossum) (40 g) appear to forage for fruit and insects both on the ground and in shrubs, vines and trees (Eisenberg 1989).

The distributions of South American opossums are given in Streilein (1982), but comparative information on the water and other nutrient requirements of this group of marsupials is meagre. Consequently, little can be said at this stage about the physiological factors limiting their various distributions. Santori, Cerqueira & da Cruz Kleske (1995) have compared digestive function in *Philander opossum* and *Didelphis aurita*. Results were consistent with the more carnivorous nature of *P. opossum*: Although the gastrointestinal tracts of both species are similar in gross anatomy to that of *D. virginiana* (Fig. 3.1), the stomach of *P. opossum* is larger and the small intestine shorter than in *D. aurita*. There were no significant differences in length of the caecum or colon between the species. *P. opossum* digested a meat diet more completely than did *D. aurita*, but digestibilities of both the meat diet and a fruit diet were in excess of 90% in both species. A diet of shrimp was digested to a much lesser extent (68–77%), reflecting the refractory nature of the chitin of the exoskeleton.

Foley, Charles-Dominique & Julien-Laferriere (1999) have examined the digestion of fruit-based diets of 0.45%, 0.90% and 2.5% total nitrogen by *Caluromys philander* (Fig. 3.2). The animals ate more of the low-nitrogen diet to compensate for the lower nitrogen content, and thereby remained in nitrogen balance. Although the caecum of the animals on the low-nitrogen diet was significantly enlarged in response to the increased food intake, the size of all other gastrointestinal tract segments was unchanged, and consequently the mean retention time (MRT) of two digesta markers (Cr EDTA as a fluid marker; $YbCl_3$ as a solid-phase marker) was less on the low-nitrogen diet (17.7–19.0 hours versus 22.6–24.8 hours on the 0.90% nitrogen diet). However, there was no significant difference in MRT between the two markers on either diet, suggesting no net selective retention of either phase of digesta in the digestive tract. All animals digested 84–87% of the dry matter of each diet, and recycled between 60% and 80% of endogenously synthesised urea to the gut. By regressing nitrogen balance against nitrogen intake, Foley *et al.* (1999) were able to estimate the maintenance nitrogen requirement from the nitrogen intake that supported zero nitrogen balance. This was 176 mg N $kg^{-0.75}$ d^{-1} on a dietary basis, or 146 mg $kg^{-0.75}$ d^{-1} on a truly digestible basis. As can be seen from Table 1.8, only the wombats, eastern pygmy-possum and sugar glider are as low. The low requirements of the wombats are related to their low basal metabolic rates. The low requirements of *Caluromys*, eastern pygmy-possum and sugar glider are related to the low fibre content of their fruit and plant exudate diets, and thus their relatively low losses of metabolic faecal nitrogen.

3.2 BANDICOOTS AND BILBIES

Bandicoots and bilbies are all terrestrial and mainly nocturnal small to medium-sized marsupials found in Australia, New Guinea and nearby islands. The family Peramelidae is predominantly Australian, with only one species, *Isoodon macrourus* (northern brown bandicoot) represented also in New Guinea. In contrast, the family Peroryctidae is predominantly New Guinean, with only one species, *Echymipera rufescens* (rufous spiny bandicoot), represented also in Australia.

3.2.1 Diets

The feeding habits of bandicoots and bilbies vary somewhat between locations, but all extant species can be described as omnivorous. Heinsohn

(1966) made a detailed study of two bandicoot species, *Perameles gunnii* (eastern barred bandicoot) and *Isoodon obesulus* (southern brown bandicoot) in farmland in north-west Tasmania. The major food items taken by both species were earthworms, adult beetles, moth larvae and pupae, and scarab larvae. The only plant products extensively used by *P. gunnii* were ripe blackberries and boxthorn berries; the stomachs of two animals collected in February were distended with berries. The only plant material taken by *I. obesulus* in large quantities was boxthorn berries, particularly in late summer. Thus both species appeared to be primarily insectivorous.

In more natural wet heath and forest habitat in southern Tasmania, Quin (1988) concluded that *I. obesulus* was an opportunistic omnivore, utilising a variety of invertebrate, plant and fungal material. Ants were consistently the main invertebrate taken, but adult beetles and beetle larvae were also prevalent. The most important plant items were grasses, seeds, clover root nodules and fungi. The bandicoots appeared to select for larger prey sizes in all seasons, suggesting that smaller, less profitable prey were bypassed. In a mixture of open pasture and eucalypt forest in southern Tasmania, *P. gunnii* consumed grasses, roots, mosses, grass seeds, fungi, earthworms, moth and beetle larvae and adult beetles (Reimer & Hindell 1996). The proportion of plant and fungal material in the faeces was highest (58%) in autumn, equal to that of insects in winter and lowest (28%) in spring. Breeding coincided with increasing proportions of insects in the faeces.

Claridge *et al.* (1991) compared the diets of *I. obesulus* and *Perameles nasuta* (long-nosed bandicoot) throughout the year in eucalypt forest in south-eastern New South Wales. Both species fed mainly on ants, cockroaches, beetle larvae and plant material in all seasons. However, there were seasonal shifts, in that cockroaches and ants declined in their relative volumetric abundance in faeces from summer–autumn to winter–early spring, while plant material and beetle larvae increased (Fig. 3.4). Additionally, *I. obesulus* included seeds in its summer and autumn diet as well as significant amounts of hypogeal (underground) fungal fruiting bodies.

The consumption of hypogeal fungi by *I. obesulus*, and to a lesser extent by *P. nasuta*, was thought by Claridge *et al.* (1991) to play a potentially important role in the ecology of eucalypt forests. Most hypogeal fungi share a symbiotic relationship with a variety of forest plants as mycorrhiza. Mycorrhiza grow in and around the roots of plants and facilitate the uptake of soil water and nutrients into the host plant. In North American coniferous forests many small to medium-sized terrestrial mammals feed extensively on hypogeal fruiting bodies. The fungal spores ingested pass through the gut of these mycophagous mammals relatively intact, are voided in the faeces and subsequently germinate. These fungus feeders are important vectors in the dispersal of otherwise immobile fungi, and are seen as a vital third link of an obligatory symbiosis between plant, animal and fungus (Maser, Trappe & Nussbaum 1978). In Australia, ground-dwelling marsu-

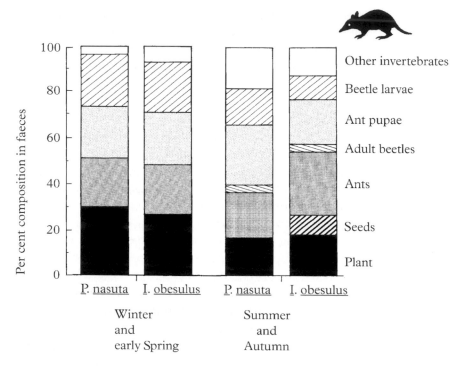

Figure 3.4 Percentage composition of invertebrate and plant remains in the faeces of long-nosed bandicoots (*Perameles nasuta*) and southern brown bandicoots (*Isoodon obesulus*) in summer/autumn and winter/early spring in south-eastern New South Wales. Modified from Claridge *et al.* (1991).

pials including bandicoots and rat-kangaroos (Chapter 8) probably play a similar role in the dispersal of hypogeal fungi in eucalypt forests.

Hypogeal fungi were also found to be an important food item for *P. nasuta* in Sydney Harbour National Park, especially in winter (Moyle, Hume and Hill 1995). The five most abundant diet categories in scats collected through summer, autumn and winter were adult beetles, fungi, monocotyledonous leaves, monocot roots and beetle larvae. Three of these five categories are obtained by digging. Disturbance of the soil surface by the foraging activities of northern brown bandicoots (*Isoodon macrourus*) was shown by Clarke, Myerscough & Skelton (1996) to be important for seedling establishment in coastal heath. The main food items of *I. macrourus* at this site were invertebrates, and plant leaves, stems and roots (Gott 1996).

Wood Jones (1924) reported on the diets of two bandicoots that are now extinct. He concluded that *Perameles eremiana* (desert bandicoot) was mainly insectivorous, but that another extinct species, *Chaeropus ecaudatus* (pig-footed bandicoot) was omnivorous, as it included grass and roots as well as grasshoppers in its diet. Earlier, Krefft (1865) had described *C. ecaudatus* caught along the lower Murray River as mainly herbivorous on the basis of the high grass content of their faeces, and on the acceptance by captive animals of grasses, leaves and bulbous roots. Some insects, particularly grasshoppers, were also eaten, but this was the extent of their carnivorous habits. Recent examination of rectal pellets in two spirit-preserved central Australian specimens of *C. ecaudatus* in the Museum of Victoria, Melbourne, by Dixon (1988) also showed them to consist almost entirely of grass. In contrast, Burbidge *et al.* (1988) reported that, according to central Australian Aborigines they interviewed, the food of *Chaeropus* was termites and ants. It could be that the pig-footed bandicoot's dietary habits were flexible according to the relative availabilities of soft grasses and insects.

A more detailed study of the contents of the digestive tract and the dentition of another central Australian specimen of *Chaeropus* from the Museum of Victoria by Wright, Sanson and McArthur (1991) has yielded a much clearer picture of the dietary strategy of the pig-footed bandicoot. The molars show a suite of characters which suggest that *C. ecaudatus* was a specialist herbivore. The occlusal cycle emphasised fine shear, as seen in grazing kangaroos (Chapter 7 and Sanson 1989), rather than the shear and grind as in *Perameles* and *Isoodon*. High crests on the molars would have provided enough shearing amplitude to reduce grasses to fine particles. Complementing these findings, the stomach contained fine particles of plant material, along with grit and small rafts of bacteria. Plant particles in the caecum were somewhat smaller than those in both the proximal and distal colon, suggestive of selective retention of small particles in the caecum. Some preparations from the proximal colon showed grass nodes as well as small amounts of dark-staining material, probably tannin, meaning that dicotyledonous plants also contributed to the diet. The presence of bacteria in all sections of the hindgut as well as the stomach supports the idea that *C. ecaudatus* was a caecum fermenter and may have been coprophagous (ingested its faeces) or even caecotrophic (selectively ingested high-nutrient faeces derived from caecal contents).

The bilby or greater bilby (*Macrotis lagotis*) (Fig. 1.2) is patchily distributed among a range of landforms in arid and semi-arid areas. Earlier workers had described the diet of *M. lagotis* to consist of grass, bulbous roots, fruits and insects. Results presented by Johnson (1980a) on the basis of microscopic analysis of scats collected from nine colonies of *M. lagotis* north-west of Alice Springs, support these observations. The food material consisted of plants (seeds, bulbs, fruits) insects (mainly ants, termites and beetles) and a small amount of fungus. There was no apparent preference for any particular food item, and Johnson (1980a) concluded that the diet appeared to reflect the seasonal and temporary availability of the various items. Insects averaged 32% of the identifiable fragments in the faeces. In two colonies living near salt lakes bulbs of the sedge *Cyperus bulbosus* formed 57–61% of food material. The annual grass *Panicum australiense* formed 67% of identifiable fragments in one colony, and seeds of the desert trigger plant *Stylidium desertorum* formed 33% in another. Later, Southgate (1990) concluded from a wider survey that in the more southern parts of its present range *M. lagotis* occupied sparse, inhospitable ridges and rises and fed predominantly on insects (mainly ants and termites). In northern sections, habitats included plains and alluvial flats as well as ridges, and the diet was broader, including seeds of *P. australiense* and bulbs of *C. bulbosus*, as well as insects. These findings suggest that *M. lagotis* is a highly flexible animal in terms of both habitat and diet, which would help to explain its very broad distribution prior to the advent of European pastoralism.

The lesser bilby (*Macrotis leucura*) was last reported alive in 1931, in northern South Australia. Johnson (1995) found large quantities of rodent skin and fur, seeds (probably of *Solanum*) and sand, but no insect remains, in the stomachs of a limited sample. Dixon (1988) found only ants and termites, as well as one other unidentifiable insect species, in the stomach of another preserved specimen from central Australia.

In contrast to members of the Peramelidae, which inhabit a wide range of habitats, including semi-arid and arid areas as well as eucalypt forests, all peroryctid species can be found in rainforests, although one New Guinean species, the spiny echimypera (*Echimypera kaluba*) is more common in grassland and abandoned garden areas (Flannery 1995). The diets of the peroryctids are poorly known, although most would seem to be insectivorous/omnivorous. Anderson *et al.* (1988) concluded that *E. kaluba* was truly omnivorous, taking fallen fruit, insect larvae and land molluscs; decayed logs were a focus for foraging. Flannery (1995) found the stomach of several *E. kaluba* to be full of seeds and fruit pulp, suggesting a more frugivorous diet. The animal is most successfully trapped beside fruiting trees of the genera *Pandanus*, *Ficus* and *Zingziber*. Van Deusen & Keith (1966) reported that the dimorphic echimypera (*E. clara*) also feeds on the fruit of *Pandanus* and *Ficus* species. Much more information is needed on diets of New Guinean species before patterns can be expected to emerge. Many peroryctids show a pronounced elongation of the snout, possibly a specialisation for extracting food from narrow crevices, and there is even a slight development of a mobile proboscis (Gordon 1995).

3.2.2 The digestive tract

Although information is not uniform among species, it appears that , as a group, bandicoots and bilbies are basically omnivorous. Differences in diet between locations and between seasons suggest a flexible dietary strategy. This, in turn, is reflected in their digestive tract anatomy and function. The gastrointestinal tract of the long-nosed bandicoot (*P. nasuta*) is shown in Fig. 3.5, and that of the bilby (*Macrotis lagotis*) in Fig. 3.6.

Tedman (1990) has provided the most detailed description of a perameloid digestive tract, that of the northern brown bandicoot (*Isoodon*

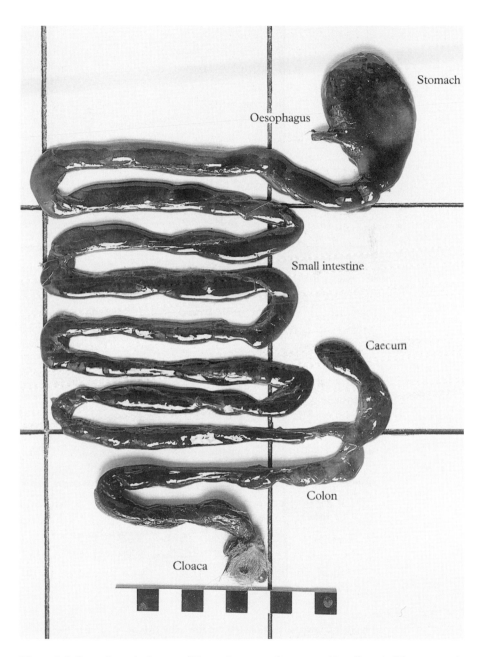

Figure 3.5 Gastrointestinal tract of *Perameles nasuta* (long-nosed bandicoot). The caecum is larger than in carnivorous marsupials. The scale bar and background grid is 10 cm.

macrourus) (Fig. 3.7). The oral cavity receives saliva from relatively small parotid but large mandibular (submaxillary) glands. Mandibular saliva is of the mucous type, while parotid saliva is a serous, highly buffered fluid. The oesophagus persists as a 4–5 mm wide tube that passes to a simple, oval stomach. The small intestine is not clearly demarcated into different re-

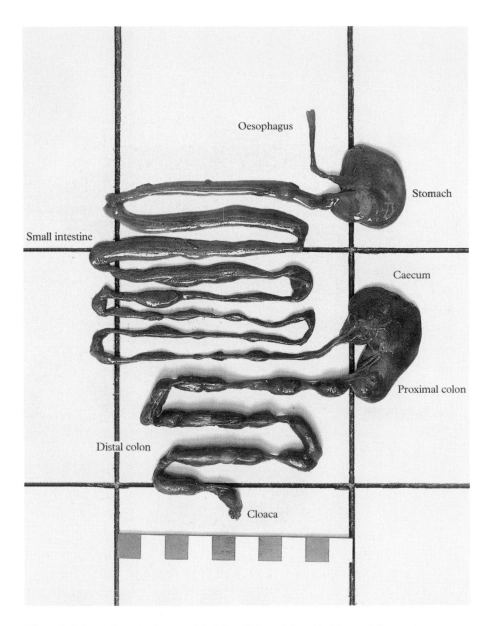

Figure 3.6 Gastrointestinal tract of the bilby (*Macrotis lagotis*). The scale bar and background grid is 10 cm. Note the long distal colon, where water is resorbed. This is an important function in arid-zone species.

gions, and is 63% of total gastrointestinal tract length. The ratio of small intestinal length to body length is 2.3:1, similar to that in *P. nasuta* (2.1:1). Brunner's glands of bandicoots form a narrow, dense band in the duodenum immediately distal to the pylorus (Krause 1972); they are restricted to a smaller area than in the Dasyuridae. The most obvious feature of the hindgut of both *I. macrourus* and *P. nasuta* is a caecum of moderate size (7%

Figure 3.7 *Isoodon macrourus*, the northern brown bandicoot. (Gordon Lyne)

of total tract length). The colon is short relative to the small intestine (26%
of total tract length).

The bilby caecum forms less of a diverticulum, but together with the
expanded proximal colon it forms a larger volume for digesta retention.
Particularly notable is the much longer distal colon in the bilby than both
the long-nosed bandicoot (Fig. 3.5) and northern brown bandicoot. As we
will see in several other groups of marsupials, a long distal colon is often
associated with arid habitats for the species. This is because the distal colon
is the principal site for net absorption of water (and electrolytes) from the
gut contents as the faeces are formed. A long distal colon provides the
absorptive surface area and the time needed for maximal water recovery.

Parsons (1903) examined a specimen of *Chaeropus* (pig-footed bandicoot) and remarked on the great size of the cardiac portion of the stomach
as well as 'the attempt at a marking off of a small secondary chamber or
pyloric antrum'. The small intestine was 180% of body length, and the
colon was equal to the length of the body. The caecum was 10 cm long and
thus relatively larger than that of *Perameles* (Fig. 3.5), supporting the idea
that plant material constituted a significant part of this animal's diet (Section 3.2.1).

The helminth parasites of peramelid marsupials are diverse, and dominated by trichostrongyloid nematodes. The nematodes, cestodes and trematodes resemble those found in dasyurid marsupials, but other helminth groups, such as the heterakoids and subuluroids, are not shared with the Dasyuridae, and were probably acquired from other host groups following the development of the caecum in the peramelids (Beveridge & Spratt 1996).

3.2.3 Digestive tract function

The rate of digesta passage through the bandicoot gut might be expected to be a little slower than in carnivores of equivalent body size, given the bandicoot's slightly longer and more complex digestive tract. Griffiths (see Waring, Moir & Tyndale-Biscoe 1966) used different species of termites as markers in *Isoodon macrourus* by examining collected faeces for undigested exoskeletons. Marker termites first appeared in the faeces 7 hours after ingestion, reached a peak concentration at 9 hours, and were completely eliminated in 29 hours.

Moyle, Hume & Hill (1995) measured rate of digesta passage in *P. nasuta* using the two chemical markers described in Chapter 2 (Co-EDTA as the solute marker and Cr-cell walls as the large particle marker). They compared passage rates on two diets, mealworm larvae (to mimic the main component of the summer diet – insects) and shredded sweet potato (to mimic a major component of the winter diet – plant roots). Results are presented, along with data on food intake and digestibility, in Table 3.1.

On the insect diet the bandicoots virtually maintained body mass at a daily digestible energy intake of $511 \, \text{kJ} \, \text{kg}^{-0.75}$, and were in strongly positive nitrogen balance. In contrast, on the plant diet bandicoots ate only one-third as much digestible energy, they lost 7% of body mass, and were in negative nitrogen balance. Clearly, plant roots alone would not sustain them in the long term. Nevertheless, the digesta passage rate data showed that they were able to maximise their use of plant material by selectively retaining solutes and small particles in the caecum and proximal colon. Mean retention time of the large particle marker on the plant diet was more than double that on the insect diet (27 hours versus 11 hours). Also, the mean retention time of the fluid marker (Co-EDTA) was significantly longer than that of the particle marker on both diets (33 and 24 hours on the plant and insect diets respectively), indicating consistent selective retention of fluid digesta irrespective of diet. It could be seen radiographically that in the mealworm-fed bandicoots the major sites of digesta retention were the distal colon and rectum, whereas in the sweet-potato-fed animals digesta spent more time exposed to microbial attack in the caecum and proximal colon. This flexibility of the bandicoot digestive tract in terms of retention time and site is probably an important factor in the ability of these

Table 3.1. *Digestive performance of long-nosed bandicoots* (Perameles nasuta) *on animal and plant diets*

	Diet	
	Mealworm	Sweet potato
Body mass		
Mean (g)	795 ± 42	710 ± 65
Change (g d^{-1})	-2 ± 1	-10 ± 1
Dry matter		
Intake (g d^{-1})	18.3 ± 0.2	13.6 ± 1.0
Apparent digestibility (%)	90 ± 1	65 ± 11
Energy		
Intake of digestible energy (kJ kg$^{-0.75}$ d^{-1})	511 ± 16	176 ± 39
Nitrogen		
Intake (g kg$^{-0.75}$ d^{-1})	2.0 ± 0.1	0.1 ± 0.2
Apparent digestibility (%)	91 ± 1	-61 ± 13
Balance (g kg$^{-0.75}$ d^{-1})	$+1.0 \pm 0.3$	-0.2 ± 0.1
Transit time (hours)		
Fluid marker	8 ± 1	21 ± 4
Large particle marker	8 ± 1	21 ± 4
Mean retention time (hours)		
Fluid marker	24 ± 3	33 ± 2
Large particle marker	11 ± 2	27 ± 2
Significance of difference between markers on each diet	$P < 0.01$	$P < 0.01$

Note: Values are means ± standard errors. All differences between diets significant except mean body mass.
Source: After Moyle, Hume & Hill 1995.

small marsupials to exploit nutritionally unpredictable habitats such as regenerating heathlands following wildfire.

A subsequent study (McClelland 1997) has confirmed the presence of selective digesta retention in a second bandicoot, *Isoodon macrourus* (northern brown bandicoot), on a diet based on a commercial small carnivore mix and agar (a binding agent) with 24% ground lucerne (alfalfa) hay included. The neutral-detergent fibre (NDF) content of the complete mix was 20% on a dry matter basis. MRTs were 10.0 hours for the particle marker, but 27.4 hours for the fluid marker. Interestingly, there was no evidence for selective digesta retention in animals on a diet of the same basal ingredients but with 50% mealworm larvae in place of the ground lucerne; MRTs were 24.7 and 30.4 hours for the particle and fluid markers respectively. This may mean that a certain minimal level of indigestible bulk (either plant cell walls or invertebrate exoskeletons) is necessary for effective digesta separation in the hindgut; the NDF content of the mealworm supplemented diet was lower than the 14% in mealworms alone. In invertebrates NDF represents the chitin of the exoskeleton.

3.2.4 Nutrition and metabolism

Bandicoots and bilbies are in the body weight range in which neither strict herbivory nor obligate insectivory is common among mammals. Herbivory is generally restricted to specialised mammals larger than 600 g (Lee & Cockburn 1985). Smaller mammals are often insectivorous. Herbivores are generally large because of the need for a large gut capacity to retain food and symbiotic microbes long enough for the slow process of fermentation to proceed. Insectivores are limited in size because energy invested in foraging for and processing of small prey cannot be fully recovered unless a large biomass is ingested relative to body mass of the insectivore. Insectivores larger than 100 g usually feed on colonial insects (ants, termites) in order to minimise foraging time and maximise rate of prey intake and processing. They also often have low rates of energy expenditure (McNab 1984). A marsupial example is the numbat (Table 1.1). Bandicoots and bilbies are therefore in a 'nutritionally critical' body mass range (Moyle 1999). Perameloid marsupials cannot afford to be either specialist insectivores or specialist herbivores, so occupy the omnivore niche. A digestive tract that is more differentiated than that of carnivores, together with selective retention of digesta in the hindgut, allows bandicoots and bilbies to switch between plant, animal and fungal material according to the relative abundance of each food type.

Rates of metabolism of bandicoots from mesic habitats are close to the 'marsupial mean' (Dawson & Hulbert 1970). The golden bandicoot and the greater bilby have lower BMRs (Table 1.1), related to their generally arid habitats, but low FMRs as a multiple of BMR have only been reported for *I. auratus* during an extended drought. Nocturnal and semi-fossorial habits may allow the bilby to escape the environmental extremes of its desert habitat, and so explain why its FMR is no different from those of bandicoots from more mesic habitats (Table 1.4).

In Chapter 2 (Section 2.8), caudal fat storage in the small dasyurid *Sminthopsis crassicaudata* was shown to be an important short-term buffer against food shortages. From a survey of tail fat storage in a range of small mammals, Morton (1980) concluded that such storage appears to occur in two broad groups. The first contains species which undergo extended torpor. The second is comprised of desert-dwelling insectivores whose habitat and diet are such that short-term energy shortages are inevitable, and food storage impossible. Storage of fat in many animals is governed by postural and locomotory factors, and fat stored in the tail does not interfere with running and manoeuvrability in the same way body fat storage may (Pond 1978).

Among marsupials, Morton (1980) listed caudal fat storage as occurring in two burramyids (*Cercartetus lepidus* and *C. nanus*), ten species of dasyurids, two didelphids (*Thylamys elegans* and *Lestodelphis halli*), the caenolestid *Rhyncholestes raphanurus*, and *Dromiciops*. More recently, Gordon & Hall (1995) found evidence of caudal fat storage in two bandicoots, both in the genus *Perameles*, viz. *P. bougainville* (western barred bandicoot) and *P. eremiana* (desert bandicoot). The latter is presumed extinct, while *P. bougainville* is extinct on the mainland but currently secure on Bernier and Dorre Islands in Shark Bay. Both were once widespread across the arid and semi-arid zones. Although both species are/were omnivores, and therefore have/had wide nutritional niches, Gordon & Hall (1995) suggested that short-term food shortages, brought about by competition for scarce resources by more specialised insectivores or more specialised granivores, may have been a sufficiently strong selective force to favour evolution of fat storage in these species. They noted that *P. bougainville* is the smallest extant bandicoot (190–250 g). Fat storage in the tail may have developed as part of an evolutionary reduction in body size in arid areas. However, this does not explain why caudal fat storage has not developed in other arid-zone bandicoots.

Of the marsupial omnivores, bandicoots have been the main species studied in relation to resistance to the plant toxin fluoroacetate (Twigg, King & Mead 1990). *Isoodon obesulus* (southern brown bandicoot) from south-western Australia, with an LD_{50} of 20 mg 1080 kg^{-1}, has the highest tolerance of any bandicoot, and is three times less sensitive than the same species from south-eastern Australia. *Perameles bougainville* and *Isoodon auratus* both have some resistance (with an LD_{50} of 9 mg 1080 kg^{-1}); although they are not exposed to the toxin on their current island refuges, their former distributions included those of fluoroacetate-bearing plants. *I. macrourus* (northern brown bandicoot) from outside the distribution of fluoroacetate-bearing vegetation in south-eastern Australia is the most sensitive peramelid species, with an LD_{50} of only 3.5 mg 1080 kg^{-1}.

3.2.5 Ascorbic acid biosynthesis in bandicoots and other marsupials

Bandicoots appear to be unique among marsupials in having retained the ability to synthesise ascorbic acid (vitamin C) in their kidneys (Table 3.2).

The final step in the pathway by which ascorbic acid is synthesised from glucose in those vertebrates capable of synthesising the vitamin is the oxidation of L-gulonolactone to L-ascorbic acid (Chatterjee 1973). All

Table 3.2. *Ascorbic acid biosynthesis (as measured by* L-*gulonolactone oxidase activity in µmol g^{-1} tissue homogenate h^{-1}) in the kidney and liver of monotremes and marsupials; values are means (with ranges)*

Species (number of animals)	Kidney	Liver
Monotremata		
Ornithorhynchus anatinus (3)	10.1 (6.6–12.6)	Nil
Tachyglossus aculeatus (3)	18.8 (15.9–23.4)	Nil
Didelphidae		
Didelphis virginiana (13)	Nil	3.3 (0.4–6.9)
Dasyuridae		
Dasycercus byrnei (2)	Nil	0.5 (0.4–0.6)
Antechinus stuartii (11)	Nil	1.2 (0.3–1.7)
Dasyurus maculatus (1)	Nil	0.9
Peramelidae		
Perameles nasuta (1)	2.8	1.7
Isoodon macrourus (3)	5.3 (4.9–6.2)	4.4 (4.3–4.4)
Pseudocheiridae		
Pseudocheirus peregrinus (7)	Nil	1.6 (0.0–3.6)
Petauriodes volans (7)	Nil	0.8 (0.5–1.0)
Burramyidae		
Cercartetus nanus (1)	Nil	2.5
Vombatidae		
Vombatus ursinus (2)	Nil	5.6 (3.7–7.4)
Phalangeridae		
Trichosurus vulpecula (9)	Nil	2.5 (0.5–4.4)
Macropodidae		
Thylogale thetis (4)	0.4 (0.3–0.4)	4.2 (1.0–5.6)
Macropus rufogriseus (6)	0.6	5.2 (3.4–7.5)
Macropus eugenii (1)	Nil	1.9
Macropus giganteus (5)	Nil	2.8 (1.7–4.4)
Macropus robustus robustus (1)	Nil	1.7
Wallabia bicolor (5)	Nil	4.0 (3.6–4.5)

Note: After Birney, Jenness & Hume 1980

species known to require dietary vitamin C lack the enzyme L-gulonolactone oxidase.

In amphibians, reptiles and some birds, L-gulonolactone oxidase is located in the kidney. All species of eutherian mammals reported to be capable of synthesising ascorbate have L-gulonolactone oxidase solely in the liver. The only marsupial investigated before 1979 was the Virginia opossum, in which the enzyme was also reported in the liver (Nakajima, Shantha & Bourne 1969). A shift in the locus of biosynthetic activities from the kidneys and other tissues to the liver during evolution seems to be a general phenomenon.

Birney, Jenness & Hume (1979, 1980) investigated the locus of L-gulonolactone oxidase in Australian marsupials, and also in monotremes because of their numerous reptilian affinities (Griffiths 1978). Their results are summarised in Table 3.2. Although the quantitative relationship between the activity of L-gulonolactone oxidase *in vitro* and the rate of synthesis of ascorbate *in vivo* is unknown, it seems safe to assume that the enzyme activities measured reflect relative rates of ascorbate biosynthesis.

Both the platypus (*Ornithorhyncus anatinus*) and short-beaked echidna (*Tachyglossus aculeatus*) contained L-gulonolactone oxidase only in the kidney, in the manner of reptiles. This was the first time that ascorbic acid biosynthesis had been demonstrated in a mammalian kidney. Bandicoots of both genera studied, *Perameles* and *Isoodon*, had similar levels of activity in liver and kidneys. Although some individuals of two macropod species, *Macropus rufogriseus* and *Thylogale thetis*, also exhibited activity in the kidney, levels were only 10–12% of liver levels. All other marsupials examined had activity exclusively in the liver, in the manner of those eutherians capable of synthesising ascorbic acid.

Ascorbate biosynthetic ability in the kidney of bandicoots is a retained reptilian feature, and is consistent with the proposed early appearance of the order Peramelina in Antarctica when Australia, Antarctica and South America were all part of Gondwana 65–34 million years ago (see Chapter 9). It can be added to the list of conservative features of modern peramelid marsupials.

3.3 OMNIVOROUS POSSUMS AND GLIDERS

This group is diverse in both body size (from 7 g to 700 g) and dietary preferences, but they have in common diets that consist of some form of plant (or insect) exudate to satisfy the bulk of their energy requirements, and arthropods (and sometimes pollen) to meet their protein requirements. When the body weights of all Australian possums and gliders are plotted on a logarithmic scale (Fig. 3.8), they fall into three distinct clusters. The largest species, together with the koala, are folivorous (although some tropical species also take some fruit). These species rely upon microbial fermentation of plant cell walls and also detanning of plant cell contents in an expanded hindgut to help meet their energy and nutrient requirements. The comparatively long time required for these processes necessitates a large body size, with its associated advantage of reduced energy requirements per unit body mass. The arboreal folivores are dealt with in Chapter 5.

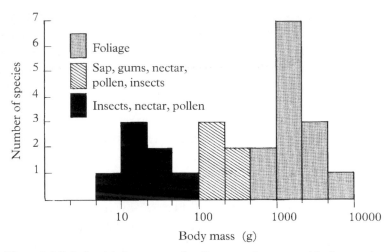

Figure 3.8 Relationship between major dietary preferences and body mass in possums, gliders and the koala. From Smith (1982a).

The other two clusters consist of omnivores. The intermediate-sized species (100–700 g) make up the family Petauridae. These are the sugar glider (*Petaurus breviceps*), mahogany glider (*P. gracilis*), squirrel glider (*P. norfolcensis*), yellow-bellied glider (*P. australis*), Leadbeater's possum (*Gymnobelidius leadbeateri*) and striped possum (*Dactylopsis trivirgata*). The first five all feed on arthropods and some form of plant exudate. *Dactylopsis* differs in that, although leaves, fruit, small vertebrates and the honey of native bees are reported to be eaten, insects provide by far the greatest part of its diet.

The smallest species (7–80 g) feed less on plant exudates, but instead concentrate on insects, seeds, fruit and pollen, as well as nectar. The Australian members include one terrestrial and four arboreal pygmy-possums (family Burramyidae), the honey possum (family Tarsipedidae) and the feathertail glider (family Acrobatidae).

3.3.1 **Diet studies**

Plant exudates that may be eaten include nectar, sap, manna, gums, kino and resins. In addition, honeydew and lerp, which are insect exudates, are derived from plant saps. All exudates have in common a low protein content.

Nectar is essentially a sugary solution produced by plants to reward pollinators, and has low concentrations of protein, vitamins and minerals. The plants visited by honey possums produce nectar containing sucrose, glucose and fructose, sometimes in similar proportions, with an average sugar concentration of 25% w/w sucrose equivalents (Richardson, Wooller & Collins 1986). Sap is also characteristically low in protein content and high in sugars. Eucalypt sap contains only 0.2% (dry weight) or less total nitrogen (Stewart *et al.* 1973). Phloem-feeding insects, mainly aphids, scale insects and lerp insects (psyllids) consume large volumes of sap in order to meet their protein requirements. Excess sugars ingested are excreted, in the same or slightly altered form, as a white encrustation called honeydew, or lerp. Many ants feed on honeydew deposited on the surface of leaves and small branches, but its availability depends on fluctuating numbers of sap-sucking insects, particularly psyllids, and so it is an unpredictable food.

Manna is also derived from sap. It is formed from sap that exudes at sites of insect damage on tree branches and leaves. The composition of the sugars in the manna is slightly changed from that of phloem sap by the action of enzymes introduced by insect saliva (Basden 1966).

Gums, or glycans, are polysaccharides of plant origin. They are hydrocolloids in that they form gels in water (Adrian 1976). Gums are produced in relatively large quantities by some Australian acacias, and are similar in composition to the commercially produced gum arabic from the African *Acacia senegal*, although they appear to be less 'soluble' in water (Mantell 1949). They are produced by trees in response to insect damage and mechanical injury to seal off injured sites and so conserve water and prevent invasion by pathogens. Lindenmayer *et al.* (1994) reported total nitrogen contents of 0.2 to 0.7% in the gums of three species of *Acacia* in alpine Victoria. These are very low, like sap. Sugar contents ranged from 24 to 68%. They suggested that differences in sugar and nitrogen content of *Acacia* gums could influence the distribution and abundance of several species of marsupials that feed on *Acacia* gum.

Kinos are produced by eucalypt trees in response to physical damage, and are often called gums. They differ from gums by containing large amounts of toxic polyphenolic compounds, are astringent in taste, and are usually red or orange in colour. Kinos are not eaten.

SUGAR GLIDER

It is clear from the results of several studies in different locations that the diet of *P. breviceps* (sugar glider) is highly correlated with patterns of resource abundance, which in turn is a reflection of habitat floristics. The most comprehensive study has been that of Smith (1982b). He used faecal scat analysis and direct observation to follow seasonal changes in the diet of sugar gliders in a strip of roadside vegetation in Victoria. This fragmented habitat was dominated by *Acacia mearnsii* and four species of *Eucalyptus* trees. Results are presented in Fig. 3.9 as a percentage of total feeding time observations devoted to six activities: feeding on *Acacia* gum , feeding on *E. bridgesiana* sap, licking branches for honeydew, peeling bark for arthropods, gleaning foliage for manna, and mixed eucalypt feeding. The latter category included random feeding and movement through eucalypts, with no con-

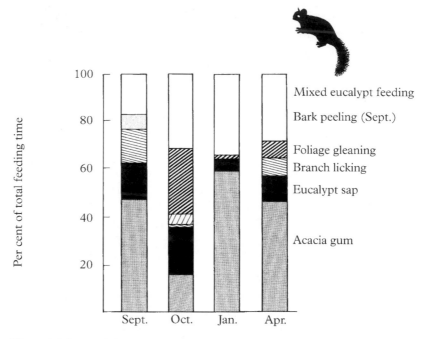

Figure 3.9 Seasonal changes in the proportion of total feeding observation time spent by sugar gliders (*Petaurus breviceps*) searching for different food resources in fragmented habitat in South Gippsland, Victoria. Modified from Smith (1982b).

sistent feeding activity evident, but when flying insects and large foliage dwelling insects were sometimes caught.

During spring and summer, sugar gliders fed mainly on insects and spiders, even though exudates were also most abundant during these seasons, probably to meet the protein requirements of reproduction. The most important insects were moths and scarabaeid beetles. The larval stages of these insects develop in pastures adjacent to the study area, while the adult stages use eucalypts for shelter. Sugar gliders obtain insects and spiders by searching eucalypt foliage, by chasing and catching insects flying amongst the foliage at night, and by stripping and searching under loose bark. They have an elongated fourth digit on the manus which they use to extract insects from crevices in the bark, and enlarged lower incisors that are used to chew into the bark of trees.

During autumn and winter, sugar gliders fed predominantly on plant exudates, mainly acacia gum (Fig. 3.10), eucalypt sap, and honeydew and manna on eucalypts. Although no fresh gum was produced in winter, some gum nodules secreted in the previous summer persisted throughout the winter. One tree produced 780 g dry weight of gum in one year, an energy equivalent of 31 kJ d^{-1}. Six trees of this type would be sufficient to satisfy the annual energy requirements of one sugar glider, based on the field metabolic rate (FMR) measurements of Nagy and Suckling (1985). Gum

Figure 3.10 A sugar glider (*Petaurus breviceps*) feeding on *Acacia* gum. (Andrew Smith)

sites generally consisted of holes made by insect borers and from which gum was exuding, or in which it had accumulated. The animals frequently prise open borer holes with their larger lower incisors to access pockets of gum. In doing so they often ingest small particles of bark.

Sugar gliders also use their incisors to strip bark and make characteristic sap feeding sites, mainly in *E. bridgesiana* trees. Smith (1982b) measured the yield of sap from one such site in late winter. The site yielded 0.98 g of dried sugars in 6 hours after it was last used by a sugar glider, equivalent to 57 kJ d^{-1}. Thus two or three sap feeding sites should be sufficient to satisfy a sugar glider's daily energy needs. Sap sites are changed or renewed frequently. Renewal is accomplished by chewing the sides of the wound, again resulting in the ingestion of small particles of bark. Change of sites is necessary because eucalypts respond to feeding damage by clogging the wound with kino after several weeks.

Honeydew was obtained by licking smooth outer branches and from beneath loose bark. Manna was harvested by gleaning new foliage and flower buds. When gleaning, sugar gliders remained in small clumps of outer new foliage for up to 45 minutes, closely examining leaves and buds by grabbing, sniffing and licking them.

Smith (1982b) concluded that in addition to enlarged incisors and the lengthened fourth digit on the manus, another important morphological feature of sugar gliders is their well-developed gliding membrane, which increases their mobility and hence food harvesting rates in open forest habitats. At his study site the most productive and least dispersed food resource available year-round was eucalypt sap. Acacia gum was also heavily utilised in all seasons, and in winter these two exudates formed the staple diet. Suckling (1984) found that differences in density of sugar glider populations were most readily explained by differences in abundance of *Acacia mearnsii*, the sole source of gum in Smith's study site. The low protein content of their staple winter diet of acacia gum and eucalypt sap may well prevent sugar gliders from breeding at this time of the year. Sugar gliders had a single breeding season beginning in early spring (Suckling 1984), coinciding with the appearance of arthropods.

Later, on the south coast of New South Wales, Howard (1989) used a similar approach with three colonies of sugar gliders in patches of eucalypt woodland surrounded by closed heath in undisturbed habitat in Jervis Bay Nature Reserve. In this area of high floristic diversity, the most observed foraging behaviour was of feeding at *Banksia* and *Eucalyptus* flowers for nectar and pollen (Fig. 3.11). All faeces collected in summer, winter and spring had pollen in them, but only 55% of faeces collected in autumn (April) had pollen. Eucalypt pollen was found mainly in summer, whereas *Banksia* pollen occurred during winter, corresponding with the main

Figure 3.11 Seasonal changes in the proportion of total feeding observation time spent by sugar gliders (*Petaurus breviceps*) in different feeding activities at Jervis Bay, New South Wales. Modified from Howard (1989).

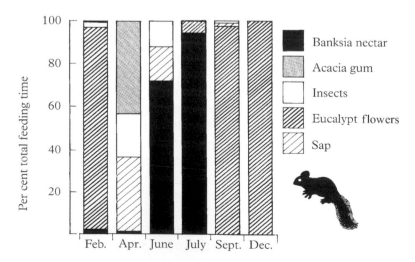

flowering periods of the two genera. When *Banksia ericifolia* was in flower (May–July) nectar would drop onto the stems and the ground.

Sap (from *E. gummifera*) was an important food resource during autumn, the trough in flower abundance. Trees incised by sugar gliders had significantly higher sap flow rates than non-incised trees. Howard (1989) measured this by hammering a '*v*' shaped aluminium plate to a depth of 1 cm and collecting all the sap that exuded out of the wound with a capillary tube (Goldingay 1987). The selection of sap-site trees is poorly understood, but Mackowski (1988) found that trees incised by *P. australis* (yellow-bellied glider) had a relatively low cambial electrical resistance. This is a measure of tree vigour. The lower the electrical resistance the more vigorous the tree, the thicker the live bark, and the wider the phloem, all of which are indicative of greater sap flow.

Arthropods (beetles, moths and spiders) were an important food item also in autumn, when pollen availability was lowest. Sugar gliders commonly sat in front of *B. serrata* cones and systematically searched them for spiders and beetles. When an insect was caught the wings and/or legs were discarded before swallowing. They also foraged on the ground during April, presumably for insects. However, overall, insects were not as important a protein source at Jervis Bay as they were at Smith's (1982b) Victorian site (Fig. 3.9). Instead, pollen, which was available most of the year, appeared to be preferred, even in summer when arthropods were presumably abundant. When visiting flowers, sugar gliders can obtain both protein and energy by foraging on both pollen and nectar. On average, 34% of *Eucalyptus* pollen and 71% of *Banksia* pollen in faeces was devoid of contents, indicating that sugar gliders could access the contents of pollen grains as a source of protein (Howard 1989). Van Tets & Whelan (1997) found a similar proportion (66%) of *Banksia* pollen grains in the faeces of sugar gliders near Wollongong, south of Sydney, to be empty.

On the north coast of New South Wales, Quin (1995) studied the population ecology of sugar gliders and squirrel gliders (*P. norfolcensis*) in an area of eucalypt forest and woodland that included *Melaleuca* swamp forest and heath understoreys. The site was relatively undisturbed habitat in Limeburners Creek Nature Reserve. Although invertebrates were taken in all seasons, sugar gliders fed heavily on nectar and pollen of *Banksia integrifolia* in winter and spring, *B. serrata* in spring and summer, *Eucalyptus pilularis* in winter and *E. gummifera* in summer, when these species flower. Honeydew was obtained from *E. pilularis* and *E. tereticornis* in summer. Recruitment of young animals into the population appeared to be most successful during years when there was heavy eucalypt and banksia flowering.

SQUIRREL GLIDER

Squirrel gliders at Limeburners Creek appeared to have very similar dietary habits to those of sugar gliders, with the same species of *Eucalyptus* and

Banksia used as sources of nectar and pollen (Quin 1995). Recruitment patterns also were most successful in years of heavy eucalypt and banksia flowering. Menkhorst & Collier (1988) also concluded that the diet of squirrel gliders in central and northern Victoria was broadly similar to that of sugar gliders, except for the preponderance of insects, particularly caterpillars. Eucalypt nectar and pollen appeared to be less important at one site but were heavily utilised at another. The availability of nectar and pollen may be both irregular and unpredictable in open eucalypt forests of low species diversity, but the high diversity and abundance of foliage invertebrates in such forests probably compensates.

A much greater diversity of dietary items was recorded by Sharpe & Goldingay (1998) in a floristically richer site on the north coast of New South Wales. Nectar and pollen were the most important foods, accounting for 59% of all feeding observations. *Banksia integrifolia* was the most important source of these foods, but four species of eucalypts were also used heavily when in flower in spring and summer. Arthropod feeding constituted 26% of all feeding observations, and was lowest in early winter when pollen ingestion was high. Other food resources used at some time(s) during the year included lichen (in winter), fruit (summer), sap (autumn, winter), *Acacia* seeds and arils (spring, summer), *Acacia* gum (autumn), and honeydew (all seasons except winter).

LEADBEATER'S POSSUM

The rare and endangered leadbeater's possum (*Gymnobelidius leadbeateri*) has the appearance of a sugar glider without its gliding membrane. It was known from only five specimens between 1867 and 1909 and had long been presumed to be extinct until its rediscovery in 1961 (Smith 1995). Leadbeater's possums are restricted to wet sclerophyll (closed eucalypt) forests dominated by mountain ash (*E. regnans*) in the Central Highlands of Victoria. Smith (1984) used both faecal analysis and direct observation to determine how Leadbeater's possums exploited the range of resources available to them. However, this fast-moving inhabitant of Australia's tallest forests is extremely difficult to observe, except briefly at dusk when it emerges from its nest hollow 6–30 m above the ground. When out of the nest the animals move rapidly through the forest canopy, often making spectacular leaps between trees (Smith 1995). A high proportion (73%) of the Leadbeater's possum's daily energy budget is used in activity (as well as the specific dynamic action of feeding) (Smith *et al.* 1982).

Analysis of diet was therefore heavily dependent on faecal scat analysis, with all its shortcomings. Dietary items included arthropods (mainly tree crickets, beetles, moths and spiders), plant exudates (*Acacia dealbata* and *A. obliquinerva* gum, *E. regnans* nectar and manna), and insect exudates (psyllid honeydew). Smith (1984) estimated that arthropods provided less than 20% of the daily energy requirement, calculated from measurement of

FMR by Smith *et al.* (1982). The most important arthropods were an undescribed tree cricket that sheltered in curled-up tips of decorticating *E. regnans* bark during the day but emerged to feed at night. Remains of lerp insects were moderately abundant in the faeces of Leadbeater's possums during spring, summer and autumn.

E. regnans, the only eucalypt in the study area, flowers only in alternate years, so eucalypt nectar availability is not assured. Leadbeater's possums do not scar *E. regnans* trees for sap, but they do notch feeding sites on the trunks and branches of understorey acacias. The gum from *Acacia dealbata* and *A. obliquinerva* is water-soluble, and so does not persist during periods of low production (spring). Leadbeater's possums were seen to lick what Smith (1984) assumed to be manna from branches of *E. regnans*.

Thus the overall dietary pattern of Leadbeater's possum was similar to that reported by Smith (1982b) for the related sugar glider, but the winter shortage of exudates was more pronounced. The presence of tree crickets in the diet throughout the year may be important in permitting Leadbeater's possums to breed in winter as well as spring (Smith 1995).

YELLOW-BELLIED GLIDER

Yellow-bellied gliders (*Petaurus australis*) are the largest of the omnivorous possums and gliders, and amongst the largest insect- and exudate feeding arboreal mammals in the world. This helps to explain their social organisation. They live in family groups consisting of a monogamous pair and one or more offspring (Henry & Craig 1984; Goldingay & Kavanagh 1990). Home ranges are large (25–85 ha), much larger than would be predicted on the basis of their body mass when compared with other exudivorous marsupials and primates (Goldingay & Kavanagh 1993). The home ranges are exclusive to each group, and likely centred on defence of food resources. Areas containing at least 150 family groups are probably needed to support viable populations (Goldingay & Possingham 1995). Ninety per cent of the time that yellow-bellied gliders spend outside their nest hollows is devoted to foraging (Goldingay 1989). Even so, they are subject to large fluctuations in body mass (Craig 1985), reflecting seasonal shortages in food supply. The well-known vocalisations of yellow-bellied gliders (Kavanagh & Rohan-Jones 1982; Goldingay 1994), together with their gliding ability, no doubt assist in the maintenance of the large home ranges.

The diet of the yellow-bellied glider has been documented in three main locations, in Victoria, southern New South Wales, and in north Queensland. Two studies in tall open eucalypt forest in the Victorian Central Highlands by Henry and Craig (1984) and Craig (1985) showed that yellow-bellied gliders fed principally on arthropods (lerps and other scale insects, beetles and their larvae, and moths), supplemented with insect and plant exudates and pollen. Eucalypt sap was accessed in two ways. Yellow-bellied gliders make well-defined '*v*'- and heart-shaped incisions deep into the cambial layers. These characteristic incisions are obvious on smooth-barked species such as *E. globulus* and *E. cypellocarpa*, but are partly obscured or completely hidden by loose bark on rough-barked species such as *E. obliqua* and *E. sideroxylon*. Most sap feeding occurred in winter and spring. The second way that yellow-bellied gliders accessed sap was to strip live bark from the limbs and outer branches of both *E. cypellocarpa* and *E. obliqua*, then to intensively lick the damaged surface. Undamaged outer branches, smooth bark under loose bark, and *E. cypellocarpa* foliage were also searched and licked, presumably for manna and honeydew. Nectar was taken from four eucalypt species, all of which were large-flowered.

In tall montane forest in south-east New South Wales the main components of the diet were eucalypt sap, arthropods and honeydew (Goldingay 1986). There was an annual cycle of food resource use that was dictated by the pattern of phenological change in the forest (Fig. 3.12). Eucalypt sap was harvested mainly in summer (from *E. viminalis*) and winter (from *E. fastigata*). Arthropods were heavily used in autumn and spring. Honeydew was used in all seasons, but most heavily in spring and early summer. Manna was only harvested during December (early summer), from *E. viminalis*. A similar pattern of seasonal food use related to the annual phenological cycle of the forest at this site was subsequently reported by Kavanagh (1987).

In coastal forest in south-eastern New South Wales, yellow-bellied gliders fed extensively (70% of total feeding observation time) on the nectar of all eucalypt species present (Goldingay 1990). Only in winter and early spring was eucalypt sap an important food component (Fig. 3.13). Such a

Figure 3.12 Seasonal changes in the proportion of total feeding observation time spent by yellow-bellied gliders (*Petaurus australis*) in different feeding activities in tall montane forest in south-eastern New South Wales. From Goldingay (1986).

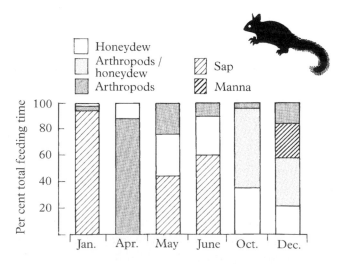

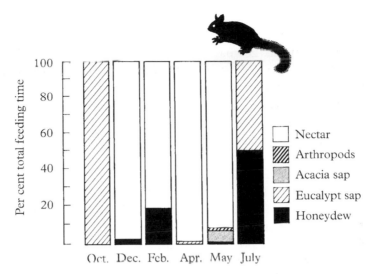

Figure 3.13 Seasonal changes in the proportion of total feeding observation time spent by yellow-bellied gliders in different feeding activities in coastal forest in south-eastern New South Wales. Modified from Goldingay (1990).

predominance of nectar in the diet of yellow-bellied gliders has not been recorded at any other site; other food types were important in the diet only in the absence of eucalypt blossom.

In north Queensland, yellow-bellied gliders were also found to rely more heavily on exudates than do the smaller sugar gliders. Sap feeding accounted for more than 80% of all feeding observations throughout the year (Quin *et al.* 1996). Smith & Russell (1982) had earlier recorded a relatively low volume of invertebrates (mainly tree crickets, adult and larval beetles, caterpillars and spiders) in yellow-bellied glider faecal pellets. They felt that the high sap and low arthropod content in the diet (and thus a relatively low protein intake) was consistent with the low fecundity of this species.

MAHOGANY GLIDER

The mahogany glider (*P. gracilis*) is one of the most endangered marsupials. It is found only in a narrow strip of coastal woodland between Ingham and Tully in north Queensland, which is under threat from clearing for agriculture and aquaculture (Van Dyck 1995). Its diet includes nectar from eucalypts and grass-trees (*Xanthorrhoea*), sap from eucalypts and acacias, gum from the flower spikes or spears of grass-trees, and lipid-rich strings of gum (arils) from *Acacia crassicarpa*. Invertebrates and lichens are also eaten. *Xanthorrhoea* gum is harvested by biting a piece out of the base of a flower spear, and returning the following night to feed on the amber mucilaginous jelly that exudes slowly from the wound. In summer, when the ripe pods of *A. crassicarpa* burst open, female mahogany gliders feed on the exposed arils to the virtual exclusion of all other foods (Van Dyck 1995).

STRIPED POSSUM

Dactylopsila trivirgata is also a north Queensland species, but is more widely distributed, from Townsville to Cape York. As in New Guinea, the striped possum's habitat in Australia includes rainforests and riverine vine-forests. Examination of the contents of two preserved specimens led Smith (1982a) to conclude that the striped possum was basically an arboreal anteater. Both animals had ingested large quantities of formicid ants, along with smaller amounts of termites and wood-boring beetle larvae. The presence of ant pupae and termite nymphs suggests that prey are obtained by breaking into nests. Striped possums have several unusual features that support Smith's (1982a) idea that it is an arboreal anteater. The procumbent lower incisors are longer than in any other possum (Kay & Hylander 1978), and the upper first incisors project forward horizontally rather than vertically. Striped possums use their lower incisors to tear out pieces of wood, with the upper incisors acting as a fulcrum (Smith 1982a). An elongated tongue and elongated fourth finger (Rand 1937) are then used to remove insects from the cavity. The only plant material Smith (1982a) found in the faeces were fragments of bark, which is consistent with their unintentional ingestion while gouging bark and searching for insects.

At 400 g, *Dactylopsila* is unusual in that it falls within the body mass range of the other petaurids, all of which feed on a mixture of exudates and arthropods (Fig. 3.8), rather than with the smaller, primarily insectivorous burramyids (see below). Like the marsupial moles (see Chapter 2), it overcomes the disadvantage of large body size by feeding on colonial ants and termites. Arthropod remains in its faeces were remarkably intact (Smith 1982a), suggesting rapid ingestion with minimal time for mastication when feeding at social ant nests. Smith concluded that *Dactylopsila* may fill an arboreal anteater niche in Australian and New Guinean rainforests comparable with that filled by myrmecophagids such as the anteaters and armadillos in South American rainforests. However, more recent field observations on *Dactylopsila* in north Queensland by Handasyde & Martin (1996) suggest that the species is a generalist insectivore that consumes some fruits and leaves as well, rather than a specialist ant-eater.

PYGMY-POSSUMS

The five species of pygmy-possums (family Burramyidae) are predominantly insectivorous and small (10–50 g). The mountain pygmy-possum (*Burramys parvus*) is the only Australian mammal with a distribution limited to alpine and sub-alpine regions. Its highly seasonal habitat is reflected in its diet. In summer (the breeding season) it feeds primarily on arthropods, mainly the bogong moth (*Agrotis infusa*). On Mt. Higginbothom in Victoria, Mansergh *et al.* (1990) found that bogong moths made up 31% of faecal samples, and 46% in females during the breeding season. Other components of the diet were other invertebrates, predominantly insects (32%) and plant material (16%). More fruits were consumed in the non-breeding season. In Mt. Kosciusko National Park, Smith & Broome (1992) found a similar pattern. The diet consisted of arthropods, seeds and berries. Arthropods (mainly bogong moths) dominated the diet in summer, with increasing occurrence of seeds and fruits of heathland shrubs (including mountain plum-pine (*Podocarpus lawrencii*) and snowbeard heath (*Leucopogon montanus*)) in autumn (Fig. 3.14). Bogong moths aestivate in great numbers in the same boulder fields used by *Burramys* as nesting and hibernation sites, and are high in fat (up to 65% of dry weight) and protein. The higher bogong moth intake by females than males in all seasons is consistent with their higher energy and protein requirements for reproduction and the need to ensure that young gain sufficient weight during the short alpine summer to survive their first winter in hibernation.

The mountain pygmy-possum's diet thus differs from those of other small omnivorous possums and gliders in the low diversity of arthropod prey, the high diversity and abundance of seeds and berries, and the absence of exudates. The use of seeds by *Burramys* is consistent with its dental morphology. Highly specialised plagiaulacoid upper premolars with serrated cutting edges and grooved sides probably help to crack and open hard seeds (Smith & Broome 1992). The highly seasonal availability of arthropods, seeds and berries in its alpine habitat limits *Burramys* to a single reproductive effort each year. It survives the winter trough in food supplies by caching seed, storing subcutaneous fat and hibernating. Torpor bouts of 3 to 17 days are interspersed with arousal periods averaging 19 hours throughout winter (Broome & Geiser 1995).

The eastern pygmy-possum (*Cercatetus nanus*) and the feathertail glider (*Acrobates pygmaeus*) (Fig. 3.15) seem to feed upon similar food resources. In a tall open eucalypt forest with an understorey of *Banksia spinulosa* in

Figure 3.14 Seasonal changes in diet (as relative surface area in faeces) of mountain pygmy-possums (*Burramys parvus*) in Mt. Kosciusko National Park. Modified from Smith & Broome (1992).

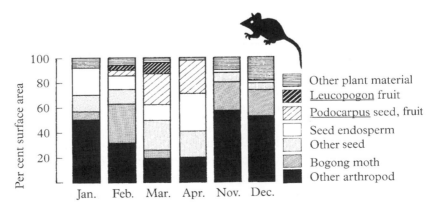

Figure 3.15 The Feathertail Glider (*Acrobates pygmaeus*). (Pavel German)

Victoria, Huang, Ward & Lee (1987) used faecal pellet analysis to conclude that both species concentrated on pollen, seeds and insects. Pollen was abundant in faecal pellets between April and September, coinciding with the flowering period of *B. spinulosa* and several species of eucalypt. Seeds were important dietary items between October and March. Insects were present throughout the year and showed no seasonal pattern in abundance in the faeces. Thus, with the exception of insects, both the eastern pygmy-possum and the feathertail glider changed their diet according to seasonal changes in availability. However, it is important to note that the importance of exudates in the diet could not be determined from this study based only on faecal pellet analysis.

Direct observations of feeding behaviour by feathertail gliders by Goldin-gay & Kavanagh (1995) in south-eastern New South Wales have provided a more complete picture of this marsupial's diet. Nectar (and pollen) appeared to be only an occasional dietary item, but another exudate, honey-dew, was thought to be very important. Most (91%) feeding observations were in live eucalypts (mainly *E. viminalis*, *E. radiata* and *E. fastigata*), and 61% of these were of feathertail gliders searching under loose shedding bark. A further 32% were of foliage gleaning. Foraging among loose bark was thought to include feeding on honeydew and arthropods, while foliage gleaning was proposed as being indicative of feeding on manna, honeydew, lerps and other arthropods. Although diet composition could not be quantified, it appears that the diet of feathertail gliders is much broader than would be indicated by faecal pellet analysis alone. Turner (1984a) had earlier recognised this, when she suggested that plant exudates (nectar, sap, manna and honeydew), which are totally digested and therefore not present in the faeces, could be the major source of energy to feathertail gliders. Their major protein source appeared to be pollen; *Eucalyptus* pollen was the most abundant food item in the faeces of feathertail gliders in dry scler-ophyll forest in Victoria, and the majority of pollen grains were empty of their contents. Arthropod fragments were not abundant in the faecal pel-lets, but caution must be exercised in the interpretation of this in the absence of direct feeding observations; many small omnivores and insec-tivores discard the hard exoskeleton and ingest mainly the highly digestible soft tissues. It is also possible that the diet of *Acrobates* differs not only seasonally but also between sites, with pollen a major source of protein in some areas, and arthropods in others.

The faeces of eastern pygmy-possums (*Cercartetus nanus*) near Wollon-gong, south of Sydney, contained 31–55% by volume of *Banksia* pollen between autumn and early spring (van Tets & Whelan 1997). Two-thirds of the pollen grains were empty, indicating that the protoplast had been extracted during passage through the digestive tract. Invertebrate material replaced the pollen during late spring and throughout summer.

The long-tailed pygmy-possum (*Cercartetus caudatus*) is a little-known burramyid from the rainforests of north Queensland and from similar habitats in New Guinea. Smith (1986) examined the contents of the stomach of seven preserved specimens that had been collected during the dry season in New Guinea. Most dietary remains consisted of finely mas-ticated arthropod fragments (mainly tree crickets, cockroaches, spiders, caterpillars and beetles). Plant structural remains consisted of a few small fragments of bark, and liverwort or moss. The presence of bark suggests that *C. caudatus* may gouge living trees to feed on sap. The relative dimen-sions of the lower central incisors are comparable to those of Leadbeater's possum, which uses its lower incisors for wood gouging (Smith 1984). The stomach of one specimen contained what could have been some type of plant exudate. One faecal sample contained pollen, none of which had been digested. From these scanty data Smith (1986) tentatively concluded that *C. caudatus* was probably predominantly insectivorous, but that the relative importance of insects and exudates in its diet had to wait for more detailed observations of the animals feeding in their natural habitats.

A species endemic to New Guinea is the feather-tailed possum (*Dis-toechurus pennatus*). It is an uncommon but widespread species in rainforest habitats (Flannery 1995). Its diet is thought to consist of invertebrates and possibly fruit as a large component. Woolley & Allison (1982) maintained several feather-tailed possums in captivity, and observed that cicadas were always eaten in preference to other insects. When feeding, they would kill large insects by bites to the head and thorax, the large upper canines and lower procumbent incisors being particularly effective in this process. The insect was then held with the paws while the contents of the thorax and abdomen were licked out, in the same way that striped possums had been observed to do by Rand (1937). The wings and usually head and exoskeleton of the thorax and abdomen were then discarded. Small insects are chewed with the premolars and molars and the hard exoskeleton spat out from the side of the mouth. The procumbent lower incisors are also used for cutting grooves in the bark of trees, and the animals then lick the sap which exudes.

The tongue of *D. pennatus* is long (21 mm from a fine tip to the epiglottis). The dorsal surface is covered with a mat of backwardly pointed filiform papillae, which Woolley & Allison (1982) thought would aid in the removal of contents from large insects. The gross morphology of the digestive tract suggests that plant material such as fruit or exudates such as gums form an important part of the diet. The hindgut is about 10 cm in length, which is quite long when compared with a small intestine of 25 cm, and there is a long (about 5 cm) caecum (Woolley & Allison 1982). These gut proportions are reminiscent of those of the sugar glider, which feeds extensively on *Acacia* gum (Smith 1982b), but as with the long-tailed pygmy-possum, direct field observations of the feeding patterns of the feather-tailed possum are needed.

HONEY POSSUM

The smallest of the exudivores are the little pygmy-possum (*Cercartetus lepidus*) and the honey possum (*Tarsipes rostratus*), both at less than 10 g. Their diets are very different. The little pygmy-possum is primarily insectivorous and feeds on a wide range of insects, spiders and even small lizards (Green 1980). It secures its prey with its forepaws and tears away edible portions with its teeth. It also takes some nectar. In contrast, the honey possum is a nectar and pollen specialist, rarely if ever taking insects (Richardson, Wooller & Collins 1986). Several features of the honey possum are correlated with its specialised diet. Its tongue has long filiform papillae at the tip and shorter compound papillae over most of the dorsal surface, like that of the feather-tailed possum (above). The long tongue, stiffened by a keratinised keel, is used to collect nectar and pollen from flowers, particularly of *Banksia* spp. The pollen is scraped from the papillae by a series of transverse combs on the roof of the mouth. The digestive tract is simple, with no hindgut caecum, but the capacious stomach is unusual in having a well defined diverticulum. This diverticulum may serve to store nectar in times of surplus. Pollen is not digested in the stomach but rather during its passage through the simple intestine (see Section 3.3.3 below). *Banksia* pollen is high in protein (36–42%) (Turner 1984b), and the pollen from only a few inflorescences would be needed to satisfy the protein requirements of *T. rostratus*. The honey possum has both a high rate of mortality (the average lifespan may only be one year) and a high reproductive potential (Wooller *et al.* 1981). These two features are probably linked to its specialised nectar and pollen diet. A high fecundity throughout the year may enable it to respond opportunistically to seasonal and stochastic changes in nectar and pollen supply resulting from fire and rain.

3.3.2 Marsupials as pollinators

Many small exudivores have been implicated as pollinators of native plants (Turner 1982). Many of them may visit flowers primarily to collect nectar, but in the process pollen attaches to their fur. For example, Goldingay, Carthew & Whelan (1987) considered that *Antechinus stuartii* visits flowers to obtain nectar rather than pollen, because pollen grains present in the faeces were mostly intact and therefore undigested. Pollen loads present on many nectarivorous animals suggest that they have the capacity to pollinate flowers. However, pollen loads on foraging animals may be much greater than those which have had time to groom in a trap. For instance, a honey possum collected by Hopper & Burbidge (1982) while it was foraging had a pollen load two orders of magnitude greater than that from trapped animals. Similarly, Goldingay *et al.* (1987) found greater pollen loads on animals which had been in traps for shorter than for longer periods.

Twenty-five species of marsupials are known to visit flowers presumably for nectar and/or pollen, but only the 11 species listed in Table 3.3 can be considered as potentially important pollinators. The greater mobility of gliding species such as *P. australis*, *P. breviceps* and *P. norfolcensis* may result in dispersal of pollen over much larger distances than by non-volant species.

Marsupials have probably been visiting flowers since the Upper Creta-ceous (Sussman & Raven 1978); some extinct small South American marsupials had features indicative of a partly nectarivorous diet. Also, flowers apparently adapted to vertebrate pollinators were present by the end of the Cretaceous. The genera most commonly visited by modern marsu-pials are *Eucalyptus* in the family Myrtaceae, and *Banksia* in the Proteaceae. Although Ford *et al.* (1979) concluded that mammal pollination probably evolved before bird pollination in *Banksia*, Turner (1982) suggested that *Banksia* may have coevolved in the Cretaceous with ancestral marsupials and parrots. The other major avian pollinators, the honeyeaters, probably entered Australia a little later, in the Miocene. Both vertebrate groups (marsupials and birds) have extant species variously specialised for feeding at flowers. Early coevolution is supported by the large number of characters in *Banksia* adapted for both marsupial and bird pollination. Table 3.4 lists the characters helpful to small marsupial pollinators.

Whatever the sequence of coevolutionary events, it is clear that small marsupials are major pollinators of a wide range of species, particularly in the families Myrtaceae and Proteaceae.

Table 3.3. *Flower-visiting marsupials in Australia and the importance of nectar and pollen in their diet*

Family and species	Body mass (g)	Nectar and/ or pollen	Plant genera visited
Dasyuridae			
Antechinus stuartii	35	+	*Banksia*
Parantechinus apicalis	70	++	*Eucalyptus, Banksia*
Burramyidae			
Cercartetus lepidus	7	+	*Eucalyptus, Banksia*
Acrobates pygmaeus	12	++	*Callistemon, Eucalyptus, Banksia*
Cercartetus concinnus	13	++	*Eucalyptus*
Cercartetus nanus	24	++	*Callistemon, Leptospermum, Eucalyptus, Banksia*
Cercartetus caudatus	30	++	*Syzygium*
Petauridae			
Petaurus breviceps	130	+	*Syzygium, Eucalyptus, Banksia, Grevillea, Hakea, Syncarpia, Xanthorrhea*
Petaurus norfolcensis	230	+	*Callistemon, Eucalyptus*
Petaurus australis	550	+	*Eucalyptus, Loranthus*
Tarsipedidae			
Tarsipes rostratus	9	+++	*Angophora, Callistemon, Calothamnus, Eucalyptus, Leptospermum, Melaleuca, Banksia, Dryandra, Grevillea, Hakea, Andersonia*

Note: Estimated proportion of diet: + small; ++ moderate; +++ large.
Source: Adapted from Turner 1982.

3.3.3 Digestive tract form and function

With the exception of the honey possum, the digestive tracts of the omnivorous arboreal marsupials are broadly similar to those of *Didelphis* and the bandicoots already discussed in this chapter, with modest enlargement of the hindgut caecum into a fermentation chamber, and a short colon. Thus the digestive tract of the feathertail glider (Fig. 3.16) is rather similar to that of *Perameles nasuta* (Fig. 3.5). The digestive tracts of the yellow-bellied glider, mountain pygmy-possum and striped possum also conform to this pattern (Smith 1982a). The diets of these four arboreal marsupials consist mainly of soluble exudates and arthropods (feathertail glider and yellow-bellied glider), arthropods and seeds (mountain pygmy-possum) or

Table 3.4. *Co-evolutionary traits of* Banksia *flowers and marsupial pollinators of* Banksia

Characteristics of *Banksia* pollinators	Characteristics of *Banksia*
Large body size relative to insects	Flowers tightly clustered, strongly attached to stems. Large stigma – nectary distance
Arboreal and scansorial foragers	Inflorescence exposed and well above ground
Scansorial and terrestrial foragers	Inflorescence hidden and close to ground
Marsupials have no colour vision but are sensitive to light tones	Mainly yellow inflorescences, but also some red and dull flowers
Marsupials sensitive to smell	Strong, sweet odour
Nectar feeders	Abundant nectar production
Pollen feeders	Abundant pollen production
Active all year	Mainly winter flowering

Source: Adapted from Turner 1982.

Figure 3.16 Gastrointestinal tract of *Acrobates pygmaeus* (feathertail glider). Redrawn from Schultz (1976).

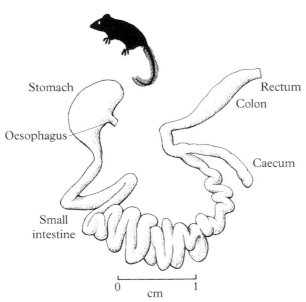

arthropods (striped possum); gums do not figure prominently. The digestive tracts of the striped possum and mountain pygmy-possum are shown in
Fig. 3.17.

In contrast, the sugar glider (Fig. 3.10) and Leadbeater's possum both
feed extensively on *Acacia* gum, which probably requires microbial fermentation to be digested (although direct experimental evidence is lacking). The
gastrointestinal tract of the sugar glider is shown in Fig. 3.18. It is note-
worthy that the caecum of both the sugar glider and Leadbeater's possum is
substantially larger than that of the other exudivores (Fig. 3.16). Similarly,
Charles-Dominique (1974) found that several species of lemurs fed on
gums, and these animals also possess an extremely long caecum (Petter,
Schilling & Pariente 1971). The mean retention time (MRT) of pollen
grains in sugar gliders fed a pollen/honey mixture was estimated by T. Leary
& A. Smith (pers. comm.) to be 29 hours. Although pollen seems to be

Figure 3.17 Gastrointestinal tracts of A, *Dactylopsila trivirgata* (striped possum) and B,
Burramys parvus (mountain pygmy-possum). Adapted from Smith (1982a).

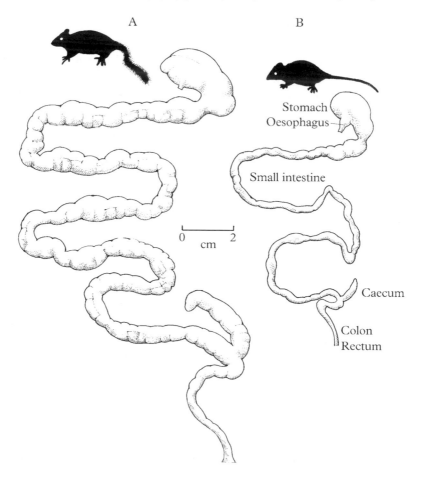

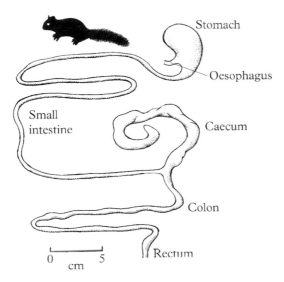

Figure 3.18 Gastrointestinal tract of *Petaurus breviceps* (sugar glider).

digested mainly in the small intestine, the long MRT was probably due to prolonged retention of empty pollen grains in the enlarged caecum. Pollen grains occupy space in the gut even after the contents have been digested.

G. Lundie-Jenkins & A. Smith (pers. comm.) measured the rate of passage of pollen grains through the gut of feathertail gliders offered diets based on sucrose or honey solutions and containing from 4% to 32% pollen. MRTs of the pollen grains decreased linearly as the content of pollen in the diet increased, from 24 hours on the 4% pollen diet to 7 hours on the 32% pollen diet. Animals on the 4% pollen diet were in negative nitrogen balance and lost body mass, but nitrogen balance increased with increased levels of pollen in the diet until 16% pollen, and then remained constant between 16% and 32% pollen inclusion. Body mass change increased linearly across the complete range of pollen inclusions. Thus in the wild it is important that feathertail gliders be able to switch from pollen to insects during periods of low pollen availability in order to meet their protein requirements.

The gross morphology of the digestive tract of the honey possum (Fig. 3.19) deviates from those so far described in this chapter, in two ways. First, it lacks a caecum, so there is little to differentiate the large from the small intestine. Second, it has a two-chambered or bilobed stomach (Schultz 1976). The main chamber is elongate with a mean length of 11 mm and a mean midlength diameter of 2.7 mm (Richardson, Wooller & Collins 1986). The second chamber is a diverticulum, about half the size of the main chamber, and connected to its medial surface by a small isthmus.

The luminal surface of the stomach is organised into simple, flat ridges running the length of the main chamber. These ridges fill most of the lumen

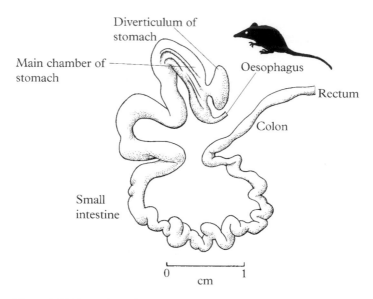

Figure 3.19 Gastrointestinal tract of *Tarsipes rostratus* (honey possum). Redrawn from Schultz (1976).

of the diverticulum. At the cardia, the stratified squamous epithelium of the oesophagus is replaced by tall columnar cells. A simple pyloric region terminates at the duodenum by a backwardly directed flange which protrudes into the duodenum. The flange is formed largely by circular muscle, which is three times thicker in the pylorus than in the main chamber and diverticulum. The mucosa of the main chamber consists of long unbranched gastric glands. Acidophilic parietal cells occupy the basal zone, while mucus neck cells occupy the apical zone. Chief cells (which secrete pepsinogen) appear to be absent (Richardson *et al.* 1986). The mucosa of the diverticulum is arranged into short simple gastric pits lined entirely by mucus-secreting columnar cells.

The distribution of endocrine cells in the gastrointestinal tract of *Tarsipes* is similar to that in the Virginia opossum (Krause *et al.* 1985), but the frequency of occurrence of several cell types, especially somatostatin- and gastrin-immunoreactive cells, was lower in *Tarsipes* (Yamada, Richardson & Wooller 1989). This pattern could reflect the much simpler diet of *Tarsipes* than of the opportunistic Virginia opossum. Endocrine cells are scarce in the cardiac gland region of both the main chamber and the diverticulum of the stomach of *Tarsipes*, but they are found in 'moderate numbers' in the fundic glands (mainly serotonin-immunoreactive cells) and pyloric glands (mainly gastrin-immunoreactive cells). Serotonin-immunoreactive cells were the dominant endocrine cell type along the small intestine, but cells immunoreactive to somatostatin, gastrin, motilin, enteroglugagon, neurotensin and GIP (gastric-inhibiting polypeptide) were also present, each with a different distribution pattern and relative frequency of occurrence. Numbers of endocrine cells in the large intestine were low (Yamada *et al.* 1989).

Richardson *et al.* (1986) found little or no pollen in the diverticulum of 12 wild-caught honey possums; 98% of gastric pollen was in the main chamber. Only a few of these pollen grains were devoid of contents, and none had pollen tubes. All 12 animals had pollen grains throughout their intestine, with an increasing proportion of them observed to be empty as they progressed through the intestine. In the most proximal quarter of the intestine most grains still had their contents; in the second quarter 1–50% had lost their contents, and in the third quarter 60–95% were empty. In the most distal quarter 50–100% were empty, a value close to faecal samples. The extent of pollen digestion differs among plant species, which accounts for the wide range of values in the wild-caught honey possums. In captive animals fed a diet containing pollen from only one species, 95–100% of grains in the faeces were empty (Richardson *et al.* 1986), a much narrower range of values.

It is clear from the wild-caught animals that pollen is not digested in the stomach of *Tarsipes* but rather during its passage through the intestine. As the pollen is not digested in the stomach, and the animals probably have no other solid protein source, Richardson *et al.* (1986) concluded that the ability to form and secrete peptides, specifically pepsinogen, may have been lost by the *Tarsipes* stomach. The stomach, particularly the diverticulum, may instead serve as a storage organ for nectar in times of surplus. Captive animals have been observed to consume nectar volumes equivalent to 10–20% of their body mass in a few minutes, and rapidly become obese if fed *ad libitum*.

The nectar taken by *Tarsipes* contains sugars at an average concentration of 25% (range 15–50%) w/w sucrose equivalents. The sugars are predominantly sucrose, glucose and fructose, sometimes in similar proportions. In preference experiments, Landwehr, Richardson & Wooller (1990) found that honey possums preferred sucrose to fructose, and preferred both of these to glucose, but there was no evidence that they could distinguish between mixtures of these three sugars. They concluded that it is more likely that other factors such as concentration and smell, rather than type of sugar, influence the foraging patterns of *Tarsipes*.

Honey possums often lick stamens to strip them of pollen before probing for nectar (Wiens, Renfree & Wooller 1979). Pollen grains typically contain 20% protein, 37% carbohydrate, 4% lipid and 3% minerals, so offer an ideal supplement to nectar. Richardson *et al.* (1986) measured the rate of passage of pollen through the digestive tract of honey possums by maintaining captive animals on a basal diet of honey, milk-based invalid health food, pollen, and nectar at either 25% or 60% w/w sucrose equivalents. The two different nectar concentrations were used because *Tarsipes* had been found to adjust the volume of nectar ingested according to its concentration, and because the volume of nectar ingested might affect the passage rate of pollen. Once acclimated to the two diets, each animal was given a pulse dose of 2 ml of the appropriate mixture containing about 2 million grains of a distinctive marker pollen type not present in the basal mixtures. Faeces were then collected at regular intervals over the next 24 hours and numbers of marker pollen grains counted on microscope slides. The pattern of pollen excretion is shown in Fig. 3.20. The first marker grains appeared after 2–4 hours, numbers peaked at 6 hours, then declined to very low values at 24 hours. However, a few marker pollen grains were still appearing after more than a week. There was a slight effect of nectar concentration. The MRT of pollen in the 60% sucrose solution can be calculated from Fig. 3.20 to be approximately 9.2 hours, versus 6.7 hours on the 25% sucrose solution. This difference indicates that the animals consumed more of the dilute nectar solution to satisfy their energy requirements, which led to faster passage of pollen through the gut.

G. Lundie-Jenkins & A. Smith (pers. comm.) found similar MRTs of pollen grains in *Tarsipes* (7.8 and 8.3 hours on diets based on sucrose and honey respectively).

In Richardson *et al.*'s (1986) study, the percentage of empty marker pollen grains in the faeces increased linearly with passage time, consistent with results from the wild-caught honey possums discussed above. The process by which the pollen is digested during its passage through the intestine is most likely to be by direct digestion of the protoplast through pores in the exine coat. All empty grains have an intact exine coat. Digestion by initiation of pollen tubes seems unlikely; Turner (1984b) found no

Figure 3.20 Pattern of faecal appearance of marked pollen grains in the honey possum (*Tarsipes rostratus*) on either 25% (filled symbols) or 60% (open symbols) sucrose solution. From Richardson, Wooller & Collins (1986).

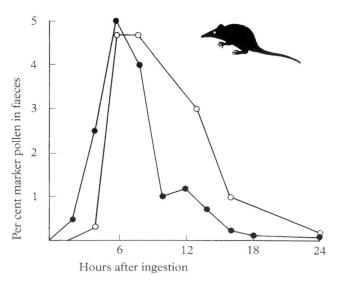

grains to have pollen tubes in the gut of *Tarsipes*. Whether exposure of pollen to acid conditions in the main chamber of the *Tarsipes* stomach is required for effective intestinal digestion is unresolved. It is interesting that *Banksia* and *Dryandra*, the two genera used most heavily by honey possums as food plants, produce large pollen grains with large germinal apertures. The intine, a relatively fragile wall inside the resistant exine coat, ruptures at the germinal apertures, allowing direct access to the protoplast. The protoplast provides the nutrients lacking in nectar, as pollen is always present where nectar is produced (Richardson *et al.* 1986).

A comparison of the honey possum, a specialist nectar/pollen feeder, with an insectivore, the hairy-footed dunnart (*Sminthopsis hirtipes*), and a more generalist omnivore, the western pygmy-possum (*Cercartetus concinnus*), showed that in all three species pollen grains lost their contents not in the stomach but progressively along the intestine (R. Wooller & K. Richardson, pers. comm.). Almost all pollen grains excreted had lost their contents 12–24 hours after ingestion, and the proportion digested was higher with slower passage through the tract. Thus the dietary specialisation of *Tarsipes* appears not to be due to any specialisation of its digestive system for pollen digestion but to specialisation in its mouth parts, particularly the tongue and palate (see 3.3.1 above), for harvesting pollen (and nectar) in amounts sufficient to yield economic returns of protein and energy.

Year-round energy costs of free existence for a 10 g honey possum were estimated by Wooller, Russell & Renfree (1984) to be 56 kJ d^{-1}. This is higher than the 34 kJ d^{-1} measured by Nagy *et al.* (1995), but the latter may be an underestimate because it was made in winter (the non-breeding season) and may also have included torpid animals. Plants most often visited yield about 40 kJ d^{-1} of nectar throughout the year, and all the nectar sources sampled by Wooller, Richardson & Collins (1993) yielded about twice this amount of energy. Comparison of mean densities of *Tarsipes* (measured as captures per hectare per day) and mean nectar production (kJ ha^{-1} d^{-1}) indicated that, for much of the time, honey possum numbers approach the limit that can be supported by the nectar available.

3.3.4 Nutrition and metabolism

As reviewed in Chapter 2, Geiser (1994) concluded that metabolic depression in marsupials fell into two categories. Shallow daily torpor, as seen in opossums, the smaller dasyurids, sugar gliders and Leadbeater's possums, is characterised by a fall in body temperature (T_b) from about 35 °C to 11–28 °C for several hours, and a fall in metabolic rate to 10–60% of BMR. Deep and prolonged torpor (hibernation), as seen in *Dromiciops australis* (Table 2.5), the pygmy-possums and the feathertail glider is characterised by falls in T_b to 1–5 °C and in metabolic rate to 2–6% of BMR, with bouts lasting 5–23 days.

The pattern of torpor in honey possums appears to differ, in that T_b falls from 37 °C to about 5 °C (characteristic of hibernation), and metabolic rate often falls by more than 90% (Collins, Wooller & Richardson 1988), but bouts do not exceed 10 hours in the laboratory (Withers, Richardson & Wooller 1990). Thus they have a pattern of deep but short-term torpor, which may be related to their nectarivorous diet and lack of extensive fat stores. It seems that torpor may be used by *Tarsipes* to survive acute shortages of food during cold weather, whenever it occurs. In the laboratory, use of torpor enabled honey possums to reduce their total daily energy expenditure by 70%, suggesting that torpor occurs when the energy reserves of animals fall below a critical value. In the field, torpor is most prevalent in animals weighing less than 7 g, and between March and September (i.e. autumn to early spring) (Collins *et al.* 1988).

Honey possums have an unusually high BMR for a marsupial (Table 1.1), at least double that expected from Dawson & Hulbert's (1970) equation. Field metabolic rates (FMR) measured in winter by Nagy *et al.* (1995) were also high, although only 2.7 times BMR (much lower than the ratio in small insectivorous marsupials, which can be as high as 6.6 – see Table 1.4). However, winter may not be the season of greatest energy expenditure by honey possums, especially with their ready use of deep daily torpor. Nagy *et al.* (1995) reported considerable variability (twice that expected from other studies) in FMR between individuals and between measurement periods. Most of the variation was due to unusually low FMRs, consistent with the use of torpor during at least some of the measurements. The three lowest values of FMR were only 38% of the three highest values, suggesting that the use of torpor could reduce daily energy expenditure by nearly two-thirds, a value consistent with the laboratory findings of Collins *et al.* (1988).

The feathertail glider also displays deep torpor, sometimes for periods of five days if ambient temperature is low enough (Jones & Geiser 1992). At higher temperatures it displays daily torpor, in groups of up to eight individuals in the laboratory (Fleming 1985). Group torpor has also been observed in the wild (Frey & Fleming 1984). Huddling in nests in groups of four or eight animals significantly reduces a feathertail glider's energy use (Fleming 1985). Although Gieser (1994) classified *Acrobates* as a deep hibernator, it lacks the characteristic pre-hibernation fattening of many hibernating species. It appears not to have a very long hibernation season, but instead uses prolonged torpor bouts during cold periods when daily torpor is not sufficient to guarantee metabolic homeostasis. The pattern of torpor use in the feathertail glider is also influenced by dietary fats. Individuals maintained on a diet rich in polyunsaturated fatty acids showed a lower minimum T_b and longer torpor bouts than animals on a diet rich in saturated fatty acids (Geiser, Stahl & Learmonth 1992). As in other heterothermic mammals, the degree of saturation of dietary fats appears to alter cell membrane composition, which in turn influences torpor patterns.

Deep and prolonged torpor is much more consistently displayed in the Burramyidae (pygmy-possums). Torpor bouts in *Cercartetus nanus* (eastern pygmy-possum) last up to five weeks (Geiser 1993), when T_b falls to as low as 1.3 °C and metabolic rate falls to less than 2% of BMR. All species of the family fatten extensively prior to hibernation, and can survive without food for up to seven months (Geiser 1994). Species of the genus *Cercartetus* can enter torpor at any time of the year in response to low ambient temperatures. In contrast, *Burramys parvus* (mountain pygmy-possum) undergoes a seasonal cycle of hibernation during winter and reproduction and growth during summer (Mansergh 1984). Entry of *Burramys* into torpor in winter seems to be controlled by body fat content (Geiser & Broome 1991). Very fat individuals entered hibernation at relatively high ambient temperatures, even when food and water were available *ad libitum*. Individuals with little fat entered hibernation at low ambient temperatures when food was withheld. Lean animals never entered torpor.

The largest marsupials that use daily torpor include the sugar glider (Fleming 1980) and Leadbeater's possum, mainly in response to food restriction. Fleming (1980) concluded that huddling appeared to be the most important mechanism for energy conservation in sugar gliders, while torpor may be used to withstand short-term reductions in winter food availability. Similarly, Smith *et al.* (1982) felt that the major responses to cold by Leadbeater's possums were behavioural. These included the construction of large, well-insulated nests of shredded bark in the hollow centres of dead trees, in which the animals spend about 74% of their time, huddling in groups of up to seven individuals, and rapid movements while foraging outside the nest.

Protein requirements of omnivorous marsupials have been published for only two species, the sugar glider and the eastern pygmy-possum. By feeding sugar gliders diets of 1.0, 3.1 and 6.5% crude protein, and by regressing nitrogen balance against nitrogen intake, Smith & Green (1987) estimated the maintenance nitrogen requirement of the sugar glider to be 87 mg kg$^{-0.75}$ d^{-1} (Table 1.8). The low maintenance estimate was related to an unusually low value for metabolic faecal nitrogen (MFN) that reflected the low roughage content of the experimental diets, and a low value for endogenous urinary nitrogen (EUN). These two avenues of nitrogen loss arise from metabolic processes in the animal, independent of nitrogen intake, and so are the two most important factors determining the maintenance nitrogen requirement of the animal (see Section 1.11). The animal must eat enough nitrogen to counter the losses of body nitrogen as MFN and EUN. Smith & Green (1987) concluded that free-living adult male sugar gliders may be able to meet their nitrogen requirements on a diet composed entirely of plant exudates (saps, nectar, gum), but that females would need to supplement this with pollen or insects to meet the additional protein demands imposed by reproduction (Fig. 1.4).

More recently, van Tets (1996) found an even lower maintenance requirement for the eastern pygmy-possum (*Cercartetus nanus*), 46 mg N kg$^{-0.75}$ d^{-1} on a dietary basis (Table 1.8), at least on a diet of pollen (predominantly from *Eucalyptus*) suspended in an agar/sugar gel. On mealworms suspended in the same gel the requirement was higher, at 147 mg N kg$^{-0.75}$ d^{-1} on a dietary basis. The large difference between the two diets was thought to be a reflection of the superior amino acid profile of eucalypt pollen (with a biological value of 72%) than of the mealworms (43%); almost two-thirds of the protein in mealworm haemolymph consists of the non-essential amino acid proline and it is deficient in three essential amino acids (ven Tets & Hulbert 1999).

Preliminary data suggest that the dietary maintenance nitrogen requirement of the honey possum (*Tarsipes*) is also likely to be low (115 mg kg$^{-0.75}$ d^{-1} (S. D. & F. J. Bradshaw, pers. comm. 1998).

The other metabolic function that has been studied in *Tarsipes* is its ability to excrete the high water loads ingested with its natural diet of nectar and pollen (Slaven & Richardson 1988). Between 50% and 90% of nectar volume may be water. Slaven & Richardson measured several parameters related to kidney function in four small marsupials. Two of these parameters are particularly informative. Relative medullary area (RMA), which is the ratio of medullary area to cortical area (Brownfield & Wunder 1976), is a measure of urine-concentrating ability. Conversely, mean cortical glomerular volume is a measure of the ability of the kidney to excrete large volumes of dilute urine. In *Tarsipes* the RMA was low (0.3) relative to two insectivorous marsupials, *Sminthopsis griseoventer* (grey-bellied dunnart) (2.8) and *Antechinus flavipes* (yellow-footed antechinus) (3.0). At the same time mean cortical glomerular volume was highest in *Tarsipes* (403 μm^3) and lowest in the two insectivores (222 and 200 μm^3). These values suggest little capacity of the honey possum's kidneys to concentrate urine. Instead, the honey possum produces copious dilute urine on its natural diet of pollen and nectar.

The fourth species included in Slaven & Richardson's (1988) comparative study was the western pygmy-possum (*Cercartetus concinnus*), which feeds on both nectar and insects. The values for its RMA (2.5) and mean cortical glomerular volume (322 μm^3) are intermediate between those of the honey possum and the insectivores. This would seem to allow the western pygmy-possum to be an opportunistic nectarivore while retaining the alternate dietary strategy of insectivory in times of nectar shortage.

Tarsipes, along with the pygmy-possums and the New Guinean feather-tailed possum (*Distoechurus pennatus*) are the first diprotodont marsupials in which we see the synthesis of the thyroxine-transporting protein transthyretin. In fish, amphibians, reptiles and dasyurid and peramelid marsupials, the main and sometimes the only plasma protein that binds thyroxine is albumin (Schreiber and Richardson 1997).

Free thyroxine is more lipid- than water-soluble. Thyroxine-binding plasma proteins counteract the leaking of thyroxine into lipid pools. During vertebrate evolution, lipid pools increased with increases in relative sizes of brains and internal organs. The concentration of plasma albumin also increased greatly during vertebrate evolution. Synthesis of transthyretin first appeared in the choroid plexus of the stem reptiles in the Paleozoic, about 300 million years ago, and independently in the liver of birds and mammals much more recently.

Within the Marsupialia, transthyretin is present in all American species so far examined, including *Dromiciops* which is in the cohort Australidelphia, and in all Australian diprotodonts (see Fig. 9.2). However, it is absent from dasyurid and peramelid marsupials, and from the monotremes. Schreiber & Richardson (1997) concluded that in herbivorous diprotodonts (wombats, folivorous possums, the koala and potoroid and macropodid marsupials), larger gastrointestinal tracts and their associated lipid pools contributed to the selective pressure favouring initiation of transthyretin gene expression in the liver. However, this does not explain why it is also present in omnivorous diprotodonts (e.g. sugar glider, striped possum) but not in omnivorous polyprotodonts (bandicoots and bilbies).

3.4 SUMMARY AND CONCLUSIONS

The nutritional niche width of omnivorous marsupials is generally greater than that of carnivores because of greater development of the hindgut caecum which houses a dense population of microbes. Although direct evidence is lacking, it is assumed that these microbes degrade the structural polysaccharides of gums in the case of the sugar glider and Leadbeater's possum, and perhaps also the chitin of insect exoskeletons in a number of species, especially the bandicoots. A feature of many omnivorous marsupials is their ability to switch between different animal and plant food resources as the availability of these foods changes between seasons. Likewise, the ability of long-nosed bandicoots and northern brown bandicoots to exploit nutritionally unpredictable habitats such as regenerating heathlands after fire owes much to their flexible dietary strategies, and equally to their flexible digestive strategies. Much more needs to be known about the function of the caecum of omnivorous marsupials, particularly whether it functions as a continuous-flow, stirred-tank reactor (see Chapter 2), or whether digesta movement in and out of this fermentative region would be more appropriately modelled as a batch reactor (with pulsed input and output) or semi-batch reactor (with pulsed input but continuous output).

Although omnivory is a convenient dietary category for the purposes of this book there is clearly not a distinct boundary between omnivory and carnivory on the one hand or between omnivory and herbivory on the other. Lee and Cockburn's (1985) classification of marsupial genera (see Table 2.1) illustrates the great diversity of food resources utilised by omnivores; seven of the 18 macroniches are occupied by species dealt with in this chapter. Another seven macroniches are occupied by herbivores, and these are dealt with in the next five chapters.

4 Hindgut fermenters – the wombats

4.1 CONCEPTS

4.1.1 The vertebrate hindgut

The term 'hindgut' is used in this book to describe a functional unit. Although different regions of the vertebrate gastrointestinal tract have different embryological origins, the term 'hindgut' is generally used to describe the entire large intestine (caecum, colon and rectum) of the adult animal. This definition is a functional one, for the hindgut serves as the final site for storage of digesta and retrieval of dietary and endogenous electrolytes and water. It is also the principal site of microbial fermentation in the digestive tract of herbivorous reptiles and most herbivorous birds and mammals (Stevens & Hume 1995).

4.1.2 Classification of herbivores

FOREGUT FERMENTERS

The other principal site of microbial fermentation in some herbivorous mammals is the foregut. The foregut consists of the oesophagus and stomach (Stevens & Hume 1995). Mammalian herbivores can therefore be conveniently divided into two main groups, the 'foregut fermenters' and the 'hindgut fermenters'. In foregut fermenters food is retained and subjected to microbial attack in the forestomach, an expanded area of the stomach proximal to the site of hydrochloric acid secretion. In ruminants, hippos, camelids, peccaries, sloths and potoroine marsupials (rat-kangaroos) (Chapter 8), the forestomach consists grossly of one or more diverticula or sacs. This sacciform morphology maximises retention of digesta for fermentation and results in high digestibilities of plant cell walls. Low rates of throughput and high extents of conversion are characteristics of continuous-flow, stirred-tank reactors (CSTRs) (Chapter 2).

In other foregut fermenters, namely the colobid monkeys and the kangaroos and wallabies (Chapter 6), the morphology of the forestomach is grossly tubiform. In chemical engineering terms, these tubiform fermentation chambers function as a number of CSTRs connected in series (Fig. 4.1). Their flow and performance characteristics are intermediate between those of a stirred-tank reactor and a plug-flow reactor (PFR). As the number of stirred-tank reactors in series increases, the reactor system becomes more and more like a plug-flow reactor (Penry & Jumars 1987). For this reason Hume (1989) termed these hybrid reactors 'modified plug-flow reactors'.

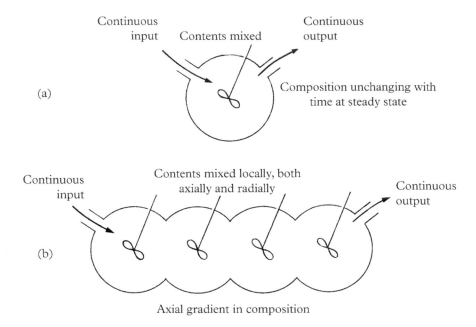

Figure 4.1 Models of (a) a continuous-flow, stirred-tank reactor, and (b) a modified plug-flow reactor (a number of continuous-flow, stirred-tank reactors in series). From Hume (1995). The wombat colon has some of the performance characteristics of a modified plug-flow reactor.

One of the main limitations of a stirred-tank reactor is that reactant concentration is diluted immediately upon entry into the vessel by material recirculating in the reactor. This dilution effect is reduced when the same total reactor volume is divided among two or more smaller stirred tanks in series. Consequently, rates of reaction and flow are higher in modified plug-flow reactors than in stirred-tank reactors, although extent of conversion to products is usually lower because of higher rates of throughput.

HINDGUT FERMENTERS

In hindgut fermenters, the principal site of microbial fermentation is either the caecum or the proximal colon. In most small hindgut fermenters (less than 10 kg body mass) the caecum is the primary and sometimes the only fermentation chamber; these animals are 'caecum fermenters' (Hume & Warner 1980). In these animals the caecum functions either as a stirred-tank reactor or a semi-batch reactor, depending on whether flow of digesta into the organ is continuous or discontinuous. A semi-batch reactor features continuous rather than pulsed output even though input is discontinuous (Penry 1993). Eutherian caecum fermenters include the rodents, lagomorphs, hyraxes and calitrichid primates (marmosets and tamarins). Marsupial caecum fermenters include the arboreal folivores, and are covered in Chapter 5.

In most large hindgut fermenters (more than 10 kg) the primary fermentation chamber is the proximal colon. In eutherian 'colon fermenters' (Hume & Warner 1980) except for the dugong a functional caecum is present, but it probably functions largely as a simple extension of the proximal colon, usually with extensive mixing of contents between the two organs. Eutherian colon fermenters include the equids (horse, donkey, zebra), tapirs, rhinos, elephants and sirenians (dugong and manatees). The marsupial colon fermenters are the wombats. In many colon fermenters, including the wombats, the proximal colon functions as a modified plug-flow reactor, much as the tubiform forestomach of the kangaroo does.

There are compelling reasons to suspect that hindgut fermentation is more primitive than foregut fermentation, and that the colon fermentation strategy is more primitive then the often more complex caecum fermentation strategy (Hume & Warner 1980). This putative sequence is followed in the organisation of this book, with the remainder of this chapter devoted to the wombats (colon fermenters), the next to the arboreal folivores (caecum fermenters), and the following three chapters devoted to the kangaroos, wallabies and rat-kangaroos (foregut fermenters).

4.2 HABITATS AND DIETS OF WOMBATS

The three living species of wombats fall into two genera. The single species of *Vombatus*, *V. ursinus* (common wombat) (Fig. 4.2) lives in forests and alpine grasslands in south-eastern Australia, including Tasmania. The hairy-nosed wombats of the genus *Lasiorhinus* live in semi-arid shrub steppe. *L. latifrons* (southern hairy-nosed wombat) is found mainly in South Australia along the Great Australian Bight and in the Murray River valley, where it burrows deep into soft limestone sediments beneath a thick surface calcrete (Wells 1978b). The endangered *L. krefftii* (northern hairy-nosed wombat) is confined to an area of 300 ha within Epping Forest National Park in central Queensland.

Wombats are the largest burrowing mammals, at up to 50 kg body mass. Apart from the wombats, the largest burrowing mammals that graze or browse are the marmots (*Marmota*), maras (*Dolichotus*) and the plains viscacha (*Lagostomus*). The average adult body mass of these burrowers is 5–8 kg. Thus there is a large gap between the wombats and their closest eutherian analogues. The combination of the three essential features of wombat ecology, viz. herbivory, large body size and burrowing, is found in no other mammalian group (Johnson 1998). Burrowing is rare among large mammals no doubt because of the energetic costs involved in this activity, as well as the problem of burrow stability at large burrow diameters. Also, the time involved in burrow construction and maintenance is incompatible with the generally long feeding times of large herbivores; for instance, eastern grey kangaroos (*Macropus giganteus*) spend 10–18 hours per day

Figure 4.2 The common wombat (*Vombatus ursinus*) from temperate environments in south-eastern Australia. (Pavel German).

feeding (Clarke, Jones & Jarman 1989). Clearly wombats must have some special features that allow them to overcome the time and energetic constraints of burrowing. These features are explored in this chapter, and many have been summarised by Johnson (1998).

The natural diet of all three wombat species is principally perennial native grasses, often of low nutritive value. Fig. 4.3 shows the chemical composition of the main grass eaten by southern hairy-nosed wombats on Brookfield Conservation Park near Blanchetown in South Australia. The high ash content is mainly non-nutritious silica. Maximum crude protein levels are no more than 11%. Even in fresh shoots soluble carbohydrates make up no more than 6% of the dry matter. The bulk of the diet is therefore cell walls, up to 70% in mature leaves. Barboza (1989) found similarly high levels of cell walls (80%) in senescent *Stipa nitida*, but higher levels of protein (19%) in the vegetative stage.

When foraging, southern hairy-nosed wombats graze closely in a circular pattern around the burrow complex to produce a 'lawn' or grazing halo of green shoots, and home ranges are small, about 4 ha (Wells 1978b). The grazed halos expand as plant growth declines. The maintenance of these halos by the wombats maximises the nutritive value of their food supply. A split upper lip allows them to use their two upper and lower incisors as efficient cutters close to the ground. All wombats have continuously growing incisors, premolars and molars, a feature not found in any other marsu-

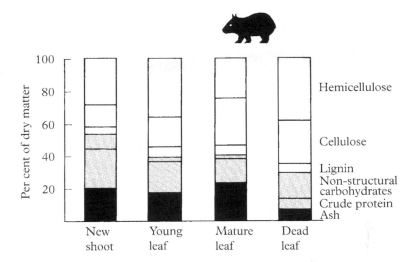

Figure 4.3 Chemical composition of speargrass (*Stipa nitida*), a major forage of the southern hairy-nosed wombat (*Lasiorhinus latifrons*) on Brookfield Conservation Park near Blanchetown, South Australia. Note the increasing proportion of cell-wall constituents (hemicellulose, cellulose and lignin) with advancing age of the leaf. From Dierenfeld (1984).

pial. Both the incisors and molars wear in such a way as to maintain extremely sharp shearing faces on the buccal (cheek) side of the lower molars and on the lingual (tongue) side of the upper molars. Consequently, tough grasses are reduced to remarkably small particles. This is seen in the comparison of particle sizes in the faeces collected from southern hairy-nosed wombats and western grey kangaroos (*Macropus fuliginosus*) grazing in the same semi-arid area on Brookfield Conservation Park (Fig. 4.4). The wombat faeces contained far fewer large particles and many more small particles that those of the kangaroos. The fine comminution of food by the wombat is probably an important factor in its ability to derive maximum benefit from plant material. Perennial grasses, especially *Stipa nida*, are the main component of the diet of *L. latifrons* (Wells 1973; Lehman 1979). However, more forbs and dicots in grassland habitats and more woody shrubs in bluebush (*Maireana*) shrubland and mallee (*Eucalyptus*) woodland habitats are eaten during severe drought (Dierenfeld 1984).

Less is known about the ecology of the endangered northern hairy-nosed wombat. Their burrows occur only on deep alluvial sands supporting an open eucalypt woodland with a grassy understorey (Crossman, Johnson & Horsup 1994). The main components of the diet of *L. krefftii* appear to be the tough perennial grasses *Heteropogon contortus* (black spear grass), *Enneapogon lindleyanus* (wiry nineawn) and *Aristida* spp. Home ranges are surprisingly small given the low nutritive value of the grasses growing around burrow systems; 70% core areas of females averaged only 5.6 ha, and those of males 6.1 ha (Johnson 1991). A significant relationship between breeding

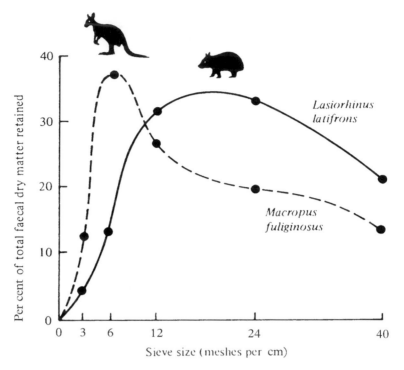

Figure 4.4 Particle size distribution in faeces from southern hairy-nosed wombats (*Lasiorhinus latifrons*) and western grey kangaroos (*Macropus fuliginosus*) grazing the same semi-arid grasslands of Brookfield Conservation Park, South Australia. Note the greater proportion of fine particles in the wombats. After Wells (1973).

rate of females and summer rainfall over a 5-year period (Fig. 4.5) suggests some nutritional constraint on breeding, although the mechanism is not clear given that the breeding cycle of most females commences before the onset of the summer rainy season (Crossman *et al*. 1994).

Common wombats usually forage in forest clearings or on pasture some distance from their burrows, and consequently they have larger home ranges, about 14 ha (McIlroy 1973). As the quality and quantity of pastures decline they move further from burrows to forage. In the highlands of New South Wales, common wombats select the leaves and roots of perennial grasses, especially *Poa*, *Danthonia* and *Themeda* spp., as well as cultivated oat crops. In periods of seasonal shortage or in areas where grasses are scarce, common wombats will feed on sedges (*Carex* spp.) and rushes (*Lomandra* spp.). They will also scratch the soil surface for small roots of grasses and other plants (McIlroy 1973; Rishworth, McIlroy & Tanton 1995).

In Coorong and Messent Conservation Parks in South Australia, the dry climate is ameliorated by a high water table which supports perennial grasses, both native and introduced (Mallett & Cooke 1986). Nevertheless, pasture growth is highly seasonal, with a flush of new growth during and after the annual winter rains. Direct feeding observations indicate that

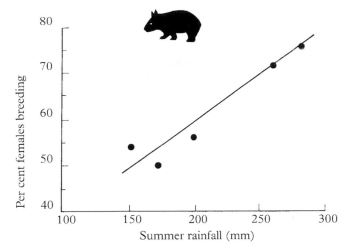

Figure 4.5 Relationship between the percentage of female northern hairy-nosed wombats (*Lasiorhinus krefftii*) breeding and summer rainfall (1985–89) at Epping Forest National Park, Queensland. From Crossman, Johnson & Horsup (1994).

common wombats in the Coorong select mainly native grasses (76–90% of total feeding time) in all seasons. However, at Messent, introduced perennial ryegrass (*Lolium perenne*) is selected along with native grasses, herbs and sedges. More sedge than grass is eaten in winter, but proportionately more grass is eaten in summer.

Barboza (1989) analysed summer and winter samples of perennial ryegrass and four species of sedge favoured by common wombats at Messent. Both the ryegrass and sedges were higher in water content during the winter rainy season than in the annual summer drought (Table 4.1) but there were no significant seasonal differences in the composition of the dry matter of either the ryegrass or the sedges. Ryegrass was consistently the better quality forage, being higher in water and protein content, and lower in all cell-wall components, in both seasons.

4.3 THE DIGESTIVE TRACT

The teeth of wombats are subject to considerable wear from silica inclusions (phytoliths) in the cells of the grasses on which they feed. In periods of low plant availability wombats also ingest considerable amounts of soil while foraging, which adds to tooth wear. Their continuously growing teeth, with a core of soft dentine and margins of hard enamel, ensure the maintenance of an efficient dental mill. The effectiveness of the dental mill of *Lasiorhinus latifrons* was seen in Fig. 4.4, which compares the particle size distribution in faeces collected from *L. latifrons* and the western grey kangaroo (*Macropus fuliginosus*) grazing in the same semi-arid area near Blanchetown in South Australia.

Table 4.1. *Nutritional analysis of perennial ryegrass and sedges selected by common wombats* (Vombatus ursinus) *at Messent Conservation Park, South Australia*

Component	Season	Ryegrass	Sedges
Water[a]	Summer	64.0	47.2
	Winter	74.0	53.2
Ash	Summer	12.1	6.2
	Winter	14.3	6.9
Crude protein	Summer	17.6	4.2
	Winter	16.8	4.6
Neutral-detergent fibre[b]	Summer	61.7	78.9
	Winter	57.8	76.2
Acid-detergent fibre[c]	Summer	32.3	48.4
	Winter	32.2	47.0
Lignin	Summer	1.9	10.2
	Winter	1.8	8.1

Note: [a] Water as a percentage of fresh matter, all other components as percent of dry matter.
[b] Total cell wall constituents.
[c] Cellulose and lignin.
Source: From Barboza 1989.

Herbivores usually have larger salivary glands than do carnivores and omnivores. Salivary function in common wombats has been examined by Beal (1991b, 1995a, b). Maximal rates of fluid secretion by the parotid gland appeared to be at the low end of the mammalian range, but levels of amylase activity were far higher than in the parotid saliva of other marsupials (Beal 1991b). This suggests that, at least for part of the year, starch levels in the diet may be considerable. In contrast, the sublingual gland was found to produce a saliva with little amylase activity. It contained high levels of protein and had sodium, potassium and calcium as the main cations and chloride as the main anion (Beal 1995a). Maximal flow rates were also relatively low for mandibular saliva, but protein levels were high and amylase activity was higher than in sublingual secretions (Beal 1995b).

The digestive tract of the Vombatidae was the subject of much discussion among the early anatomists. The tracts of the common wombat and the southern hairy-nosed wombat are shown in Fig. 4.6.

The most striking feature is the great development of the colon compared with the omnivorous marsupials described in Chapter 3. However, it is the stomach and the caecum which, until recently, have attracted most attention. The first reported observations on the stomach of a wombat were those of Home (1808). The stomach is small and simple, except for the thickening of the wall of the lesser curvature near the cardia (opening of the oesophagus into the stomach) (Fig. 4.6). Internally, the mucosa of the lesser curvature is organised into a specialised cardiogastric

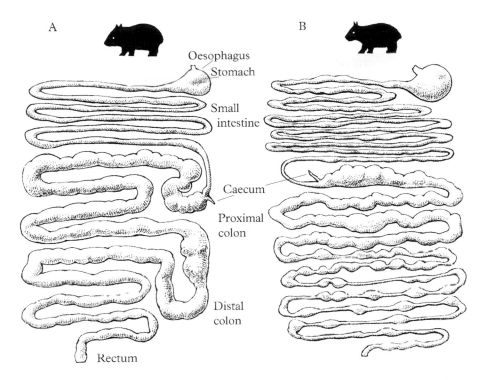

A

B

Oesophagus
Stomach
Small
intestine
Caecum
Proximal
colon
Distal
colon
Rectum

Figure 4.6 The gastrointestinal tracts of the common wombat (*Vombatus ursinus*) (A) and the southern hairy-nosed wombat (*Lasiorhinus latifrons*) (B). From Barboza & Hume (1992a).

gland region, similar in morphology and histology to that of *Caenolestes* (Chapter 2).

The cardiogastric gland of the wombats is a complex group of mucosal sacculations which open into the stomach lumen via 25–30 large, crater-like ostia (Hingson & Milton 1968). As in *Caenolestes*, there is no histological or cytological difference between the cardiogastric gland region and the rest of the stomach mucosa (Oppel 1896; Johnstone 1899). Chief cells are present in the basal region of the gastric glands, parietal cells are present mainly in the middle 60% of the glands, and the neck cells produce mucus. The cardiogastric gland secretions are of similar composition and concentration to those of the body of the stomach, but the cardiogastric gland must contribute a large proportion of the total gastric juice (Milton, Hingson & George 1968). The functional significance of this anatomical specialisation is still obscure. Oppel (1896) considered that the glandular apparatus of the stomach of both the wombat and koala developed into the cardiogastric gland in order to facilitate the assimilation of large amounts of food, but we now know that voluntary food intakes of both wombats (Barboza 1993a) and the koala (Cork, Hume & Dawson 1983) are lower than those of most other marsupials.

The small intestine of both *Vombatus* (36% of total gastrointestinal tract length) and *Lasiorhinus* (40%) is relatively much shorter than that of carnivorous marsupials such as *Caenolestes* (87%), and of omnivores such as the northern brown bandicoot (*Isoodon macrourus*) (63%). The Brunner's glands in the wombats are not organised into such large lobes as in the carnivorous dasyurids or the omnivorous bandicoots but they extend further along the duodenum (Krause 1972).

The wombat caecum is extremely small (Fig. 4.6). It was referred to by Owen (1868) as 'extremely short, but wide', and 'provided with a vermiform appendix'. Lönnberg (1902) and Mitchell (1916), however, regarded the narrow projection at the end of the ileum as the true caecum, the greater part of which has become transformed into a solid vermiform appendix. The wide caecal pouch of Owen (1868) is then merely one of the haustrations of the proximal colon.

There are two important differences in the digestive tracts of the common and southern hairy-nosed wombats. The first is that the total length of the gastrointestinal tract of *Lasiorhinus* is greater than that of *Vombatus*, both in absolute terms (13% longer) and in relation to body length (30% longer). The second difference lies in the relative importance of the proximal and distal regions of the colon.

The great size of the colon is functionally the most important feature of the wombat gut (Fig. 4.6). It is here that we see the second important difference between the two southern species. Although in both species the total colon is about 60% of total tract length, in *Vombatus* the proximal region is the more important (68% of total colon length versus 41% in *Lasiorhinus*), whereas the distal colon is longer in *Lasiorhinus* (Barboza & Hume 1992a). These differences in relative lengths are reflected also in mucosal surface areas in the two species (Fig. 4.7); *Vombatus* has a greater surface area in the proximal colon and *Lasiorhinus* has a greater surface area in the distal colon.

The proximal colon is haustrated throughout its length in both species. Haustra are non-permanent sacculations formed when the outer longitudinal muscle layer is concentrated into one or more bands (taeniae); contraction of the circular muscle layer between these bands draws the included segment of gut wall into pockets, or haustra. Haustra can help delay digesta passage, while at the same time waves of contraction of the circular muscle serve to mix the digesta and help to propel it either caudally (peristalsis) or orally (antiperistalsis). Haustra continue along part of the distal colon, but they are much weaker than those of the proximal colon. In both *Vombatus* and *Lasiorhinus* there is a pair of permanent sacculations about midway along the length of the proximal colon. They mark the site of the gastrocolic ligature, and in the human intestine the bend between the transverse and descending colon. The function of these permanent sacculations, if any, is unknown.

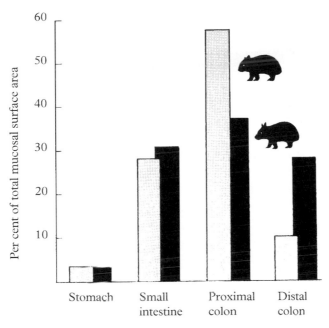

Figure 4.7 Distribution of mucosal nominal surface area of the stomach, small intestine, proximal colon and distal colon of captive common wombats (grey columns) and southern hairy-nosed wombats (black columns). From Barboza & Hume (1992a).

4.4 DIGESTIVE FUNCTION

The pH of digesta in the wombat stomach is low (2.9–3.2), but it increases to 7.0–7.1 in the small intestine, then decreases again to 6.5–6.9 in the proximal colon. The distal colon of *Vombatus* is more acidic (pH 6.3) than that of *Lasiorhinus* (7.1) (Barboza & Hume 1992b). Direct counts of bacteria in the digesta from the common wombat by Gowland (1973) yielded numbers in the order of 10^{10} g^{-1} in all regions of the gut, but numbers were three to eight times higher in the colon than in the stomach and small intestine. There are no reports on the composition of the colonic bacterial population. The helminth fauna in the gut of the two southern wombats is restricted in its range of species, being limited to anophocephalid cestodes in the small intestine and strongyloid nematodes in the colon. The latter are similar to nematodes found in the forestomach of kangaroos, but whether they are ancestral to the kangaroo nematodes or are derived from them is uncertain (Beveridge & Spratt 1996).

The markedly higher numbers of bacteria in the colon compared with the stomach and small intestine suggest that microbial fermentation of plant cell walls in wombats is confined to the colon. This suggestion is confirmed by measurements of the concentration of the short-chain fatty acids (SCFA), the main end-products of microbial breakdown of carbohydrates in the anaerobic environment of the gut. Levels of SCFA are negligible in the stomach and small intestine, but in the proximal colon of wild *Vombatus* they occur at 67–87 mmol L^{-1}, and in wild *Lasiorhinus* at 104–127 mmol L^{-1} (Barboza & Hume 1992b). The main SCFA produced in the fermentation of carbohydrates in the gut are acetic, propionic and butyric acids. Other, more minor SCFA (isobutyric and isovaleric acids) are produced from the fermentation of proteins. The three main SCFA are found in the hindgut of wild wombats in the molar proportions of 66–74% acetate, 17–26% propionate and 6–8% butyrate.

Barboza & Hume (1992b) measured the rates at which SCFA were produced in the wombat hindgut by an *in vitro* fermentation technique. This involves removal of digesta from the animal immediately after death. These digesta are then incubated anaerobically in duplicate at the temperature of the gut (37 °C) without addition of extra substrate or buffer, and samples are removed from the glass incubation vessels at 15 or 30 minute intervals for 150 minutes. SCFA concentrations in the digesta samples are plotted against time, and the rate of SCFA production at zero time (i.e. the time of removal from the animal) is estimated by fitting a line of best fit to the data points and extrapolating back to zero time. Results from one common wombat are shown in Fig. 4.8.

Figure 4.8 Change in the concentration of acetic, propionic, butyric and valeric acids with time during incubation of proximal colon digesta from a captive common wombat (*Vombatus ursinus*). From Barboza & Hume (1992b).

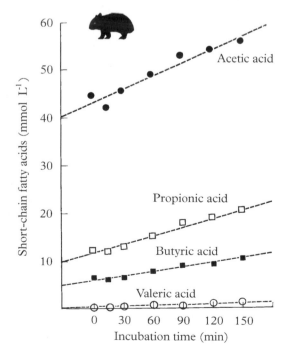

In the wombats the line of best fit was found to be linear in all cases. In contrast, similar plots of SCFA concentrations against time using contents of the forestomach of kangaroos (Prince 1976) and wallabies (Hume 1977a) and of ruminants (Carrol & Hungate 1954; Whitelaw *et al.* 1970; Hume 1977a) generally yield curvilinear lines. This is presumably because the forestomach contains more readily fermentable substrates than does the hindgut. As these substrates disappear from the incubation vessel the rate of fermentation tends to decline. Fitting a straight line to these data points would underestimate zero-time production rate.

Some readily fermentable substrates disappear extremely rapidly, and may even disappear in the interval between removal of digesta from the animal and commencement of the incubation. This is thought to be a principal reason why SCFA production rates measured *in vitro* are usually lower (by as much as 50%) than estimates made *in vivo* by isotope dilution (Whitelaw *et al.* 1970). However, isotope dilution techniques can only be used satisfactorily when mixing of the injected or infused isotopically labelled SCFA with a single large pool of SCFA is instantaneous. This situation is approached in the forestomach of sheep (Leng & Leonard 1965), but not in tubiform fermentation organs such as the wombat colon, which is best modelled as a number of stirred tanks arranged in series (a modified plug-flow reactor, see Fig. 4.1). Thus estimates of SCFA production are restricted to *in vitro* procedures. Fortunately though, it seems that agreement between *in vitro* and *in vivo* estimates of SCFA production is likely to be closer in the hindgut than in the forestomach because of the much lower concentrations of rapidly fermentable substrates distal to the small intestine.

It has not been feasible to test this conclusion in the wombats, but Faichney (1969) reported similar estimates of SCFA production rate in the sheep caecum both *in vitro* and *in vivo*. No other studies have been reported in which *in vivo* and *in vitro* estimates have been made simultaneously in the hindgut. However, on the basis of information available, it seems reasonable to assume that *in vitro* estimates of SCFA production in the hindgut are probably close to actual rates.

When SCFA production rates are expressed as a proportion of the animal's intake of digestible energy, a second assumption is involved, that the rate of fermentation measured at one point in time is representative of the rate throughout the remainder of the 24 hours. Again, this has not been tested in the wombats. Certainly, in the forestomach both total SCFA concentration and the fermentation rate vary markedly throughout the day unless the animal feeds at frequent and regular intervals throughout the entire 24 hours. Effect of feeding pattern is likely to be much less in the hindgut. Foley, Hume & Cork (1989) found that in the caecum of captive greater gliders the zero-time production rate of propionic acid was higher when the animals were actively feeding than at other times of the day, but estimates of total SCFA production based on zero-time rates measured before, during and after the nightly feeding period were similar (38, 37 and 34 mmol d^{-1} respectively). Similarly, in the caecum of wild rabbits and captive rabbits fed *ad libitum* there is little variation in SCFA concentrations over the 24 hours (Henning & Hird 1972; Parker & McMillan 1976).

In wild wombats in winter, SCFA production was faster in the proximal colon of *Lasiorhinus* (28 mmol L^{-1} digesta fluid h^{-1}) than in *Vombatus* (16 mmol L^{-1} h^{-1}). In both species production rate declined along the length of the colon, to only 15 and 6 mmol L^{-1} h^{-1} in the distal colon of the two species respectively. The higher SCFA production rate in *Lasiorhinus* is consistent with a higher quality diet selected by hairy-nosed wombats while grazing the 'lawn' they maintain around their burrow complex, compared with the diet of *Vombatus*. When captive animals were fed a common diet of pelleted straw and corn and containing 65% plant cell walls, SCFA production rate was 12 mmol L^{-1} h^{-1} in both species. This rate is lower than in the wild, suggesting that the natural diet selected was of much higher quality than the captive diet.

The decline in SCFA production rates along the length of the colon reflects the gradual change in the substrate fermented to one which is progressively higher in fibre content as the more readily fermented fractions disappear. This is illustrated in Fig. 4.9, which shows the sequence of digestion of dry matter, neutral-detergent fibre (plant cell walls) and protein along the digestive tract of captive wombats fed a 65% fibre diet. These trends are consistent with the modelling of the colon as a modifed plug-flow reactor; rates of reaction and concentrations of substrates decline along the length of plug-flow reactors (Fig. 2.2).

Figure 4.9 The sequence of digestion of digestible dry matter (closed circles), total nitrogen (open circles) and neutral-detergent fibre (squares) along the gastrointestinal tracts of (a) the common wombat (*Vombatus ursinus*) and (b) the southern hairy-nosed wombat (*Lasiorhinus latifrons*). From Barboza & Hume (1992a).

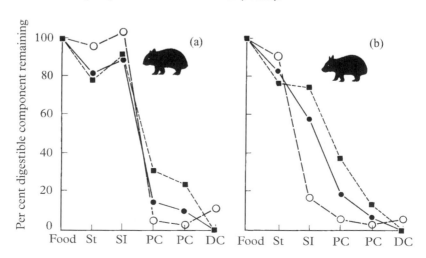

In the common wombat, relatively little digestion of this rather low-quality diet took place in the stomach or small intestine compared with the proximal colon. In contrast, in the southern hairy-nosed wombat, relatively more digestion took place in the stomach and especially the small intestine, and less in the hindgut. The proximal colon is a larger proportion (86%) of a larger gut (4.8 kg of contents) in the common wombat than in *Lasiorhinus* (76% of 3.6 kg of contents). It can also be seen in Fig. 4.9 that less of the fibre than of the protein disappeared in the small intestine, and more in the colon, the site of microbial fermentation.

Along the length of the colon the molar proportion of propionic acid declined, which also reflects a fermentation based less on non-structural carbohydrates and more on fibre towards the distal colon. At the same time the molar proportions of isobutyric and isovaleric acids increased, from a total of less than 1% in the upper proximal colon to 6% in the distal colon. This indicates increasing fermentation of protein, perhaps including microbial cells, towards the lower end of the colon.

Barboza & Hume (1992b) calculated the contribution made by the SCFA produced in the colon to the energy economy of wombats. In captive animals of both species, 30–33% of the total energy assimilated was in the form of SCFA. This is similar to the eutherian pig, another colon fermenter (Stevens & Hume 1995).

Barboza (1993a) measured rates of digesta passage through the two southern species of wombats fed pelleted diets. He used Cr-EDTA as the solute marker and Ru-Phenanthroline as a particle-associated marker. Mean retention times were much longer for the particle marker (52–62 h on the high-fibre diet, 69–75 h on the low-fibre diet) than for the solute marker (30–36 and 49–50 h respectively). There were no differences between the species. These results indicate longer retention of food particles than of solutes, primarily in the proximal colon, the main site of microbial fermentation of plant cell walls.

This pattern of digesta retention is characteristic of a modified PFR. In the proximal colon, contractions of the haustra express the fluid and small particles through a matrix of large particles, resulting in faster flow (shorter mean retention times) of the fluid or solute fraction. Digesta retention times, especially of the large food particles, were longer in the wombats than in horses, even though their body size and digestive tract capacity are much smaller than those of the equids. This is the main reason for the relatively high fibre digestibilities recorded by Barboza (1993a) in the wombats (Table 4.2), similar to those seen in donkeys fed grass hay of similar fibre content.

The slow passage of digesta in the two wombat species is a direct consequence of their low food intakes. Intakes of digestible dry matter were only 33% of those of the donkeys mentioned above, yet they maintained body mass on both diets throughout the study. These low food intakes are a

Table 4.2. *Intake and digestion in wombats fed chopped lucerne (alfalfa) hay*

Measure	*Vombatus ursinus*	*Lasiorhinus latifrons*
Body mass (kg)	27.2 ± 2.7	25.7 ± 1.2
Intake of dry matter		
(g d⁻¹)	210 ± 13	353 ± 79
(g kg⁻⁰·⁷⁵ d⁻¹)	18 ± 1	39 ± 5
Apparent digestibility of dry matter (%)	68 ± 4	60 ± 2
Digestibility of neutral-detergent fibre (%)	52 ± 3	42 ± 3

Source: From Barboza 1993a.

direct reflection of the low BMR recorded for *Lasiorhinus latifrons* (Table 1.1) and form the basis for the low maintenance energy requirements of captive *Lashiorhinus* and *Vombatus* shown in Table 1.2.

The inverse relationship between passage rate of digesta and food intake within a wombat species (*Lasiorhinus latifrons*) is shown in Fig. 4.10. This is a common phenomenon that has been observed in several vertebrate species. The only exceptions are seen when gut capacity expands to accommodate additional digesta, in which case mean retention times of digesta may remain constant (Karasov & Hume 1997).

4.5 METABOLISM AND NUTRITION

The BMR of *Lasiorhinus latifrons* (Table 1.1) is one of the lowest recorded for any marsupial. It is only 44% of the value predicted from Kleiber's (1961) equation for eutherians. The maintenance energy requirements of *Lasiorhinus* and *Vombatus* (Table 1.2) are the lowest recorded for any marsupial. A low rate of metabolism in the wombats is also consistent with the plasma concentrations of thyroid hormones reported by Barboza, Hume & Nolan (1993), which are the lowest for any mammal (Hulbert & Augee 1982). Although plasma concentrations of the hormones do not directly indicate their rates of secretion by the thyroid, they have been correlated with basal rates of metabolism among rodents by Scott, Yousef & Johnson (1976). Low rates of metabolism and low rates of nutrient turnover provide animals with enormous advantages in terms of their ability to survive long periods of food shortage under adverse environmental conditions.

Consistent with the low energy requirements of wombats, maintenance requirements for nitrogen have also been found to be low. On a dietary basis, the requirement for *Vombatus* was 158 and for *Lasriorhinus* it was 201 mg N kg⁻⁰·⁷⁵ d⁻¹. On a truly digestible basis the values were 71 and 116 respectively (Barboza *et al.* 1993). These are at the lower end of the range for marsupial herbivores (Table 1.8), and thus much lower than any

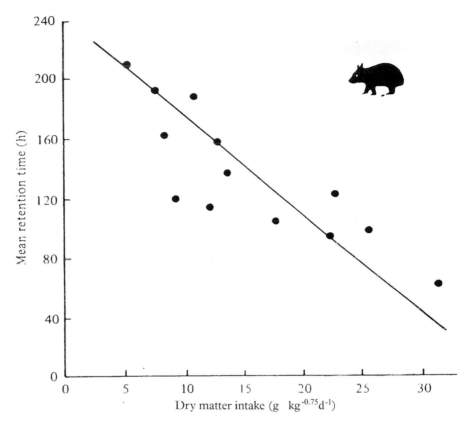

Figure 4.10 The inverse relationship between mean retention time of stained hay particles and dry matter intake in southern hairy-nosed wombats (*Lasiorhinus latifrons*) observed by Wells (1973).

eutherian herbivore. Among the marsupials only the omnivorous sugar glider, eastern pygmy-possum and honey possum are lower. The low maintenance nitrogen requirements of the wombats are partly explained by low rates of nitrogen excretion in the urine and high rates of recycling of endogenous urea to the hindgut. On a low protein diet, 78% of urea synthesised in the liver was recycled and degraded in the gut, and 49–60% of this was incorporated into microbial protein. The other principal explanation lies in the low rates of protein synthesis in the wombats. These and other parameters of nitrogen metabolism in wombats are shown in the model depicted in Fig. 4.11.

The model consists of three pools of nitrogen in the body: free amino acids (mainly in the blood), urea (mainly in the liver) and protein (mainly in muscle). Flux of nitrogen between these pools can be estimated using isotope tracer techniques. Nitrogen enters the body as dietary protein, which is digested in the stomach and small intestine and absorbed into the amino acid pool. Amino acids are being continuously exchanged with the

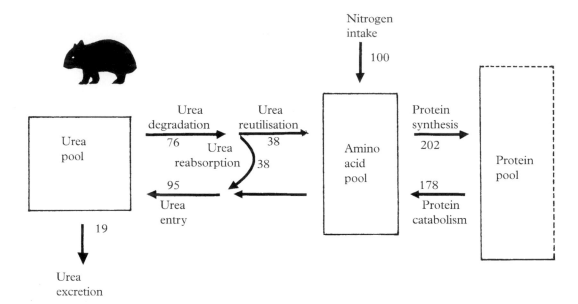

Figure 4.11 Quantitative model of nitrogen metabolism in the common wombat (*Vombatus ursinus*). Values as per cent of nitrogen intake. From Barboza, Hume & Nolan (1993).

protein pool during synthesis and catabolism of proteins. Amino acids are also catabolised, the nitrogen appearing as ammonia which is incorporated into urea in the liver. In non-herbivores most of this urea is excreted in the urine, but in herbivores such as the wombats a proportion of the urea synthesised in the liver recycles to the gut where the resident bacteria degrade it to ammonia again. Bacteria in the gut use part of this ammonia for their own protein synthesis. This recycling of nitrogen to the gut of herbivores reduces renal loss.

Rates of whole-body protein turnover calculated from the model ranged between 4.1 and 6.7 g crude protein $kg^{-0.75}$ d^{-1}. These rates are only 26–43% of the mean value of 15.4 $g\,kg^{-0.75}$ d^{-1} for six eutherian species calculated by Waterlow (1984).

A comparison of the kidneys of the two southern wombats suggest differences that reflect the more arid habitat of *Lasiorhinus* compared with that of *Vombatus*. Although both species had glomerular filtration rates (Barboza 1993b) that were only 26–34% of those predicted from the allometric equation of Yokota, Benyajati & Dantzler (1985), *Vombatus* had a relative medullary thickness (RMT) of 3.6, less than that of *Lasiorhinus* (5.4), suggesting greater urine concentrating ability in the latter species (McAllan, Roberts & Barboza 1995). Twice the number of collecting ducts per unit area in the medulla of the *Lasiorhinus* kidney indicates that the nephrons join the collecting ducts deep in the medulla, another feature associated with greater urinary concentrating ability. Faeces were also drier in *Lasiorhinus* (45% dry matter) than in *Vombatus* (36%) when

Table 4.3. *Sodium content of soils and grasses from four environments*

	Sodium concentration (mEq kg^{-1} dry weight)			
	Snowy Mountains	Canberra	Western New South Wales	Coastal Victoria
Soil	0.5	0.8	3.6	3.3
Grass (mixed species)	0.2–8.5	5–10	96–200	68–309
Saltbush (*Atriplex*)	–	–	2200	–

Note: Compared with coastal Victoria and western New South Wales, the Snowy Mountains and Canberra grasslands are severely sodium-deficient.
Source: After Blair-West *et al.* 1968.

the water intake of both was restricted to 50% of *ad libitum* levels (Barboza 1993b).

Rather than a shortage of water, the common wombat has been shown to live in at least one sodium deficient area, namely the Snowy Mountains in south-eastern New South Wales, and to a lesser extent the grasslands surrounding Canberra nearby. It is here that Blair-West *et al.* (1968) examined the adaptations to a constant sodium deficiency in *Vombatus*, and also the eastern grey kangaroo (*Macropus giganteus*), by comparing a number of physiological parameters in animals from the Snowy Mountains with those in the same species near the coast in Victoria. The low sodium status of soils and grasses from the Snowy Mountains and Canberra is compared with soils and plants from coastal Victoria and western New South Wales in Table 4.3.

As can be seen from Table 4.4, sodium content of urine from both herbivores in the Snowy Mountains was virtually zero. The value in coastal *M. giganteus* was similar to that reported by Dawson & Denny (1969) in the euro (*M. robustus erubescens*) in western New South Wales. The low value in coastal *Vombatus* was thought to reflect differences in diet; grass roots, which are relatively low in sodium, rather than tops, appeared to be eaten.

Only limited observations on blood electrolytes were made by Blair-West *et al.* (1968), but the sodium content of plasma of alpine *M. giganteus* was reduced relative to the coastal animals. The adrenal glands of both *Vombatus* and *M. giganteus* examined in the Snowy Mountains were approximately double the weight of those in coastal animals, and the area of the zona glomerulosa, the site of synthesis and secretion of aldosterone, was greater in adrenal glands of both species from the sodium-deficient Snowy Mountains. Correlated with this was a higher concentration of circulating aldosterone in both species in the alpine environment. There was little difference in the levels of the glucocorticoids cortisol and corticosterone in

Table 4.4. *Physiological parameters in common wombats* (Vombatus ursinus) *and eastern grey kangaroos* (Macropus giganteus) *from a sodium-deficient alpine environment (the Snowy Mountains) and a coastal environment in Victoria*

	Snowy Mountains		Coastal Victoria	
	V. ursinus	*M. giganteus*	*V. ursinus*	*M. giganteus*
Urine sodium (mEq L^{-1})	0	0	23	215
Plasma sodium (mEq L^{-1})	n.a.	139	n.a.	148
Weight of adrenal glands (g)	0.7	1.1	0.4	0.6
Blood aldosterone (μg dL^{-1})	12	40	6	9
Area of zona glomerulosa of left adrenal (mm^2)	14	15	6	6

Note: n.a. – not available
Source: After Blair-West *et al.* 1968

sodium-deficient and sodium-replete animals. Thus the several differences described are clearly adaptations specifically to achieve sodium homeostasis in a sodium-deficient environment.

Blair-West *et al.* (1968) also examined the salivary glands of *M. giganteus* from the two environments because of the high salivary flow rates in forestomach fermenters such as ruminants and macropodid marsupials. Maintenance of microbial fermentation in the forestomach involves circulation of large volumes of fluid of high sodium content. The capacity to replace sodium with potassium in this cycle in sodium deficiency reduces the haemodynamic stress of sodium deficiency and thus is pivotal to adaptation. There were marked structural differences in the salivary glands between the two populations of *M. giganteus*. Animals from the Snowy Mountains had a much more extensive duct system of both the parotid and mandibular glands, and blood vessels were extraordinarily abundant around the striated (secretory) ducts. These structural differences are adaptations for conserving sodium by the salivary glands. Although salivary concentrations of sodium and potassium were not measured by Blair-West *et al.* (1968), there would undoubtedly be a much lower sodium/potassium ratio in the saliva of alpine *M. giganteus* than in that of coastal animals. Salivary glands from *Vombatus*, a hindgut rather than a foregut fermenter, showed similar structural differences between the two environments but not to the same degree.

Both southern wombats are extremely sensitive to the plant toxin fluoroacetate, with LD_{50} values of only 0.2 mg 1080 (sodium monofluoroacetate) kg^{-1} (Twigg & King 1991). This is a small fraction of the LD_{50} reported for macropods such as *Macropus fuliginosus* (western grey kangaroo) of 40 mg 1080 kg^{-1}. Such a large difference is consistent with a western origin of the western grey, where it coevolved with fluororacetate-containing plants, but an eastern origin for the wombats, where there is no natural exposure to fluoroacetate in the environment.

A metabolic difference seen within the wombats appears to be most closely related to habitat aridity. Agar *et al.* (1996) reported much lower levels of the enzyme glucose-6-phosphate dehydrogenase in the red blood cells of *Vombatus* (5 units g^{-1} haemoglobin) than in both species of *Lasiorhinus* (30–36 units g^{-1} haemoglobin). Glucose-6-phosphate dehydrogenase protects against oxidation of haemoglobin to methaemoglobin (which cannot transport oxygen). Rates of methaemoglobin formation were significantly greater in *Vombatus* erythrocytes treated with acetyl phenylhydrazine, a strong oxidant. A higher degree of defence against methaemoglobin formation in the two *Lasiorhinus* species may be related not to aridity *per se*, but to the consumption of more woody shrubs during drought (see Section 4.2) and the increased intake of plant secondary metabolites, some of which are strong oxidants, that it may entail.

4.6 DIGESTIVE AND FORAGING STRATEGIES

From the information now available on the two southern wombats, it is possible to compare and contrast their digestive and foraging strategies in relation to their respective habitats. Contrary to what might be expected, the semi arid-zone *Lasiorhinus* eats a higher quality diet than the mesic *Vombatus*. It does this by maintaining grazing 'lawns' around the burrow complex in which it lives. The higher quality of the *Lasiorhinus* diet is reflected in its relatively large small intestine and small proximal colon. Size of the proximal colon in *Lasiorhinus* may be constrained by the need to maximise water resorption in a long distal colon. The primary limiting factor to this foraging strategy is the intermittent nature of rain in the Murraylands and Nullabor Plain, and lack of plant growth for long periods. Low rates of energy metabolism, water turnover and protein synthesis assist *Lasiorhinus* to survive these periods of food shortage.

In contrast, *Vombatus* is not often subjected to severe water shortage in its natural habitat. This is reflected in a higher water content of its faeces when water intake is restricted, and in a shorter distal colon. A larger proximal colon provides greater capacity for digesta retention and fermentation. This in turn permits the more solitary common wombat to forage on tussock grasses which are relatively abundant in its mesic forest habitat but which are often of low quality. A relatively short small intestine reflects the generally lower quality of the diet of *Vombatus*.

The combination of the several features of wombats that lead to energy conservation occurs in no other radiation of mammals. This allows them to spend most of their time underground, which contributes further to energy and water conservation. Johnson (1991) noted that northern hairy-nosed wombats spent only 2–6 hours per day above ground, depending on seasonal conditions. The combination of low food requirements with the energy-saving advantages of living below ground results in extremely conservative ranging behaviour, as reflected in the small home ranges noted above. It also suits them for survival in habitats of very low productivity (Johnson 1998). For example, common wombats persist in alpine environments up to and above the snow line. At lower altitudes they often are found in sandy coastal environments. The two species of hairy-nosed wombats are restricted to semi-arid habitats of low fertility. But in all these environments they may reach high population densities, suggesting a competitive advantage over other large herbivores.

4.7 RESPONSES TO DROUGHT

A study of the condition of *Lasiorhinus latifrons* in the Brookfield Conservation Park during a prolonged drought by Gaughwin *et al.* (1984) demonstrated the remarkable survival ability of the southern hairy-nosed wombat. Plant growth and body mass of the wombats over a three-year period are shown in Fig. 4.12. There is a close relationship between periods of net primary production after rain and increases in body mass of the wombats.

The effects of the drought on various physiological parameters of animals caught early and late in the drought and then after a period of significant rain are shown in Table 4.5.

Six months of severe drought resulted in a 10% loss in body mass. Associated with this, body condition (fat index) and haematocrit (packed

Figure 4.12 Plant growth and body mass changes in southern hairy-nosed wombats (*Lasiorhinus latifrons*) during drought. From Gaughwin *et al.* (1984).

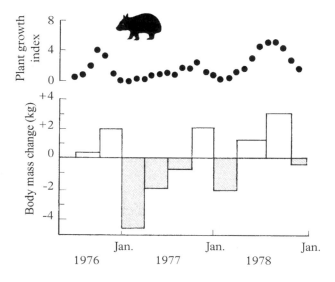

Table 4.5. *Effect of drought on various physiological parameters of southern hairy-nosed wombats* (Lasiorhinus latifrons)

	Aug. 1977 Early drought	Feb. 1978 Late drought	Aug. 1978 After drought
Sample size	10	4	5
Body mass (kg)	23.6 ± 0.8	21.2 ± 1.0	25.2 ± 0.7
Digestive tract mass (kg)	5.9 ± 0.2	5.3 ± 0.3	4.9 ± 0.1
Fat index (scale 0–5)	2.0 ± 0.3	0.9 ± 0.5	1.8 ± 0.4
Packed cell volume (%)	40.6 ± 1.5	32.0 ± 1.9	39.5 ± 1.5
Plasma			
Chloride (mmol L^{-1})	101 ± 3	107 ± 1	97 ± 1
Glucose (mmol L^{-1})	6.6 ± 0.6	5.0 ± 0.3	7.3 ± 0.4
Urea (mmol L^{-1})	$13.5 + 0.8$	15.4 ± 0.8	12.6 ± 1.7
Creatinine (mmol L^{-1})	0.15 ± 0.01	0.19 ± 0.02	0.13 ± 0.01
Liver			
Iron (mmol kg^{-1} dry)	30 ± 3	48 ± 7	5 ± 1
Copper (mmol kg^{-1} dry)	0.33 ± 0.02	0.92 ± 0.25	0.21 ± 0.01
Zinc (mmol kg^{-1} dry)	1.71 ± 0.09	2.14 ± 0.15	1.50 ± 0.08
Stomach contents			
Ash (% of dry matter)	26.8	29.6	17.8
Nitrogen (% of organic matter)	3.4 ± 0.2	3.3 ± 0.2	3.1 ± 0.2
Cell walls (% organic matter)	65.5	78.0	47.1
Faeces			
Water (% of wet matter)	60.3 ± 1.7	52.2 ± 1.8	74.8 ± 1.5
Ash (% of dry matter)	36.5	35.8	29.7
Nitrogen (% of organic matter)	2.9 ± 0.1	2.2 ± 0.1	3.1 ± 0.2

Note: From Gaughwin *et al.* 1984.

cell volume) declined, and there were changes in blood plasma composition. The most notable is the 24% fall in glucose concentration (a direct result of reduced carbohydrate intake) and the 15% rise in urea and 26% rise in creatinine concentrations (both a result of body protein catabolism). The significant increases in liver concentrations of iron, copper and zinc reflect increasing intakes of soil as the drought progressed. Stomach contents consisted of 30% ash, and 78% of the organic matter consisted of cell walls (fibre). Similarly, the falls in faecal water and nitrogen contents reflect the deteriorating quality of the plant material available.

Nevertheless, few animals died, and most physiological parameters returned to normal soon after the drought broke.

4.8 **SUMMARY AND CONCLUSIONS**

The wombats are the world's largest burrowing mammals. The broad nutritional niche of wombats, as demonstrated by their ability to survive extended droughts, is a consequence of three main factors. The first of these are their low metabolic rates and as a consequence low maintenance requirements for energy, protein and water. The second is their efficient masticatory system with its open-rooted teeth. The third is their digestive strategy, which is based on a colon fermentation system. This system, which has performance characteristics of a modified plug-flow reactor, means that highly fibrous grasses and sedges can be processed and passed through the hindgut at a rate sufficient to avoid excessive gut fill and depression of food intake resulting therefrom.

These metabolic and digestive features, together with their burrowing habits, fit them for life in low-productivity environments. Their capacity to reach high population densities in poor nutritional environments suggests a competitive advantage over the other large herbivores of Australia, the kangaroos.

5 Hindgut fermenters – the arboreal folivores

5.1 CONCEPTS

Chapter 4 introduced the concept of a classification of herbivores based on the location of the principal site of microbial fermentation. The first division was into foregut fermenters and hindgut fermenters. The latter were further divided into colon fermenters and caecum fermenters. Wombats, the group dealt with in Chapter 4, are colon fermenters. In these terrestrial herbivores the proximal colon can be modelled as a modified plug-flow reactor (PFR) (i.e. a linear series of continuous-flow, stirred-tank reactors (CSTR)) (see Fig. 4.1).

In contrast, with the exception of the tree kangaroos (see Chapter 6), the marsupial arboreal folivores (i.e. animals that live in trees and eat tree leaves) are caecum fermenters. This means that most plant cell wall digestion takes place in a greatly expanded caecum. In some species such as the koala (*Phascolarctos cinereus*) there is also some microbial fermentation in the proximal colon as well, but in the majority of caecum fermenters microbial fermentation is virtually confined to the caecum. Although Hume (1989) suggested that in chemical reactor terms the mammalian caecum could be modelled as a stirred-tank reactor, this is an oversimplification of a complex system. Depending on the patterns of filling and emptying, the caecum may be better modelled as a batch reactor (Chapter 2) or a semi-batch reactor. In a batch reactor, reactants are added and products are removed in discrete batches, which disrupts flow and the fermentation process. In a semi-batch reactor, materials may be added in batches, but the flow of materials out of the reactor vessel is continuous (though not necessarily constant), and the vessel never completely empties (Penry 1993). Some avian caeca fill and empty in a pulsatile fashion (Fenna & Boag 1974) and thus can be modelled as batch reactors, but the caecum of most mammals is closer to a semi-batch reactor. Caecum fermenters exhibit a range of patterns for filling and emptying the caecum (Stevens & Hume 1995). Some of these are described in this chapter.

Most caecum fermenters are smaller than colon fermenters. At 40–60 kg the capybara (*Hydrochoerus hydrochaeris*) is an obvious exception, but generally caecum fermenters are less than 10 kg body mass. There are advantages for arboreal folivores in being small. These are related to the high energy costs of vertical movement in a three-dimensional habitat, and to the ability to reach young leaves at the end of small branches without breaking the branch. But there are also disadvantages, the main one being their small digestive tract in absolute terms, which limits the extent of cell wall digestion. Thus in most arboreal folivores we see foraging strategies that involve selective feeding behaviour against old leaves with their highly lignified cell walls, and digestive strategies to eliminate ingested lignified cell walls as rapidly as possible. The smallest arboreal folivores are the ringtail possums. Many species have an adult body mass in the range 0.5–1.2 kg, but the smallest species, *Pseudocheirus mayeri* (pygmy ringtail possum) from New Guinea is only 150 g. The largest arboreal folivores are koalas, at 6–12 kg.

5.2 THE ARBOREAL FOLIVORES

Australia has two species of tree-kangaroos, and New Guinea has six. All tree-kangaroos are rainforest or vine forest inhabitants and, importantly, do not feed on the foliage of *Eucalyptus*, even if available, probably because the essential oils in eucalypt leaves are toxic to the forestomach bacteria. Tree-kangaroos are dealt with in more detail in Chapters 6 and 7.

The arboreal hindgut fermenters include the families Phalangeridae, Pseudocheiridae and Phascolarctidae (the koala). The Phalangeridae includes two brushtail possums, two cuscuses and the scaly-tailed possum in Australia, and eight cuscus species in New Guinea. The Pseudocheiridae includes seven ringtail possums and the greater glider in Australia, and eight ringtail possums in New Guinea (Flannery 1994a; Strahan 1995). The koala is found only in Australia.

5.3 DIETS

The highly specialised dietary preferences of the koala (*Phascolarctos cinereus*) (Fig. 5.1) for the foliage of *Eucalyptus* are legendary: almost as fussy is the greater glider (*Petauroides volans*). Although the list of eucalypt species preferred by koalas varies among locations, about 35 species can be described as highly preferred. This is a small proportion of the 600 or more species so far described, although it must be pointed out that many species do not grow within the koala's present range. Also, many other eucalypt species in addition to the 35 or so highly preferred species are eaten to a lesser extent. Numerous reports of koalas eating non-eucalypt species such as *Melaleuca* spp. and introduced *Pinus radiata* are not enough to alter the popular image of the koala as a eucalypt specialist. It is an animal of open woodland habitats, although it can also be found in coastal forests along the eastern seaboard and in remnant River Red Gum (*E. camaldulensis*) forests that skirt numerous inland river systems in eastern Australia. Its present range is a small fraction of what it was before the advent of European agriculture, due principally to the loss and fragmentation of habitat.

Figure 5.1 The koala (*Phascolarctos cinereus*), which feeds almost exclusively on the foliage of *Eucalyptus* species. (Pavel German)

The greater glider lives in a variety of eucalypt-dominated habitats, from low open forests on the east coast to tall forests of the Great Dividing Range and low woodland on the western slopes of the Great Divide. Its diet in south-eastern Australia consists principally of the leaves, buds and flowers of certain species of *Eucalyptus* (Marples 1973; Kavanagh & Lambert 1990), but not necessarily the same species as those favoured by the koala. Comport, Ward & Foley (1996) found similar dietary patterns for greater gliders in north Queensland. Like that of the koala, the reproductive rate of greater gliders is low and they have a slow rate of population turnover, features which are adaptations to a stable environment (Tyndale-Biscoe & Calaby 1975).

Of the ringtail possums, the diet of the common ringtail (*Pseudocheirus peregrinus*) has been studied in most detail. *P. peregrinus* is widespread throughout south-eastern Australia, in habitats ranging from coastal shrublands to the lower edges of snow gum (*E. pauciflora*) communities of the highlands. Thomson & Owen (1964) reported that common ringtails in Victoria were strictly herbivorous, the diet consisting almost entirely of leaves, shoots or flowers, often of only one or two species. In a climax

151

association of *Eucalyptus* and *Kunzea*, 60% of stomach samples contained eucalypt leaf, 56% *Kunzea*, 20% *Acacia dealbata* and 4% *Rubus fruticosa* (the introduced blackberry). A single species only was found in 56% of stomachs, 40% contained two species and only 3% contained three species. No seasonal shifts in species preference were observed. The possums appeared to prefer young leaves, shoots and buds over the mature leaves of *Kunzea*, *Melaleuca* and *Leptospermum*, and young phyllodes over mature phyllodes of *A. dealbata*. With *Eucalyptus* the mature leaf was eaten and young foliage avoided. In contrast, Pahl (1984) reported that for common ringtails eucalypt foliage was the major food source, and young foliage was preferred to mature foliage for most of the eucalypts and for all shrub species used as food. However, young foliage is not always available, and so mature foliage is consumed over much of the year. The possums preferred the mature leaves of some eucalypt species over others, and the highest population densities were associated with these preferred species. In areas in which the dominant eucalypt was not a preferred species the foliage of understorey shrubs was a more important component of the diet.

Common ringtails have a high reproductive rate and rapid population turnover. In unstable situations this provides the potential for rapid colonisation or recovery after a population crash (Tyndale-Biscoe & Calaby 1975). It is thus a more resilient species than the greater glider.

The green ringtail possum (*Pseudochirops archeri*), a species of dense upland rainforests, is also strictly folivorous, with only 1.3% of total feeding time spent on non-leaf plant parts (Proctor-Gray 1984). Goudberg (1990) studied three sympatric species, the green ringtail, the lemuroid ringtail (*Hemibelideus lemuroides*) and the Herbert River ringtail (*Pseudochirulus herbertensis*), in a patch of vine forest on the Atherton Tableland in north Queensland. Based on direct feeding observations, all three species had a diet containing over 90% leaves, the remainder being mainly fruits and flowers. Dietary partitioning occurred because each species concentrated on a different plant species. *P. archeri* appeared to be a fig specialist, eating *Ficus* leaves at all times of the year and fruits when they were available, and supplementing this with vines (*Cissus* spp.) and a few lauraceous species. *H. lemuroides* fed on large amounts of *Sloanea langii*, especially during the dry season, when there was a flush of new growth. *P. herbertensis* fed throughout the year on the new and mature leaves of *Alphitonia petriei*, a common regrowth species along rainforest margins.

In a study of museum specimens of six species of New Guinean ringtail possums that range in body size from 150 to 2000 g, Hume, Jazwinski & Flannery (1993) found that although the stomach contents of all species were composed principally of leaves, those of the two smallest species (*Pseudocheirus canescens*, the lowland ringtail possum (226 g) and *P. mayeri*, the pygmy ringtail possum (148 g)) also contained tissues of mosses, ferns and lichens, and *P. mayeri* contained fungal material and pollen as well.

The diets of these small species are thus more diverse and, on the basis of higher total nitrogen levels in their gastric contents, of higher quality than those of the four largest species. Selection of a higher-quality diet is consistent with the relatively high mass-specific nutrient requirements of small endotherms.

The results of numerous studies indicate that, in contrast to the Pseudocheiridae, members of the Phalangeridae are less strictly folivorous and more frugivorous in their dietary habits. The common brushtail possum (*Trichosurus vulpecula*) is perhaps the most adaptable of all Australian marsupials. It has the widest distribution of any Australian marsupial. It has successfully adapted to the presence of Europeans, in contrast to many other marsupials which have greatly diminished in numbers, several to extinction, since the arrival of Europeans in 1788. It has successfully adapted to different conditions in New Zealand since its introduction there in 1858, quickly reaching pest status. Its success has been attributed to its high fecundity and the early dispersal of young, which allow rapid colonisation of new or regenerating habitat, and the ability to exploit the most abundant plant types, which means that a wide variety of habitat types can be exploited (Tyndale-Biscoe & Calaby 1975).

Numerous studies on the feeding ecology of the common brushtail were collated and used by Kerle (1984) to indicate the great dietary flexibility of this species. Grasses and herbs contributed 40–60% of the diet of common brushtails in dry sclerophyll forest adjacent to pasture in Tasmania, based on faecal analysis (Fitzgerald 1984); *Eucalyptus* and *Acacia* leaves made up the bulk of the remainder. In mature wet sclerophyll forest, leaves of *Nothofagus* and *Phebalium* were the most important food items (38–73% of the faeces), with ferns and mosses making up a further 7–19%; *Eucalyptus* leaves made up no more than 12% of faeces. In open eucalypt woodland in south-eastern Queensland, the diet of common brushtails consists of approximately 66% eucalypt leaves (Freeland & Winter 1975). In arid central Australia, Evans (1992a) found that common brushtails prefer the foliage of a mistletoe (*Amyema*), the foliage and flowers of *Acacia* spp., the foliage and fruits of *Rhagodia* spp. and fruits of *Solanum* spp. from the shrub layer. They ate little grass or eucalypt foliage, even in areas where *E. camaldulensis* (river red gum), often a favoured food tree in other parts of Australia, was present.

Its ability to utilise a wide range of plant materials is probably the main reason why *T. vulpecula* is more widely distributed through various eucalypt forests in south-eastern Australia than other folivorous possums (Braithwaite, Dudzinski & Turner 1983). In Western Australia, many plants contain high concentrations of sodium monofluoroacetate (compound 1080), a metabolic poison which blocks the Kreb's tricarboxylic acid cycle at the citrate stage (see Section 2.9). Common brushtails living within the range of fluoroacetate-containing plants possess an unusually high tolerance for the toxin; Mead, Oliver & King (1979) demonstrated that the Western

Australian common brushtail was nearly 150 times more resistant to fluoroacetate intoxication than was the South Australian common brushtail. The high tolerance of the common brushtails in Western Australia has enabled them to exploit the foliage of a wider range of plant species than would otherwise be possible. Thus they have an expanded nutritional niche.

The mountain brushtail possum (*T. caninus*) is anatomically rather similar to *T. vulpecula*, but its ecological requirements are very different. In contrast to the wide distribution of *T. vulpecula* in all wooded habitats in Australia, *T. caninus* is restricted to the tall open forests of eastern Australia and closed forest (rainforest) in the northern part of its range. Where the two species are sympatric, *T. vulpecula* is the less abundant. Within the tall open forests of Victoria, *T. caninus* selects a wide range of plant species, many of which are found on the forest floor or in the lower storey, whereas *T. vulpecula* in the same forest subsists on the foliage of eucalypts (Owen & Thomson 1965). Items consumed by *T. caninus* include foliage, principally of *Acacia dealbata* but also of several other non-eucalypts, as well as fungi and small amounts of grass, seeds, moss and invertebrates (Seebeck, Warneke & Baxter 1984); staminate cones of the introduced *Pinus radiata* were important in winter when other food items were less available.

An explanation for the much more limited distribution of the mountain brushtail compared with the common brushtail is not readily apparent. Tyndale-Biscoe & Calaby (1975) thought that the lower fecundity of *T. caninus* and its high site attachment made it less resilient and adaptable than *T. vulpecula*, but whether there are metabolic differences between the two *Trichosurus* species, as well, is not known. Barnett, How & Humphreys (1979a) found that several blood parameters, including plasma glucose and protein concentrations, haemoglobin concentration and red blood cell count, showed greater seasonal variation in *T. caninus* than in sympatric *T. vulpecula* in north-eastern New South Wales. They suggested that the lower seasonal variation in *T. vulpecula* may reflect its ability to ameliorate environmental stress and so occupy more diverse habitats, although the effect of dietary differences cannot be ruled out.

Further study of blood parameters of the two species from preferred and peripheral habitats (Barnett, How & Humphreys 1979b) showed that habitat had a large effect on *T. caninus* but little effect on *T. vulpecula*, again indicative of a more adaptable physiology in the latter species. However, an investigation of the metabolism and nutritional requirements of the two species is still needed before any metabolic differences can be evaluated (Hume 1982).

Proctor-Gray (1984) found that the diet of the coppery brushtail possum (*T. vulpecula johnstoni*) in north Queensland rainforest included a substantial proportion of non-leaf plant parts (fruits and flowers) and a high proportion of introduced plant species. The stomach contents of four New Guinean phalangerids also contained a large proportion of non-leaf materials (fruits and bark) as well as leaves of *Syzygium*, *Ficus* and *Nothofagus* spp. (Hume *et al.* 1993). The stomach of one species (*Phalanger vestitus*, Stein's cuscus) also contained flowers, flower buds and pollen. M. J. Runcie (pers. comm. 1998) found that the scaly-tailed possum (*Wyulda squamicaudata*) in the Kimberley region of northern Western Australia fed on the leaves of the trees *Xanthostemon paradoxus*, *X. eucalyptoides*, *Planchonia careya* (mangaloo or cocky apple) and an unidentified species of *Eucalyptus*, but also on the stems, flowers and seeds of the perennial herb *Trachymene didiscoides*. Together these studies indicate a wider range of non-foliage foods in the dietary spectrum of phalangerid than of pseudocheirid marsupials.

5.4 *EUCALYPTUS* FOLIAGE AS FOOD

Although *Eucalyptus* foliage is an abundant food resource, and features in the diets of several Australian folivorous possums, in addition to the koala, its utilisation is potentially limited by two sets of factors. The first of these is a low concentration of nutrients; the second is the presence of plant secondary metabolites (also referred to as allelochemicals or antinutrients). Plant secondary metabolites found in eucalypt leaves are of two types. One type interferes with the digestive process. Examples are lignin, which interferes with microbial digestion of plant cell-wall polysaccharides, and condensed tannins, which form complexes with proteins, making them less available to the animal. The other type are toxic to the animal's tissues. Examples of this type are hydrolysable tannins and other low-molecular weight phenolic compounds, cyanogenic glycosides and the essential oils (terpenoids). Essential oils are toxic to bacteria also, so they potentially have an effect at both the gut and tissue levels.

The relative importance of a low concentration of nutrients and the presence of antinutrients (plant secondary metabolites) as limitations to the utilisation of *Eucalyptus* foliage is still imperfectly understood, but it is becoming clear that rarely is a single component the limiting factor. More often it is probably the ratio of several nutrients to one or more plant secondary metabolites that determines the usefulness of certain leaves as food (Cork 1992; Hume and Esson 1993), although, as will be seen below, one class of plant secondary metabolites found in eucalypts, the diformyl-phloroglucinols, have such strong effects that they may override other factors in some instances. How widespread these instances are is not yet known.

Eucalypts evolved on the old, highly weathered, nutrient-depleted soils that characterise much of the Australian continent (Attiwill & Leeper 1987). Consequently eucalypt leaves generally have a high degree of sclerophylly and contain low levels of nutrients. The chemical composition of *Eucalyptus* foliage is compared with domesticated grasses, legumes and vegetables in Fig. 5.2. Relative to one or more of the non-eucalypt ma-

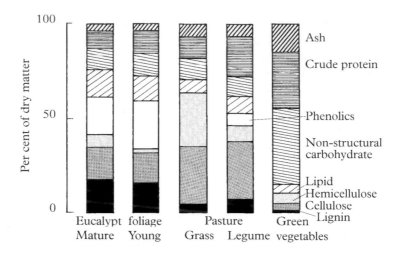

Figure 5.2 The chemical composition of young and mature foliage of *Eucalyptus punctata* compared with typical analyses for domesticated grasses, legumes and vegetables. From Cork & Sanson (1990).

terials, eucalypt foliage is low in ash (i.e. minerals), protein and available carbohydrate (sugars and starches), but high in lipid and phenolic compounds, including lignin. The crude lipid fraction includes the essential oils and other lipophilic plant secondary metabolites. New growth is slightly higher in protein and available carbohydrate and lower in fibre (i.e. cell walls), but is also higher in phenolics.

Although the total fibre content of eucalypt leaves is not especially high, the fibre is highly lignified and for this reason has a low absolute digestibility.

The low protein content is exacerbated by the presence of condensed tannins. Eucalypt foliage is much higher in total phenolics and condensed tannins than foliage in other temperate and all types of tropical forests (Cork 1996). The proteins rendered inactive by complexing with condensed tannins include dietary proteins, digestive enzymes secreted into the gut and enzymes of microbial origin. Thus the effect of tannins is seen in a reduced digestibility of several dietary components, especially cell-wall carbohydrates, not just protein.

It is against this background survey of the several factors that make eucalypt leaves a difficult resource to exploit that food choice by the four main eucalypt folivores can now be discussed.

5.4.1 Diet selection

Interest in this area began with koalas, but more recent studies have also used other folivorous marsupials in attempts to elucidate common principles that influence feeding choice and patterns, and thus the factors that determine habitat quality for these animals. Two approaches have been used. The first is based on correlations between leaf intake and the composition of selected versus rejected foliage. This approach is not likely to yield unambiguous answers involving a single component of the foliage, but it has provided clues to various combinations of nutrients and plant secondary metabolites that are correlated with the persistence of particular folivores in particular forests.

The second approach is based on a combination of correlations and bioassay-guided fractionation of the chemical components of foliage. Comparisons of foliage from browser-susceptible and browser-resistant trees are used to identify the plant secondary metabolites responsible for browser resistance. Likely fractions separated from foliage are added to plant secondary metabolite-free basal diets and short-term intake responses are recorded. Although the number of factors that can be tested in combination by this means is necessarily limited, short-term intake controls can be unambiguously identified.

Both approaches are useful, but the combination of correlation and bioassay-guided fractionation has greater potential to explain differences in the width of the nutritional and dietary niches between species of arboreal folivores.

The correlative approach has been used with both captive and free-living animals. For instance, at San Diego Zoo in California, Ullrey, Robinson & Whetter (1981a) found that foliage preferred by koalas had significantly higher concentrations of crude protein, non-structural carbohydrates, phosphorus and potassium and lower concentrations of lipid, cell-wall polysaccharides, lignin, calcium, iron and selenium compared to rejected foliage. The koalas selected the younger leaves when offered a mixture of foliage from 11 *Eucalyptus* species, and the younger leaves in general contained higher levels of several important nutrients such as crude protein and phosphorus and lower levels of cell-wall polysaccharides and lignin. More recently, Zoidis & Markowitz (1992) at San Francisco Zoo found that leaf consumption by koalas was positively correlated with leaf nitrogen content and negatively with fibre content, but there was no consistent relationship with either lipid or essential oil content. Osawa (1993) at Saitama Children's Zoo in Japan reported a negative correlation between total sugar content and leaf intake. *Eucalyptus* species eaten readily contained 0.6% total sugars on average compared with 1.7% in species seldom eaten. Whether this correlation has any causative basis is unknown.

Crude protein (nitrogen) has also been suggested to be a determining factor in leaf choice by free-living koalas by Degabriele (1981, 1983), Martin (1985) and Hindell, Handasyde & Lee (1985), although no direct evidence was presented. However, some support for this suggestion comes from a study with captive koalas by Pahl & Hume (1990). They found that koalas consistently selected younger foliage when available, and were strongly selective in their choice among 18 species of *Eucalyptus* offered. Although the amount of leaf eaten was significantly and positively correlated with the levels of both nitrogen and water in the leaves, they explained these relationships in terms of the operation of minimum threshold levels of nitrogen (1.8% of dry matter) and water (65% of fresh weight), below which foliage was avoided and above which it was eaten (Fig. 5.3). The factors determining the amounts eaten above these thresholds appeared to be more than just nitrogen and water. No significant relationships were found with the contents of ash, non-structural carbohydrates, cellulose, hemicellulose or lignin. Essential oil and phenolic contents were not determined.

In a more comprehensive study, Hume & Esson (1993) enlisted the help of 13 wildlife parks and zoos in New South Wales and compared the preferences of koalas for *Eucalyptus* foliage as ranked by their keepers with the content of a number of nutrients and plant secondary metabolites in samples of the leaves offered to koalas in each park. No single factor separated the four preference ranking groups used, but koalas selected foliage that was above minimum threshold levels of water (approximately 55%) and essential oils (approximately 2% of dry matter) (Table 5.1).

Koalas usually smell unfamiliar leaves before they eat or reject them, suggesting that some volatile constituent such as the essential oils may provide a proximate cue in foliage choice. Also, there has long been interest in the supposed role of *Eucalyptus* essential oils in limiting koala populations (Eberhard *et al.* 1975), yet no consistent relationships between koala

Figure 5.3 Relationships between intake by koalas (*Phascolarctos cinereus*) of *Eucalyptus* foliage and (a) per cent nitrogen on a dry matter basis, and (b) per cent water in the leaves offered. From Pahl & Hume (1990).

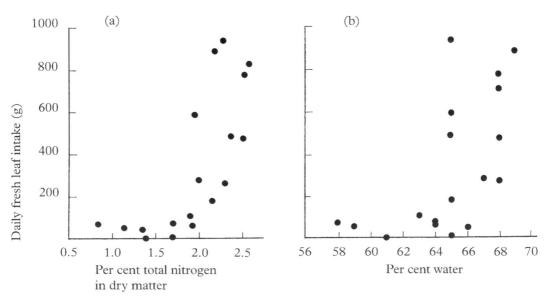

Table 5.1. *The chemical components (% of dry matter) of* Eucalyptus *foliage from four preference groups determined by offering to captive koalas*

Component	Preference group*			
	1 (Highest)	2	3	4 (Lowest)
Water (% of fresh leaves)	55[a]	56[a]	55[a]	50[b]
Total essential oils	2.7[a]	2.6[a]	2.8[a]	1.9[b]
Oil components (% of total)				
Volatile monoterpenes	74[a]	76[a]	59[b]	55[b]
Less volatile monoterpenes	13	11	16	14
Sesquiterpenes	13[a]	13[a]	26[b]	31[b]
Nitrogen (N)	1.7[a]	1.7[a]	1.5[b]	1.6[a]
Fibre	34	34	35	35
Total phenolics	12	13	10	12
Condensed tannins	4[a]	7[b]	8[b]	11[b]
(N:Fibre) × 10	0.53[a]	0.54[a]	0.47[b]	0.48[b]
N: Total phenolics	0.14	0.13	0.15	0.14
N: Condensed tannins	0.46[a]	0.23[b]	0.19[b]	0.14[b]

Note: *1 = most preferred; 4 least preferred. Statistically different values on any one line carry different superscripts.
Source: After Hume & Esson 1993.

leaf preferences and total oil yield had previously been reported. Nevertheless, in data in Southwell (1978), the mean oil yield of eucalypts that were browsed or heavily browsed (1.3 and 1.4% respectively) was significantly greater than the yield of trees not browsed (0.6%). These data, together with those in Table 5.1, suggest that, at least within a certain range, koala leaf preferences are related to oil yield in a positive rather than a negative manner.

That eucalypt oils may be acting as a positive feeding cue to koalas may at first appear counter-intuitive, given the cytotoxic nature of terpenes and the metabolic cost in their detoxification, conjugation with glucuronic acid and urinary excretion. However, it may be of selective advantage to the koala if this metabolic capacity is not shared by other folivores.

The other finding related to the volatile constituents that Hume & Esson (1993) reported was that above the threshold oil content, or perhaps in addition to it, the composition of the oil also plays a role in koala leaf preference. Leaves containing a higher proportion of volatile monoterpenes and/or a lower proportion of sesquiterpenes were more attractive (Table 5.1). This is also suggested by Betts' (1978) report of a significant, though weak, negative relationship between leaf intake by captive koalas and the level of sesquiterpenes in the leaves. Recent work by Pass, Foley & Bowden

(1998a) indicates that this relationship may be explained not by the sesquiterpenes themselves but by a group of polar phenolic compounds (diformylphloroglucinols) that are complexed with sesquiterpenes, and even with some monoterpenes. These compounds, with 18 to 28 carbons, must be solvent extracted as they are not volatile during steam distillation. Several appear to have strong antifeedant properties. These include Macrocarpal G (Yamakoshi *et al.* 1992). Pass *et al.* (1998) concluded that Macrocarpal G was at least partially responsible for the selective feeding behaviour of common ringtail possums offered *E. ovata* foliage.

There is also reason to believe that sesquiterpenes may be more highly toxic and more difficult to detoxify than monoterpenes. For instance, M. L. Baker *et al.* (1995) found that in both the koala and the common brushtail possum haemolysis of erythrocytes was greater when treated with a mixture of sesquiterpenes extracted from *E. haemastoma* leaves by steam distillation than a mixture of monoterpenes extracted from the same leaves. That the erythrocytes of folivorous marsupials are subject to oxidative stress from plant secondary metabolites is suggested by their relatively high levels of catalase activity. Catalase protects against the oxidation of haemoglobin to methaemoglobin, which cannot bind oxygen. Koalas and brushtails had catalase activities that were 9–20 times higher than activities in the erythrocytes of two macropodid grazers, the whiptail wallaby (*Macropus parryi*) and the black-striped wallaby (*M. dorsalis*) (Agar & Baker 1996).

The strong negative effects of some plant secondary metabolites such as the diformylphloroglucinol compounds (DFPCs) on feeding by marsupial arboreal folivores may be ultimately due to their cytotoxic properties. However, animals may be expected to limit their intake of particular plant secondary metabolites before overt illness is observed. This could be achieved if the metabolite stimulated the emetic (vomiting) system of the mid brain (Provenza 1995). Animals feeding on the metabolite would then develop conditioned food aversions as a result (Provenza 1996). Provenza, Pfister & Cheney (1992) emphasised that this emetic stimulation need not result in overt illness and that the animal need not even be aware of the event that triggers intake modulation.

Lawler *et al.* (1998a) studied the effects of an aromatic ketone, jensenone, a naturally occurring phenolic constituent of *E. jensenii* foliage, on the intake of a non-eucalypt based diet by common ringtail possums and common brushtail possums. When offered diets containing graded levels of jensenone, both species regulated their intake of the diet so as not to exceed a ceiling intake of jensenone. The ceiling for ringtails was twice that for brushtails. If this finding can be extrapolated to other plant secondary metabolites it may partly explain why ringtails routinely include more eucalypt foliage in their diet than do brushtails. Other possible reasons for the greater intake of eucalypt foliage by ringtails are explored below.

When Lawler *et al.* (1998a) injected brushtails with the antiemetic drug ondansetron the animals ate significantly more jensenone than others injected with water alone, but ondansetron had no effect on the intake of food that did not contain jensenone. These findings suggest that emetic stimulation could be an important mechanism limiting the intake of plant secondary metabolites by marsupial arboreal folivores.

Lawler *et al.* (1998b) recorded short-term intake responses of both koalas and ringtails to foliage from several *E. ovata* and *E. viminalis* trees that were highly negatively correlated with the total essential oil content, and more specifically the diformylphloroglucinol content of the foliage of individual trees. On *E. ovata*, average daily dry matter intakes varied from 3 to 50 g kg$^{-0.75}$ for the ringtails compared with 22 to 36 g kg$^{-0.75}$ for the koalas. Intakes were lower on the *E. viminalis*, ranging from 1 to 6 g kg$^{-0.75}$ for the ringtails and from 14 to 46 g kg$^{-0.75}$ for the koalas. These results demonstrate not only large differences in resistance to ingestion between individual trees of *Eucalyptus*, as has been reported in rainforest trees utilised by folivorous primates (Glander 1978), but also that the koala has a higher ceiling to diformylphloroglucinol compounds (DFPCs) than the common ringtail possum. Again, if this is extrapolated to other plant secondary metabolites, it provides an explanation for the greater ability of koalas to use eucalypt foliage as a sole source of nutrients than either the common ringtail possum or the common brushtail possum. On this basis the greater glider would also be expected to exhibit a high ceiling to its intake of DFPCs, as it is almost as specialised on *Eucalyptus* foliage as the koala. More information is needed on the metabolic fate of DFPCs and thus the basis for differences among species of arboreal folivores in their tolerance to them.

The findings of minimum threshold levels of nutrients (water and nitrogen) and even one class of secondary plant metabolites (essential oils in Hume & Esson's (1993) study with koalas) suggest that while one set of factors may determine whether or not herbivores eat a plant, another set of factors may determine how much is eaten in the long term (Braithwaite, Turner & Kelly 1984; Marquis & Batzli 1989). Whether a plant is eaten or not may be determined by the presence of single components, such as the DFPCs. In plants that are eaten, the amount actually consumed is probably a net effect of the balance between nutrients and other (perhaps non-DFPC) plant secondary metabolites (Cork & Foley 1991).

The water content of eucalypt leaves has been found to be significantly linked to the occurrence of koalas in the semi-arid woodlands of north Queensland (Munks, Corkrey & Foley 1996). No other nutrients were related to koala occurrence, but this could have been because of the paramount importance of water in this environment. Mean water contents ranged between 38 and 50%, well below the threshold levels established previously with captive animals (55 to 65%). Similarly, differences in water content of foliage may have been the dominant factor in the differential mortality of koalas during drought in south-western Queensland (Gordon, Brown & Pulsford 1988). Foliage water content was not measured, but around large permanent water holes foliage showed little evidence of browning or leaf fall, and koalas had good body condition and mortality was low. In contrast, along stretches of dry river bed there was extensive leaf fall and/or browning of the foliage of food trees, and koalas were in poor condition, with high tick loads, anaemia and dehydration, and mortality was high.

The correlative findings of Hume & Esson (1993) with captive koalas were partly confirmed with free-living animals by Krockenberger (1993). The most preferred species on Krockenberger's study site, *E. acaciiformis* (wattle-leafed peppermint) was comparatively high in volatile monoterpenes, nitrogen, and the ratios of nitrogen to fibre, total phenolics and condensed tannins. It was low in sesquiterpenes and fibre. The least preferred species, *E. stellulata* (black sallee), was the lowest of the species in water, total oils, volatile monoterpenes and nitrogen, and ratios of nitrogen to fibre, total phenolics and condensed tannins. It was high in fibre, total phenolics and condensed tannins.

On a broader scale, Braithwaite, Turner & Kelly (1984) showed that in the forests of south-eastern New South Wales, a threshold in leaf quality, correlated with the concentrations of nitrogen, potassium and phosphorus in foliage, is a major discriminator between eucalypt communities that support permanent populations of arboreal marsupials, including koalas, and communities that support only transient, non-breeding individuals at best. This pattern has been explained by Cork (1992) in terms of the resource-availability hypothesis of Coley, Bryant & Chapin (1985). This hypothesis holds that the relative availabilities of nutrients versus carbon determine the amounts and types of chemical defences employed by plants against herbivores. When availability of nutrients from soil is low, natural selection should favour plants that grow slowly and defend themselves strongly to maximise leaf longevity. The defensive compounds should be carbon-based, such as phenolics and terpenes, because carbon is available in excess due to nutrient limitations on plant growth. Carbon-based defences should decrease when nutrient availability increases.

Consistent with predictions from the resource-availability hypothesis, Cork (1992) found significant negative relationships between the total phenolic content of foliage and the concentration of each of nitrogen, phosphorus and potassium (Fig. 5.4).

Even stronger relationships were found when concentrations of nitrogen and phosphorus were expressed as ratios of phenolic contents (Table 5.2). The use of these ratios is consistent with the general conclusion that, above minimum thresholds of certain nutrients, foliage acceptability is likely to be determined by the relative proportions of nutrients and plant secondary metabolites, not just the concentrations of one or the other. The use of the

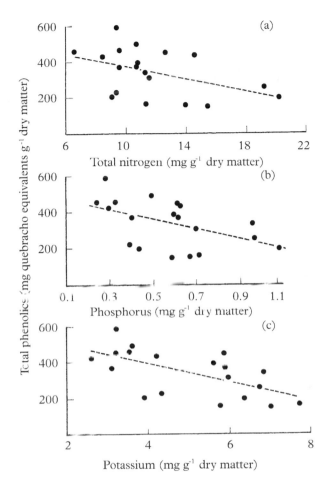

Figure 5.4 Relationships between the concentration of total phenolics and levels of (a) nitrogen (b) phosphorus, and (c) potassium among 18 species of *Eucalyptus* and one of *Angophora* in forests in south-eastern New South Wales. From Cork (1992).

ratio of phosphorus to phenolics (other than condensed tannins), and the equivalent ratio for nitrogen, reflects recent findings that when phenolics (and terpenes) are absorbed from the gut they are conjugated in the liver to form strong organic acids that challenge the acid–base equilibrium of folivorous marsupials (Foley & Hume 1987c; Foley 1992). Dietary phosphorus helps to buffer these acids. Excretion of ammonia in the urine also helps by conserving bicarbonate, which also buffers these acids, but this mechanism elevates the animal's nitrogen requirement. This is the main reason for the high maintenance nitrogen requirement of greater gliders fed *E. radiata* foliage (Table 1.8), which is high in total phenolics and particularly high in terpenes (Foley & Hume 1987c). Thus phosphorus and nitrogen are important nutrients because, among other things, they counteract the effects of plant secondary metabolites.

Table 5.2. *Concentrations of three nutrients and their ratios to phenolics and cell walls (neutral-detergent fibre) in* Eucalyptus *foliage in preferred and non-preferred communities in forests in south-eastern New South Wales*

Nutrient/Ratio	Community group	
	Preferred ($n = 9$)	Non-preferred ($n = 13$)
Nitrogen (% of dry matter)	1.23	0.96
Phosphorus (mg g^{-1} dry matter)	0.69	0.39
Potassium (mg g^{-1} dry matter)	5.21	3.73
P: non-condensed tannin phenols	49.5	31.8
N: non-condensed tannin phenols	8.9	7.4
N: non-condensed tannin phenols plus cell walls	8.7	7.2
N: condensed tannins	347	269

Note: All comparisons statistically significant.
Source: After Cork 1992.

The ratios of nitrogen to tannins and cell walls are also nutritionally important because both condensed tannins and fibre increase nitrogen losses in the faeces (Cork 1986). Phenolics have also been implicated in leaf and habitat selection by folivorous primates in tropical African forests (Oates *et al.* 1990). Manipulation of the ratios of nutrients to phenolics in forests by management practices in order to improve their value as wildlife habitat is an area where more research is needed.

Conversely, Lawler *et al.* (1997) demonstrated experimentally that when seedlings of *E. tereticornis* (forest red gum) were grown to 1.5 m saplings under controlled conditions, elevated carbon dioxide, low nutrient availability or high light intensity increased the C:N ratio of leaves, and led to lower leaf nitrogen concentrations and higher levels of total phenolics and condensed tannins. When these leaves were fed to the larvae of the chrysomelid beetle *Chrysophtharta flaveola*, there were reductions in digestibility of leaf and in pupal body size, and increased mortality. There appeared to be a threshold level of leaf nitrogen (about 1.0% of dry matter) below which reductions in performance were most severe.

The low-nutrient, high-light treatment was designed to mimic present conditions in the forests of south-eastern Australia. Extrapolation of their results to marsupial folivores led Lawler *et al.* (1997) to conclude that any significant elevation in atmospheric carbon dioxide concentration could easily reduce eucalypt leaf quality below the threshold levels demonstrated by Braithwaite *et al.* (1984) and Cork (1992). That is, a large proportion of forests presently suitable for folivorous marsupials would become unsuitable. The potential for substantial reductions in population densities and ranges is such that local extinctions could well be widespread. The effects of projected increases in atmospheric carbon dioxide over the next century will be most pronounced in forests growing on poor soils; it is these forests that dominate conservation reserves in Australia.

5.5 DIGESTION AND METABOLISM IN THE ARBOREAL FOLIVORES

The four marsupials that use eucalypt foliage as either their principal food or as a major component of their diet (the koala, greater glider, common ringtail possum and common brushtail possum) have now been studied sufficiently to allow for meaningful comparisons of their digestive strategies.

5.5.1 Koala

DIGESTIVE TRACT

The koala is remarkable not only for its virtually sole diet of *Eucalyptus* foliage but also for its large and complex digestive system (Fig. 5.5). Although the enormous size of the hindgut is the most obvious feature of the gross morphology of the koala gastrointestinal tract, the stomach has also received attention from anatomists since, like that of its nearest relatives, the wombats, it contains a cardiogastric gland (Oppel 1896: Johnstone 1898). The gland measures about 4 cm in diameter and contains about 25 distinct openings (Krause & Leeson 1973; Harrop & Degabriele 1976). The composition of the gland is similar to that of the wombats, but the tubules, which in the wombats are straight and unbranched, in the koala are branched and thus more complex (Oppel 1896). Yamada *et al.* (1987) found that the mucosa adjacent to, overlying and forming the tubules consisted of gastric (oxyntic) glands whose tubular structure and constituent cell types were similar to those found in other mammals. Four types of endocrine cells were identified. Somatostatin-immunoreactive cells were the most numerous, primarily in the lower two-thirds of the gastric glands. Bovine pancreatic polypeptide-immunoreactive cells were confined to the lower two-thirds and increased in number towards the basal portion of the glands. Some 5-hydroxytryptamine-immunoreactive cells were scattered throughout the gastric glands, and a few glucagon-immunoreactive cells were seen in the basal region of the glands. Exogenous pancreatic glucagon inhibits gastric acid secretion and stimulates mucus secretion in humans, but the role of glucagon-immunoreactive cells in the koala gastric glands is not known (Yamada *et al.* 1987).

The cardiogastric gland is the only remarkable feature of the koala's stomach, which in gross morphology is small and simple. Most digestion takes place in the small intestine; Cork, Hume & Dawson (1983) showed that 91% of the koala's daily energy requirement was met by digestion of

165

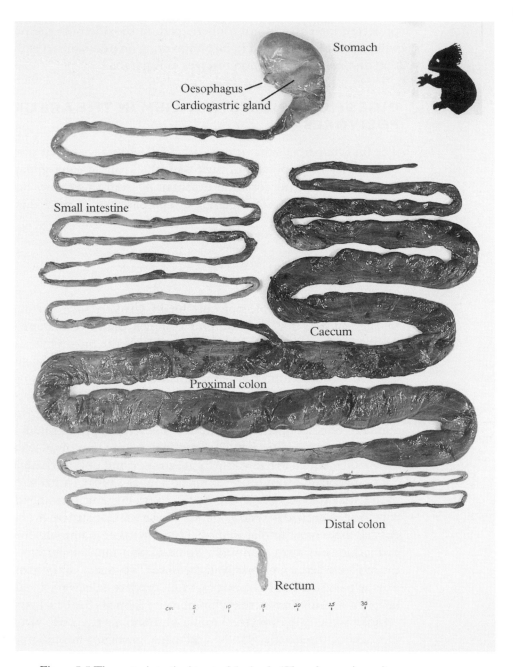

Figure 5.5 The gastrointestinal tract of the koala (*Phascolarctos cinereus*).

the cell content fraction of eucalypt leaves in the small intestine (see Table 5.5). Nutrient availability in the small intestine is affected by the particle size of digesta leaving the stomach. In the koala, with its simple stomach, this depends on the efficiency of its masticatory system. The dentition of the koala is shown in Fig. 5.6. There is a large gap or diastema between the

166

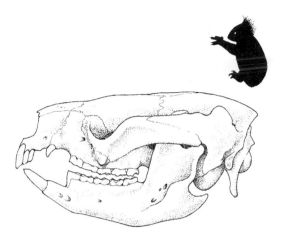

Figure 5.6 The skull of the koala (*Phascolarctos cinereus*) to show its dentition.

incisors and the single pair of premolars. The single pair of canines on the upper jaw are vestigial. When feeding, koalas grasp a small branch with one of their forepaws, then sniff the leaves. If acceptable, leaves are stripped using the sharp longitudinal blades of the premolars. Mastication then commences with the leaf lying diagonally across the mouth.

The leaf enters the occlusal plane of four pairs of molars anteriolingually. One of the roles of the diastema is to keep the leaf steady and oriented so that the occluding cheektooth row spans the leaf (Lanyon & Sanson 1986a). The subselenodont molar structure emphasises a cutting-shearing action. This is achieved by having high, tightly interlocking cusps with opposing long, curved blades and additional accessory crests which provide extra cutting edges. In animals with relatively unworn teeth the leaf is cut into pieces that have evenly serrated edges that can be matched to the cutting edges of the molar tooth row. The koala does not chew the leaf again once it has been through this single size reduction process. There is no crushing component at the end of the occlusal stroke (Lanyon & Sanson 1986a). Thus the efficiency of size reduction of ingested leaves is dependent on the maintenance of the cutting edges on the molars.

With age, distinctive wear patterns result from the progressive attrition of the accessory crests of the molars and the exposure of the underlying dentine. Martin (1981) used seven classes of tooth wear on the premolar from the right upper jaw as a means of assigning koalas to age classes. In a study of tooth wear, Lanyon & Sanson (1986b) assigned molars in the upper right tooth row of 43 koala skulls to one of 28 composite wear classes and related these classes to the total lengths of the cutting edges of the entire molar row. The relationship is shown in Fig. 5.7. They also determined the distribution of particle sizes of leaf fragments in the stomach, caecum and proximal colon of 13 wild koalas obtained as roadkills by wet

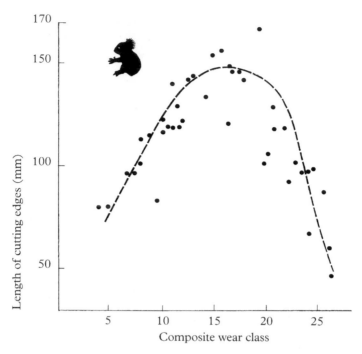

Figure 5.7 Length of cutting edges in each upper right molar row of koalas (*Phascolarctos cinereus*) plotted against composite wear class. From Lanyon & Sanson (1986b).

sieving gut segment contents through a nest of sieves of five mesh sizes. Table 5.3 shows a partial summary for five composite wear classes and for particles retained by the two smallest mesh sizes (small particles) and by the two largest mesh sizes (large particles).

Fig. 5.7 indicates that some wear is necessary in order to maintain maximal total length of cutting edges. Very young and very old animals are thus less efficient in comminuting eucalypt leaves. This is illustrated by Table 5.3. The ratios of small to large particles in the stomach of composite wear classes 5 (young animals) and 25 or greater (very old animals) are both less than 1, compared with ratios of about 2.5 in wear classes between 10 and 20. Aged koalas bred in captivity have minimal chances of surviving if released into the wild partly because of their inability to comminute *Eucalyptus* leaves finely enough for adequate nutrition.

In contrast to the stomach, ratios of small to large particles in the caecum and proximal colon do not reflect differences in tooth wear, except perhaps in very old animals. This is because particle size distributions in the koala hindgut are modified by a colon separation mechanism which results in selective retention of solutes and fine particles in the caecum and proximal colon (see below).

Secretions from the parotid salivary glands of koalas are surprisingly highly buffered for a hindgut fermenter, being at least 50% of that of foregut fermenters such as the red kangaroo (*Macropus rufus*) and sheep

Table 5.3. *Particle size distribution (% of total) in the stomach, caecum and proximal colon of wild koalas in five different composite molar tooth wear classes*

Composite Wear Class	Stomach			Caecum			Proximal colon		
	Small	Large	Small Large	Small	Large	Small Large	Small	Large	Small Large
5	32	35	0.9	74	9	8.2	69	11	6.3
10	48	20	2.4	62	11	5.6	61	15	4.1
15	43	17	2.5	64	8	8.0	61	10	6.1
20	45	17	2.7	62	11	5.6	62	12	5.2
25+	22	56	0.4	71	16	4.4	59	24	2.5

Note: Small particles 75–230 µm; large particles > 1015 µm.
Source: After Lanyon & Sanson 1986b.

(Beal 1990a). The high buffering capacity is a consequence of high levels of bicarbonate and to a lesser extent phosphate ions. Amylase activity is low, despite high maximal protein secretion rates.

Amylase activity is also lacking in sublingual (Beal 1990b) and mandibular saliva (Beal 1991a). Beal (1990a) related this to the low starch levels of *Eucalyptus* foliage. Flow rates of sublingual saliva are low, but those of the mandibular gland are quite high, about four times the rates per g gland tissue of the parotid glands. Beal (1991a) concluded that the koala mandibular gland, because of its relatively high secretory capacity, plus its responsiveness to acute changes in circulating levels of mineralocorticoid hormones (which produce a markedly hypotonic saliva), is better adapted than the parotid gland to be the primary source of saliva for evaporative cooling. That koalas use saliva for this purpose was demonstrated by Degabriele & Dawson (1979) when they observed koalas to pant and lick and spread saliva in very hot conditions.

The saliva of many mammalian browsers contains proteins that are rich in proline (i.e. contain 20% or more proline) (McArthur, Sanson & Beal 1995). Proline-rich proteins may have evolved in saliva to maintain oral homeostasis, but in many browsers they also protect the animal against dietary tannins by preferentially binding with them. However, whether proline-rich salivary proteins protect against the protein-binding properties of eucalypt tannins is questionable. Mole, Butler & Iason (1990) recorded only weak affinity of the salivary proteins of koalas and common ringtail possums, even though they both contain 31% proline.

The small intestine of the koala is relatively short (Fig. 5.5), only 29% of the total intestine (Snipes, Snipes & Carrick 1993). As in the wombats, Brunner's glands, which secrete mucus, extend along the duodenum further than in all other marsupial groups (Krause 1972), but the reason for this is obscure. The small intestine houses the only gut helminth reported in koalas, the anoplocephalid cestode *Bertiella obesa*, which is highly host specific (Beveridge & Spratt 1996). Thus the koala harbours an extremely limited helminth fauna, possibly because of its adoption of arboreal habits and evolutionary loss of parasites which are transmitted via terrestrial arthropods or whose larval stages are ingested with ground-level herbage. The essential oils of eucalypts may also be toxic to most gut helminths (Beveridge & Spratt 1996).

Several early anatomists remarked on the enormous size of the koala caecum. Owen (1868) wrote: 'in the koala the caecum and large intestines arrive at their maximum of development', and Mackenzie (1918) stated that in the koala 'we meet with the greatest instance of caecal development in the Mammalia'. At approximately 1.3 m, the koala caecum is 23% of total intestinal length, and 35% of total intestinal volume (Snipes *et al.* 1993). The presence of 8–14 longitudinal folds in the caecum and proximal colon of the koala augment the surface area of both regions enormously. The importance of the koala caecum and proximal colon as a site of digesta retention and microbial fermentation has now been established (Cork & Warner 1983; Cork & Hume 1983).

McKenzie (1978) reported that although the koala's caecal epithelium was basically similar to that of other mammals, distended intercellular spaces and numerous interdigitations of the lateral membrane of epithelial cells gave the ultrastructural appearance of an epithelium designed for transporting water in much the same way as the rabbit gallbladder. He concluded that the caecal epithelium absorbed water and nutrients from the lumen and secreted mucus to lubricate the passage of digesta. The other feature of the epithelium was the strong attachment of bacteria with its luminal surface to form a tall palisade fringe composed of Gram-positive and Gram-negative bacilli, cocci and actinomycete-like organisms. No spirochaetes or fusiform bacilli were seen. The palisade arrangement of bacteria on the epithelium would allow the maximum number of bacteria to have a close association with the caecal wall.

One of the organisms colonising the caecal (and proximal colon) wall of the koala was identified by Osawa *et al.* (1993) as a facultatively anaerobic Gram-negative pleomorphic rod that degrades tannin-protein complexes. The same organism isolated from koala faeces was found to be capable of degrading protein when complexed with gallotannin (a hydrolysable tannin) but not when complexed with quebracho (a condensed tannin) (Osawa, Walsh & Cork 1993). It is probable that some hydrolysable tannins escape acid hydrolysis in the stomach and reach the hindgut. The tannin-protein complex degrading bacteria described by Osawa *et al.* (1993) may therefore provide a means of salvaging any dietary or endogenous proteins complexed with hydrolysable tannins in the intestine. However, there is no evidence for active transport of amino acids across the wall of the mammalian hindgut (Stevens & Hume 1995). Thus any protein

Table 5.4. *Total concentrations and molar proportions of short-chain fatty acids (SCFA) along the gut of captive koalas*

Region	Total SCFA (mmol L^{-1})	Molar proportion (%)			
		Acetic	Propionic	Butyric	Valeric
Stomach	15.1	93.1	6.5	0.3	0.1
Small intestine	12.6	93.0	1.9	4.7	0.4
Caecum	25.9	86.3	8.8	4.7	0.2
Proximal colon	28.1	86.1	10.3	3.7	0.1
Distal colon	36.3	91.7	5.6	2.5	0.2
Faeces	62.1	95.7	2.8	1.6	0.0

Note: Butyric acid contained less than 0.1% of iso-butyric acid, and valeric acid contained less that 0.2% of iso-valeric acid.
Source: After Cork & Hume 1983.

released from complexes with tannins can only be useful if it is degraded to ammonia in the hindgut. It could then serve as a nitrogen source for bacterial growth in the hindgut, or be absorbed and incorporated into non-essential amino acids in the liver.

Bacteria in the lumen of the caecum and proximal colon of koalas have been characterised by London (1981). Viable counts of anaerobes averaged 1.1×10^{10} per g wet contents in the caecum and 3.0×10^9 in the proximal colon. Viable counts of aerobes were much lower, 9.7×10^6 and 2.3×10^7 per g wet contents in the caecum and proximal colon respectively. Thus the ratio of anaerobic to aerobic bacteria in the caecum was 1150:1. Viable counts of caecal anaerobic and aerobic organisms represented 45% of total cell numbers. The bacteria were predominantly Gram-positive rods (61% of cells isolated). The most common bacteria isolated were presumptively grouped into the genera *Bacteroides, Eubacterium, Peptococcus, Peptostreptococcus, Propionibacterium* and an unidentified group of Gram-positive rods.

A close-to-neutral pH (6.5–6.6 in the caecum and 6.6 in the proximal colon) (Cork, Hume & Dawson 1983) and a constant temperature of 36–38 °C (London 1981) provides conditions favourable for bacterial growth and fermentation in the koala hindgut. The concentrations of total SCFA, the main end-products of the gut fermentation, along the gastrointestinal tract are shown in Table 5.4. Levels in the stomach and small intestine are within the range found in other simple-stomached herbivores (Stevens & Hume 1995). Concentrations increase in the caecum and proximal colon, but they are a good deal lower than in the hindgut of a range of other small hindgut fermenters. This could be due to either faster absorption of SCFA in the koala or slower production. Studies on the absorptive epithelia of the gut in both foregut- and hindgut-fermenting eutherians indicate little difference in the potential rate of SCFA absorption per unit surface area (Stevens & Hume 1995). Thus the low concentration of SCFA in the hindgut of the koala suggests not faster absorption but slower fermentation rates than in the caecum and colon of other hindgut fermenters.

This suggestion is supported by rates of production of SCFA measured in the hindgut of captive koalas by Cork & Hume (1983) using the same *in vitro* incubation procedure described for the wombats (Chapter 4). Zero-time rates of production were similar between the caecum and proximal colon (11 mmol L^{-1} fluid h^{-1}), and comparable with rates in the proximal colon of captive wombats fed a high-fibre diet (12 mmol $L^{-1} h^{-1}$). However, they are low compared with rates in the hindgut of most other mammalian herbivores, probably as a consequence of the refractory nature of the cell walls of eucalypt foliage. Acetate formed 85% of the SCFA present (Table 5.4) and 62% of that produced, which is typical of microbial fermentation of high-fibre diets; in the wombats acetate accounted for 74% of the SCFA present in the proximal colon and 52% of that produced. The extent of reductive acetogenesis in the hindgut of koalas is unknown; this is an additional source of acetate in the hindgut of humans, some rodents and termites (Drake 1994).

Despite the huge size of the koala hindgut, daily SCFA production represented only 9% of digestible energy intake (or energy absorbed). This is much lower than the 30–33% calculated for captive wombats, and barely accounts for digestion of the cell-wall carbohydrates of the eucalypt leaves. Thus fermentation of other substrates must be minimal and quantitatively unimportant in the koala's energy economy.

Table 5.5 summarises digestibilities and contributions of several euca-lypt leaf components to the daily energy absorbed by captive koalas. In line with the contribution of 9% made by SCFA, digestion of cell-wall carbohy-drates (cellulose and hemicellulose) was also low (8–11% of digestible energy intake). The value for lignin is dubious as it is based on lignin digestibilities of 19%, which are out of line with the notion that lignin breakdown in the anaerobic conditions of the mammalian gut is unlikely. Lignin breakdown by fungi living on the forest floor is both aerobic and slow. More likely, dissolution and absorption of less-polymerised phenolic compounds from the small intestine and their urinary excretion could have occurred. Alternatively, these small phenolic compounds may be degraded by microbial action in the hindgut. If these compounds are in the small particle fraction they may be retained in the caecum and proximal colon for extended periods (see below), which would maximise the opportunity for their microbial degradation.

As can be seen from Table 5.5, the cell contents are the major source of energy to the koala. The relative contributions of dietary constituents to the koala's metabolism in this table are based on three considerations (Cork & Sanson 1990). First, failure to account for faecal excretion of endogenous

Table 5.5. *Digestibility of various components of* Eucalyptus punctata *foliage and their contributions to the energy absorbed (i.e. digestible energy intake) of captive koalas*

	Winter			Spring		
Constituent	digested (%)	percent of absorbed energy	percent of metabolised energy	digested (%)	percent of absorbed energy	percent of metabolised energy
Cell contents						
Crude protein	45	7	25	45	9	33
Starch, sugars	92	15	30	92	14	28
Lipid	43	30	28	43	20	18
Total phenolics	91	33	2	91	35	2
Cell walls						
Hemicellulose	24	11	20	24	8	15
Cellulose	31			31		
'Lignin'	19	10	0	19	10	0
Digestible energy intake (kJ kg$^{-0.75}$ d^{-1})	533			494		

Source: After Cork, Hume & Dawson 1983 and Cork & Sanson 1990.

crude protein, carbohydrate and lipid underestimates the amounts of these components of leaf cell contents that are absorbed. Second, most absorbed phenolics and essential oils (part of the lipid fraction) do not contribute to the animal's energy economy, as they must be detoxified and excreted in the urine; their energy value may indeed be negative. Third, any components of lignin that are absorbed from the gut are unlikely to be useful, as discussed above. Thus the crude protein, starches and sugars, and some of the lipid in the cell contents, which together make up only 36% of the leaf dry matter at most, probably supply more than 80% of the useful energy derived from eucalypt leaves by koalas. This conclusion is consistent with both the small contribution made by microbial fermentation in the hindgut and the high activity of several disaccharidases (except trehalase) in the small intestinal mucosa of the koala (Table 2.3); the small intestine is by far the main site of energy absorption in this arboreal folivore .

The negative value of absorbed essential oils is partly because of the urinary loss of the glucuronic acid used to conjugate modified terpenes in the liver. Eberhard *et al.* (1975) estimated that koalas normally excrete 1–3 g of glucuronic acid daily. The glucose required to synthesise this glucuronic acid has been calculated by Cork (1981) to be equivalent to as much as 20% of the koala's fasting glucose requirement, a considerable energetic cost. However, *Eucalyptus* foliage is high in ascorbic acid (Dash & Jenness 1985), which is reflected in high circulating levels of ascorbate in the blood of common ringtail possums maintained on *E. andrewsii* foliage (Dash, Jenness & Hume 1984). If some of this absorbed ascorbate can be used to synthesise glucuronic acid, as postulated by Dash (1988), this would spare glucose for more essential metabolic functions and would represent a significant energy saving.

The long retention times in the hindgut of koalas alluded to above were established in captive animals by Cork & Warner (1983). They used ^{51}Cr-EDTA as a solute marker and ^{103}Ru-Phenanthroline as a particle-associated marker, as Barboza (1993a) did with wombats (Chapter 4). Mean retention times were extremely long, averaging 99 hours for the particle marker and even longer, 213 hours, for the solute marker. Among the mammals, only the sloth, dugong and manatee have been found to have longer particle retention times (Stevens & Hume 1995). The patterns of excretion of the two markers in the faeces of the koala are shown in Fig. 5.8. The common brushtail possum is included for comparison and is discussed later (see Section 5.5.5).

From the concentrations of the two markers at various sites along the digestive tracts of three koalas killed 3, 12 and 24 hours after a single oral dose of the markers, Cork & Warner (1983) were able to conclude that while the retention time of the particle marker in the stomach (9.0 hours) was longer than that of the solute marker (4.5 hours), retention in this organ contributed little to total MRT. It was clear that both markers

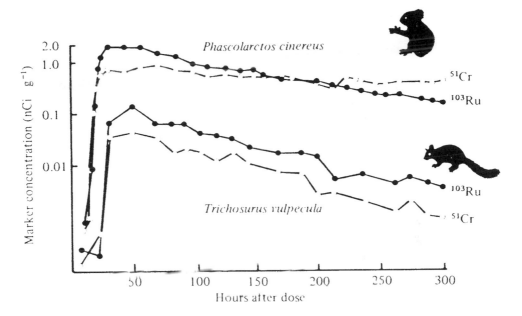

Figure 5.8 Patterns of appearance of a solute marker (^{51}Cr-EDTA) and a particle-associated marker (^{103}Ru-Phenanthroline) in the faeces of the koala (*Phascolarctos cinereus*) and the common brushtail possum (*Trichosurus vulpecula*) following a pulse oral dose. After Cork & Warner (1983) and Wellard & Hume (1981a).

accumulated in the caecum and proximal colon, from which, after 24 hours, the particle marker left at a faster rate. This explains the difference in MRT between the solute and particle markers. This difference is shown clearly by the slopes of the two excretion curves in Fig. 5.8. The caecum and proximal colon contained notably higher proportions of small particles than either the stomach or the distal colon, suggesting that along with solutes, small digesta particles were selectively retained in the caecum and proximal colon.

Selective retention of solutes and small particles has several important nutritional consequences for the koala. First, digestibility of nutrients in the solutes and small particles is likely to be increased. Second, because small particles include bacterial cells, faecal loss of microbial protein may be reduced. Third, the facilitated elimination of fibre in the large particles should reduce the gut-filling effect of the diet, thus relaxing the upper limit to food intake. This mechanism, together with low energy and nutrient requirements (see Chapter 1), are important features of the koala's physiology that allow it to exploit a low-quality diet which few other small mammals can utilise.

Despite the relatively rapid elimination of large particles, their MRT is still unusually long. The low digestibilities of fibre reported for koalas (Table 5.5) are therefore not a consequence of the separation mechanism in the koala's hindgut. More likely they reflect the absolute limit to the

175

digestion of eucalypt foliage cell walls by mammalian herbivores, which may be independent of differences in MRT beyond a certain number of hours. The time suggested by Cork and Warner's (1983) results is 100 hours. However, this was recorded with captive animals. More recently, Krockenberger (1993) measured rate of passage in free-living koalas using Co-EDTA as a solute marker and Cr-mordanted particles. MRTs were much shorter, being 99 hours for the solute marker and 32 hours for particles. Fibre digestibility was not measured, but perhaps the retention time beyond which little further digestion of eucalypt leaf cell walls is achieved is considerably less than 100 hours. A study of the kinetics of digestion of *Eucalyptus* foliage by the use of *in vitro* incubation would be necessary to resolve this question.

Despite the much shorter MRTs in the free-living koalas, selective retention of the solute marker was still clearly evident. The shorter MRT for the particle marker is partly due to lack of migration of the mordant from larger to smaller particles, in contrast to [103]Ru-Phenanthroline, resulting in a truer picture of large particle passage. The shorter MRTs of both the solute and particle markers in the wild koalas is no doubt due to the greater food intakes required to meet the additional energy costs of free existence, such as greater activity and thermoregulation. Non-lactating females ate 35% more dry matter than did captive animals, and lactating females ate 69% more. Rate of passage of particles is generally increased when food intake increases (Warner 1981a), unless there is a concomitant increase in gut capacity (Karasov & Hume 1997).

NITROGEN METABOLISM

The first nutritional study on koalas was by Harrop & Degabriele (1976). They maintained captive koalas on a sole diet of *E. punctata* (grey gum) foliage throughout the year. Seasonal differences in most nutritional parameters were slight. For instance, dry matter intake averaged 41 $g\,kg^{-0.75}$ d^{-1} in summer and 49 in winter, but at the same time apparent digestibility of dry matter fell from 59% in summer to 52% in winter, so that intake of digestible dry matter remained unchanged with season at 25 $g\,kg^{-0.75}\,d^{-1}$. Apparent digestibility of nitrogen followed the same trend as that for dry matter, with the result that nitrogen balance of the animals remained low but positive throughout the year.

Harrop & Degabriele (1976) were not able to estimate the maintenance nitrogen requirement of koalas because of the narrow range of leaf nitrogen content (0.9–1.2%). In a more detailed study, Cork (1986) used a slightly wider range of nitrogen contents (1.1–1.5%) and was able to conclude that on a sole diet of *E. punctata* foliage adult, non-reproductive koalas required 283 $mg\,kg^{-0.75}$ of dietary nitrogen or 271 $mg\,kg^{-0.75}$ of truly digestible nitrogen daily to remain in nitrogen balance. These values are in the same range as estimated requirements for common brushtail possums on a sole diet of *E. melliodora* (yellow box) foliage and for common ringtail possums on a sole diet of *E. andrewsii* (New England blackbutt) foliage (Table 1.8), but below those of greater gliders on a sole diet of *E. radiata* (narrow-leafed peppermint) foliage. Reasons for the surprisingly high values for the greater glider are explored below.

The relatively low requirements for the koala are partly explained by low urinary losses, particularly of urea. Urea is the normal end-product of mammalian nitrogen metabolism. It is synthesised in the liver from ammonia that is absorbed from the gut or released during catabolism of body proteins. The UR ratio is the ratio between the concentrations of urea-nitrogen and total nitrogen in the urine, and can be used as an index of recycling of urea to the gut (Kinnear & Main 1975). In the gut the urea can be hydrolysed to ammonia by ureolytic bacteria and some of the ammonia used as a source of nitrogen for microbial protein synthesis. A low UR ratio is indicative of substantial urea recycling. In captive tammar wallabies (*Macropus eugenii*), Chilcott, Moore & Hume (1985) found that a UR ratio of 0.37 related to a urea recycling rate of 79%. Wild tammars had a UR ratio of 0.74 during the winter wet season when the protein content of available forage was high (23.6%) but a ratio of 0.32 during the annual summer drought when plant protein levels were minimal at 6.8% (Kinnear & Main 1975). Cork's (1986) koalas had a UR ratio of 0.36 throughout the year, and the crude protein content of the *E. punctata* leaf eaten was also consistently low, 6.9–9.2%, suggesting that urea recycling was high in all seasons of the year.

This suggestion was confirmed by Cork (1981) when he measured urea recycling directly using an isotope dilution technique. Of the urea synthesised in the liver each day, 79% was recycled to the gut in winter, 78% in spring, and 76% in summer.

In contrast to the low urinary loss of nitrogen, Cork (1986) found that faecal loss was higher than expected. This is probably because of the presence in *E. punctata* leaves of significant amounts of tannins (Fig. 5.2). Tannins reduce the digestibility of proteins by complexing with them. This affects endogenous as well as dietary proteins, leading to increased metabolic faecal nitrogen (non-dietary faecal nitrogen). In Cork's (1986) koalas 83% of faecal nitrogen was of non-dietary origin. Thus, a diet of mature eucalypt foliage can be expensive in terms of nitrogen loss in the faeces, but in koalas fed *E. punctata* this is partly compensated by extensive urea recycling and thus a low level of nitrogen loss in the urine. Unfortunately, comparable data for koalas on other diets are not available. As will be seen, urinary nitrogen losses in other marsupial folivores can sometimes be very high.

WATER METABOLISM

Daily water turnover rates (WTR) in captive koalas measured by Degabriele, Harrop & Dawson (1978) were quite low in both summer and winter

Table 5.6. *Partitioning water intake and loss in captive Koalas under various conditions; values are percentages of total intake or loss*

	Summer Water *ad lib.*	Winter Water *ad lib.*	Winter Water deprived
Intake			
Water drunk	26	26	0
Preformed water in leaves	45	44	58
Metabolic water	29	30	42
Loss			
Urine	20	15	19
Preformed water in faeces	21	25	11
Metabolic water	10	13	9
Evaporative water loss	49	47	61
Total water turnover (mL kg$^{-0.80}$ d^{-1})	80.0	92.3	63.3

Source: After Degabriele, Harrop & Dawson 1978.

(80 and 92 mL kg$^{-0.80}$ respectively). These values are only 65–75% of the predicted eutherian level under similar conditions (Richmond *et al.* 1962). The higher winter value may be related to the higher dry matter intakes in that season (Harrop & Degabriele 1976), and thus a greater contribution of metabolic water to the koalas' water budget. The animals drank free water in both seasons. When deprived of drinking water in winter, water turnover rate dropped to 63 mL kg$^{-0.80}$ d^{-1}. Under these conditions preformed water in the leaves contributed 58% of the total water input of the animals, and metabolic water the remainder (Table 5.6). The main avenue of water loss in all cases was evaporation, but the greatest reduction in water loss when deprived of drinking water was seen in faecal output.

Krockenberger (1993) measured water turnover rates in free-living female koalas on the Northern Tablelands of New South Wales. Daily turnover was 88 mL kg$^{-0.71}$ in both summer and winter, similar to the value of 83 mL kg$^{-0.71}$ reported by Nagy & Martin (1985) in Victoria in spring (see Table 1.7). However, lactating females turned water over faster than non-lactating females in both seasons (91 vs. 84 mL kg$^{-0.71}$ d^{-1}), and at peak lactation in spring the difference was more marked (82 vs. 71 mL kg$^{-0.71}$ d^{-1}). This corresponded to a 27% greater daily food intake in the lactating animals at this time of the year (Krockenberger 1993).

Note that different scaling factors are used for comparing water turnover rates in captive and free-living marsupials. This is explained in Chapter 1.

Higher water turnover rates were recorded in central Queensland by Ellis *et al.* (1995). Daily turnover rates in male koalas were 107 mL kg$^{-0.71}$ in summer and 91 $mL\,kg^{-0.71}$ in winter. More water may have been used in summer for evaporative cooling, while the water content of the foliage of selected eucalypt species averaged only 47% in winter (vs. 62% in summer). Because FMR was higher in winter (501 vs. 462 $kJ\,kg^{-0.58}\,d^{-1}$ in summer), the authors concluded that changes in diet selection reflected increased energy requirements in winter and increased water requirements in summer.

The low rates of water turnover recorded in koalas are not associated with great concentrating power of the koala's kidneys. In their laboratory study Degabriele *et al.* (1978) recorded a maximum urine osmolality of 1692 $mOsmol\,kg^{-1}$ water in winter with no drinking water available. Administration of vasopressin (an antidiuretic hormone) induced maximally concentrated urine of 1843 $mOsmol\,kg^{-1}$. This is little more than would be predicted from Beuchat (1990a, b) on the basis of the koala's body size. The lack of great concentrating power in their kidneys may explain why koalas reject foliage that contains less than a minimum (threshold) water content.

NUTRITION OF YOUNG KOALAS

Given the poor nutritional quality of eucalypt leaves, a pertinent question is how do koalas meet the additional energy demands of reproduction on such a diet? Krockenberger (1993) studied this question in free-living koalas on the Northern Tablelands of New South Wales. He concluded that females met the extra energy demands for reproduction in a number of ways. First, they spread lactation over a long period (nine months), which is 58% longer than expected for a mammal of that size, eutherian or marsupial (Krockenberger, Hume & Cork 1998). This minimises daily energetic demands; at the time of peak lactation there was no measurable difference in field metabolic rate between lactating and non-lactating females. Second, they produce milk which does not increase in lipid concentration around the time of pouch exit by the young (Krockenberger 1996). This also reduces the energetic costs of lactation at the time of peak demand. In this they differ from terrestrial marsupials, carnivores, omnivores and herbivores alike (Fig. 5.9), but it is a feature they share with two other arboreal folivorous marsupials, the common ringtail and common brushtail possums.

Lactating female koalas meet at least 90% of the additional demands of lactation by increasing food intake, by 27% over that of non-lactating females at the time of peak lactation. This was accommodated without any increase in passage rate by a 33% increase in fluid capacity of the digestive tract. This increase was probably mainly in the caecum and proximal colon, the site of selective retention of solutes and small particles (Cork & Warner 1983). Digestive efficiency of all but the cell walls was therefore probably maintained despite the greater amount of leaves being processed by the gut.

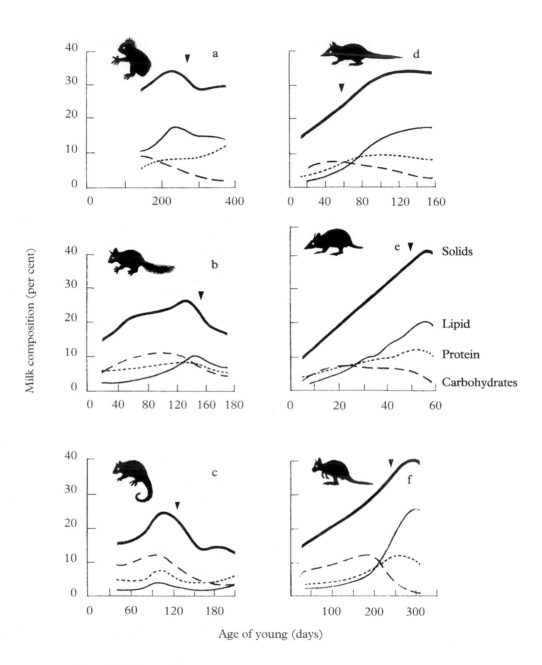

Figure 5.9 Composition of the milk of three marsupial arboreal folivores, the koala, *Phascolarctos cinereus* (a), common brushtail possum, *Trichosurus vulpecula* (b) and common ringtail possum, *Pseudocheirus peregrinus* (c) compared with that of a marsupial carnivore, the eastern quoll, *Dasyurus viverrinus* (d), an omnivore, the northern brown bandicoot, *Isoodon macrourus* (e) and a terrestrial herbivore, the tammar wallaby, *Macropus eugenii* (f). Note the decline in milk solids and lack of an increase in milk lipid content at the time of pouch exit (arrow head) in all three arboreal folivores but in none of the other species. Sources: koala, Krockenberger (1996); common brushtail possum, Cowan (1989); common ringtail possum, Munks *et al.* (1991); eastern quoll, northern brown bandicoot and tammar wallaby, Green & Merchant (1988).

However, the additional energy assimilated would be expected to show up as a higher field metabolic rate. That it didn't suggests that lactating koalas compensate by reducing other aspects of their energy budget. When Krockenberger (1993) measured resting metabolic rate at peak lactation it increased by only 42% of that predicted, a considerable saving. Also, lactating females reduced the size of their home ranges; core areas (defined by 50% harmonic mean isopleths) of lactating females were only 0.2 ha compared with 0.4 for non-lactating females. Although this would have saved minimal energy it would reduce time spent on the ground moving between trees, when back young are most vulnerable to attack by terrestrial predators. It may also indicate that lactating females concentrate feeding activity on trees of the highest nutritional quality at peak lactation; trees browsed by lactating koalas at this time were significantly lower in total phenolics and higher in the ratio of nitrogen to total phenolics compared with those utilised by non-lactating females. Some energy saving may also result from reduced thermoregulatory expenditure; Krockenberger (1993) calculated that by carrying her young on her belly or her back the lactating female would effectively reduce the ratio of surface area to mass of combined mother/young by about 10%, thereby saving up to 108 kJ per day, equivalent to 26–40% of the expected increase in resting metabolic rate during lactation. Thus it seems that koalas use a range of strategies to meet the additional costs of reproduction from their unpromising diet of eucalypt leaves.

5.5.2 Greater glider

Almost as highly specialised on *Eucalyptus* foliage as the koala, the greater glider is an ideal animal to use for testing the universality of some of the digestive and metabolic features of koalas discussed above.

The greater glider and other members of the family Pseudocheiridae have more complex shearing teeth than members of the family Phalangeridae (which includes the brushtail possums), all of which are less folivorous and more frugivorous in their dietary habits. Shearing teeth triturate the highly resilient foliage of *Eucalyptus* and thereby expose a large surface area of leaf for attachment by bacteria in the enormous caecum of the greater glider (Fig. 5.10). Within the Pseudocheiridae, the upper molars of the greater glider have more cutting edges than those of the common ringtail possum (Gipps 1980).

The digestive tract of the greater glider includes a simple stomach and short small intestine. Brunner's glands, immediately distal to the pyloric sphincter, form only a narrow collar and consist of groups of small lobules rather than lobes as seen in the koala (Krause 1972). The small intestine carries cestodes of a single genus, *Bertiella* (Beveridge & Spratt 1996). As mentioned above, the caecum is enormous. Microbial fermentation is

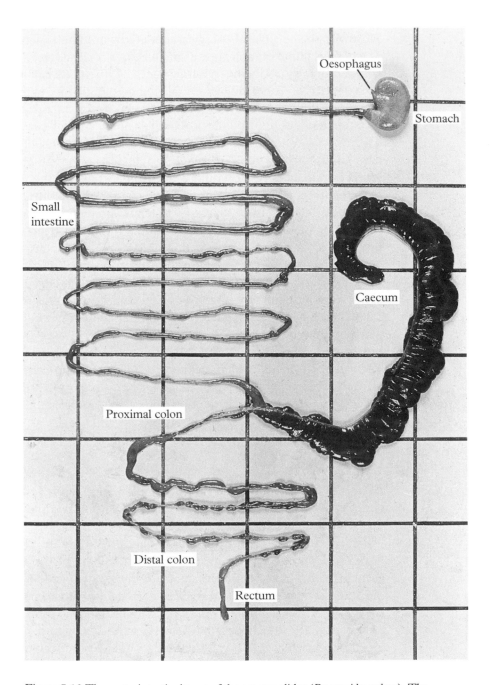

Figure 5.10 The gastrointestinal tract of the greater glider (*Petauroides volans*). The background grid is 10 cm.

confined to this organ, with no parallel development of the proximal colon into a fermentative region as in the koala. The caecum contains a dense oxyuroid nematode fauna, with as many as 1 million nematodes per greater glider (Beveridge & Spratt 1996). No information is available on the microbial population of the caecum, however.

Foley & Hume (1987a) showed that, as in koalas, solutes and small particles of digesta are selectively retained in the greater glider caecum, with mean retention times (MRT) of 51 and 23 hours for a solute marker (^{51}Cr-EDTA) and ^{51}Cr-mordanted large particles respectively on a sole diet of *E. radiata* foliage. (When ^{103}Ruthenium-Phenanthroline was used as the particulate marker, its MRT was 50 hours, showing that this marker was biased towards selectively retained small particles). Concentrations of SCFA were less than 5 mmol L^{-1} in the stomach and small intestine, but they increased to 36 mmol L^{-1} in the caecum of captive animals on *E. radiata* foliage, and to 70 mmol L^{-1} in the caecum of wild animals (Foley, Hume & Cork 1989). Acetic acid made up 63%, propionic acid 22%, butyric acid 13% and valeric acid 2% of the total SCFA in the wild animals. Fermentation rates measured *in vitro* in the caecum of captive greater gliders during feeding were 25 mmol SCFA produced L^{-1} h^{-1}, which yielded 37 mmol d^{-1}, only 7% of the energy absorbed (digestible energy intake). In wild greater gliders shot while feeding, SCFA production was slower, 18 mmol L^{-1} per hour, but the fermentation yielded 50 mmol d^{-1} because the caecum of the wild animals contained twice the amount of digesta (150 g versus 75 g in the captive animals). Foley *et al.* (1989) concluded that, as in the koala, the low rate of fermentation in the greater glider caecum was largely due to the highly lignified nature of the fibre of eucalypt leaves, and possibly the inhibitory effects of leaf phenolics as well.

SCFA are absorbed throughout the caecum and colon of the greater glider, with the highest rates measured in the proximal colon (Rübsamen *et al.* 1983). There was no relationship between SCFA absorption and uptake of sodium or water, because there was net secretion of both sodium and water into the proximal colon but net uptake in the caecum and distal colon (Fig. 5.11). The net inflow of water into the proximal colon is probably part of the separation mechanism whereby solutes and small digesta particles are pushed back into the caecum while large particles continue caudally (see below). The driving force for the inflow of water was probably the net secretion of sodium ions, as it is in other vertebrates (Stevens & Hume 1995).

Although *E. radiata* foliage contains only a moderate level of phenolics (18% of dry matter), it contains a high level of essential oils (11% of dry matter). This was thought to be the main reason for the high nitrogen requirement that Foley & Hume (1987b) established for greater gliders on this diet. For maintenance of zero nitrogen balance they needed 700 mg kg$^{-0.75}$ daily on a dietary basis, or 560 mg kg$^{-0.75}$ on a truly digestible

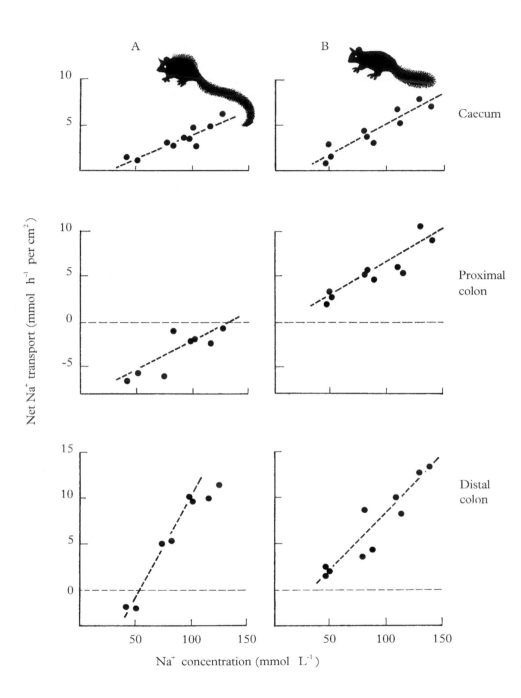

Figure 5.11 Relationship between net sodium transport and luminal sodium concentration in the caecum, proximal colon and distal colon of (A) greater glider (*Petauroides volans*), and (B) common brushtail possum (*Trichosurus vulpecula*). Net sodium uptake was positive in all cases except in the proximal colon of the greater glider. Net secretion of sodium in the greater glider proximal colon is indicative of net water secretion during selective retention of fluid and fine particles in the caecum of this species. From Rübsamen, Hume, Foley & Rübsamen (1983).

basis. These are more than double the requirements established by Cork (1986) for the koala on *E. punctata* foliage (283 and 271 respectively). Comparison of the two diets reveals that, compared with *E. radiata*, *E. punctata* foliage is much lower in essential oils (only 1%).

The principal avenue of nitrogen loss from the greater gliders on *E. radiata* foliage was found to be in the urine, mainly as ammonia rather than urea as would be expected in a mammal. When the diet was switched to *Angophora floribunda* (rough-barked Angophora) foliage, which contains virtually no essential oils, the greater gliders reverted to urea excretion (Fig. 5.12).

Foley (1992) subsequently demonstrated the same effect of *E. radiata* foliage in the common ringtail possum. Animals fed this species produced acid urines (pH 5.7) due to more titratable acid and glucuronic acid excreted. Glucuronic acid is the principal conjugate used in the detoxification and excretion of absorbed terpenes (essential oils) in the common ringtail (McLean *et al.*1993). It seems that ammonium ions are excreted in order to maintain acid–base balance in the animal.

The role of ammonium ion excretion in the acid–base metabolism of mammals was reviewed by Foley, McLean & Cork (1995). Detoxification of plant secondary metabolites often results in the production of organic acids that threaten acid–base balance. Since these acids are largely ionised at physiological pH, disposal of the hydrogen ion and the anion may

Figure 5.12 Effect of changing the diet of a captive greater glider (*Petauroides volans*) from *Eucalyptus radiata* (high in plant secondary metabolites) to *Angophora floribunda* foliage (low in plant secondary metabolites) on the proportion of urinary nitrogen excreted as urea (white bars) or ammonia (grey bars). Values are means ± standard errors. From Foley & Hume (1987b).

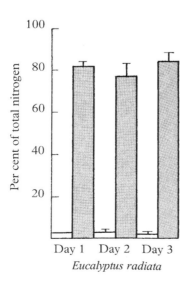
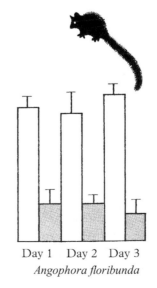

proceed independently. Most hydrogen ions are eliminated in one or more of four ways: (1) By direct excretion of hydrogen ions; but this can only ever account for a small percentage of the daily hydrogen ion production. (2) By retention in the skeleton; but this may lead to excessive calcium excretion and demineralisation of bone. (3) By reaction with dibasic phosphate (HPO_4^{2-}) to form monobasic phosphate ($H_2PO_4^-$) which is excreted in the urine; but this is expensive in terms of phosphorus. (4) By reaction with bicarbonate to form carbon dioxide which is exhaled; but the bicarbonate ions must be replaced to maintain acid–base homeostasis.

Reports of negative calcium balances in mammals fed browse diets (e.g. in hares (*Lepus* spp.) by Pehrson (1983) and Reichardt *et al.* (1984)) suggest that demineralisation of bone may occur in some browsers and thus limit their nutritional niche width. Similarly, the finding of Braithwaite *et al.* (1984) of a threshold level of phosphorus in terms of the quality of eucalypt foliage in the forests of south-eastern New South Wales suggests that the availability of phosphorus may play a major role in dietary selection by arboreal folivores in the phosphorus-poor forests of Australia.

However, leaf nitrogen concentration is likely to be a more general limiting factor in browse diets. Urinary excretion of ammonium ions is one of the most widespread responses when mammals are acid loaded. The main proximate source of the ammonium ions is glutamine extracted from the plasma of circulating blood by the kidneys and metabolised in the cells of the renal proximal tubules to NH_4^+ and to α-ketoglutarate which is then metabolised to generate HCO_3^-. Because this process occurs in the kidney rather than the liver, and the ammonium ions are excreted directly into the renal collecting ducts, each ammonium ion appearing in the urine means the net synthesis of a bicarbonate ion which is available to titrate part of the acid load. The cost to the animal is seen in increased nitrogen requirements to make good the loss of nitrogen as urinary NH_4^+.

Thus greater gliders are able to utilise *E. radiata* foliage despite its high content of terpenes no doubt because of its relatively high nitrogen content (Foley & Hume 1987b). Similarly, Kavanagh & Lambert (1990) found that the pattern of foliage selection by free-living greater gliders was related to foliar concentrations of nitrogen, and more particularly to the ratio of nitrogen to acid-detergent fibre (i.e. cellulose and lignin). The most preferred species in the study site on the south coast of New South Wales was *E. viminalis*, especially the young leaves. The ratio of nitrogen to fibre in young leaves of *E. viminalis* (0.21) was three times the ratio in the four other eucalypt species present (mean 0.07).

That the essential oils are readily absorbed from the digestive tract of greater gliders was shown by Foley, Lassak & Brophy (1987). The apparent digestibility of the oils of *E. radiata* foliage was 97%. Similarly, Eberhard *et al.* (1975) found that koalas apparently digested up to 97% of the essential oils of *E. punctata*. Only minor losses of leaf oils occurred during mastication, and most absorption took place from the stomach. Further absorption took place in the small intestine, so that only 1% of ingested oils reached the caecum. Thus there was little chance of major interaction with the microbes; terpenes are cytotoxic because they are readily soluble in cell membranes but when incorporated into membranes they disrupt membrane function. However, oils absorbed from the digestive tract are also potentially toxic to the animal, and must be detoxified and excreted. Absorbed terpenes are detoxified in the liver by a two-stage process consisting of the structural modification of the molecule by mixed-function oxidases, which renders the molecule more polar and thus more water soluble, followed by conjugation with a small water-soluble molecule, which in the marsupial arboreal folivores is glucuronic acid (Hinks & Bolliger 1957; McLean *et al.* 1993). The conjugated terpene is then excreted, largely in the urine. The metabolic cost in terms of nitrogen excretion has been discussed above. The other metabolic cost is the energy required to maintain the enzymatic machinery involved in the liver, and the energy excreted in the form of glucuronic acid. Eberhard *et al.* (1975) estimated that the koala normally excretes 1–3 g glucuronic acid daily. The glucose required to synthesise this glucuronic acid was calculated by Cork (1981) to be equivalent to as much as 20% of the koala's fasting glucose requirement, a considerable energetic cost.

The low energy availability of the eucalypt foliage on which the greater glider feeds is matched by this folivore's low field metabolic rate. Foley *et al.* (1990) found that although the greater glider's BMR was no lower than expected for a marsupial (Table 1.1), its FMR was only 2.5 times BMR. This is similar to the koala (1.7–2.4), but much lower than ratios in carnivorous and omnivorous marsupials (Table 1.4). By assuming that 43% of the animal's energy intake was lost in the faeces and 15% was excreted in the urine, Foley *et al.* (1990) were able to calculate from measured FMRs that the gross energy intake of the greater gliders at their field site in coastal eucalypt forest near Maryborough in Queensland was about 1130 kJ d^{-1}. Animals would need to eat 45–50 g dry matter daily to meet that requirement. This is little more than the 44 g dry matter d^{-1} consumed by captive greater gliders on *E. radiata* foliage (Foley 1987).

The average nitrogen content of the foliage consumed by the free-living greater gliders was about 1.0%, substantially below the 1.7% nitrogen in the *E. radiata* leaves fed to the captive animals. That the animals maintained body mass and condition during the field study attests to the adequacy of the low nitrogen intake. The reason for the lower dietary nitrogen level required was the much lower level of terpenes in the foliage selected by greater gliders at the field site (1–3% versus 9–13% of dry matter in *E. radiata*). This is the basis for assuming that only 15% of ingested energy was lost in the urine of the free-living animals, compared with 25% in the captive animals feeding on *E. radiata* (Foley 1987).

Water turnover rates in the free-living greater gliders averaged 87 mL kg$^{-0.71}$ d^{-1}, similar to that of free-living koalas in central Queensland at the same time of year (winter) (Table 1.6). At 51% water, consumed leaves provided about 58% of the animals' needs. The bulk of the remainder came from metabolic water, but unlike koalas, greater gliders probably have to drink free water (about 21 mL d^{-1} in Foley *et al.*'s (1990) study) to remain in water balance. This could come from animals licking dew and rainwater from the surface of leaves or grooming water from their fur.

5.5.3 Common ringtail possum

The most researched and well-known ringtail possum is the common ringtail (*Pseudocheirus peregrinus*) (Fig. 5.13). This species is equally as folivorous as the greater glider, but includes more non-eucalypt species in

Figure 5.13. A common ringtail possum (*Pseudocheirus peregrinus*) from the Sydney region. (Pavel German)

its diet. Nevertheless, it can easily be maintained on a single-eucalypt species diet for extended periods.

The highly folivorous nature of the common ringtail is reflected in its dentition. The incisors are not particularly large, but the molars have long shearing blades and high cusps. A shearing action is accentuated by many accessory crests, and crushing has been minimised by restriction of the size of the trigonid and talonid basins (Kay & Hylander 1978). These features result in the trituration of their highly resilient *Eucalyptus* foliage diet. Gipps & Sanson (1984) showed that as animals aged and their teeth became more worn the ratio of small particles (< 280 μm) to large particles (> 560 μm) in the stomach decreased from 0.85 to 0.55, digestibilities of dry matter and neutral-detergent fibre decreased (by 7% and 46% respectively), and animals lost body mass.

The digestive tract of the common ringtail (Fig. 5.14) is remarkably similar to that of the greater glider. A small, simple stomach and short small

Figure 5.14 The gastrointestinal tract of the common ringtail possum (*Pseudocheirus peregrinus*).

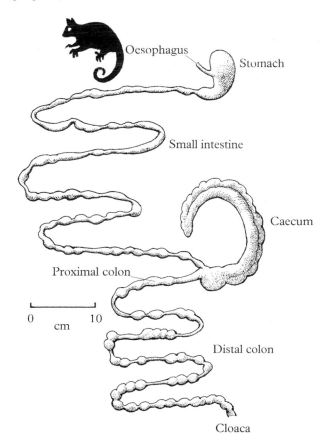

intestine are followed by a large, strongly haustrated caecum. Brunner's glands form a diffuse area of lobules at the beginning of the duodenum in the same way they are found in the greater glider, rather than a distinct lobular collar as in carnivorous and omnivorous marsupial groups (Krause 1972). As in the greater glider, there is no development of the proximal colon for retention of large amounts of digesta, and microbial fermentation is therefore confined to the caecum. In the common ringtail, 33% of the nominal mucosal surface area of the intestinal tract is contributed by the small intestine, 40% by the caecum and 27% by the colon (Kingdon 1990). In the greater glider the values are 31%, 46% and 23% for the small intestine, caecum and colon respectively, reflecting the slightly larger caecum in the latter species. The bacteria inhabiting the common ringtail caecum have not been investigated. Beveridge & Spratt (1996) reported no oxyuroid nematodes in the caecum, unlike the greater glider, but cestodes of the genus *Bertiella* were found throughout the small intestine.

Measurement of digesta passage rates by Chilcott & Hume (1985) showed that solutes and small digesta particles are selectively retained in the caecum of the common ringtail, as they are in the greater glider; mean retention times of the solute marker ^{51}Cr-EDTA (63 h) were almost double those of the particle-associated marker ^{103}Ru-Phenanthroline (35 h). However, unlike the greater glider, Chilcott (1984) demonstrated conclusively that common ringtails are caecotrophic, which means that they ingest material derived from their caecum. This material is excreted as soft faeces (caecotrophes) at regular intervals during daylight hours, when common ringtails are resting in tree hollows or dreys (nests made of twigs and leaves in the foliage of trees). Caecotrophes are taken directly from the cloaca (Fig. 5.15) and are chewed before swallowing.

Hard faeces are excreted during the foraging phase, at night, and are not eaten. The chemical composition of hard and soft faeces is compared in Table 5.7. Samples of caecotrophs were obtained by fitting animals with a plastic collar for short periods. These collars prevented them from reaching their cloaca. The soft faeces were higher in water and nitrogen, and lower in cell-wall polysaccharides and lignin. Soft faeces were also similar to caecal contents in their high proportion (approximately 60%) of particles less than 75 μm, in contrast to gastric contents that contained only 35% of these small particles.

Fig. 5.16 shows how the content of dry matter and nitrogen in the faeces of captive common ringtails varies over the 24 hours. The animals were maintained on a sole diet of *E. andrewsii* (New England blackbutt) on 12 hours of dark and 12 hours of light each day. There was a rapid change in composition as soon as the animals entered their nest boxes at 'dawn', when the nitrogen content increased from 1.2% to a maximum of 6.0%, while dry matter content declined from about 54% to a minimum of 22%. Faecal composition reverted to that of hard faeces toward the end of the light phase.

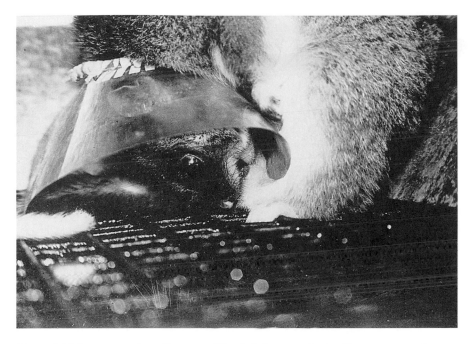

Figure 5.15 A common ringtail possum (*Pseudocheirus peregrinus*) taking a caecotrophe from its cloaca. The animal is wearing a control collar during an experiment in which complete collars were used on other animals to prevent caecotrophy. From Chilcott (1984).

Table 5.7. *Composition of soft and hard faeces produced by common ringtail possums (*Pseudocheirus peregrinus*) fed* Eucalyptus andrewsii *foliage*

	Soft faeces	Hard faeces
Dry matter (%)	23	53–57
Gross energy (kJ g^{-1})	21.4	24.9
Total nitrogen (%)	4.7	2.1
Neutral-detergent fibre (%)	38.7	67.3
Acid-detergent fibre (%)	31.1	57.7
Lignin (%)	28.0	36.0

Source: After Chilcott & Hume 1985.

By direct observation through the light phase the total number of caecotrophes could be counted. The nutritional importance of caecotrophy was then calculated by multiplying caecotrophe production by average caecotroph composition, and then comparing this with dietary intakes of energy and nitrogen. Caecotrophy contributed 254 kJ kg$^{-0.75}$ d^{-1} to total energy intake, equivalent to 58% of digestible energy intake. The contribution to

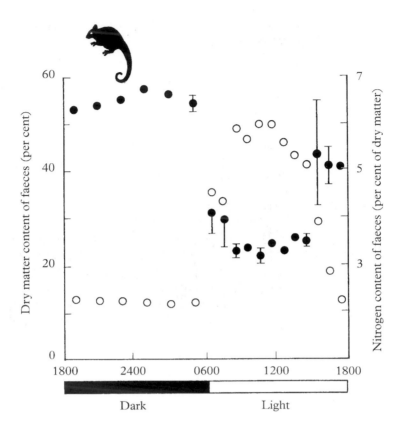

Figure 5.16 Changes in the dry matter (●) and nitrogen content (○) of the faeces of common ringtail possums (*Pseudocheirus peregrinus*) over 24 hours. Caecotrophes (soft faeces) were collected from animals wearing complete plastic collars of the type shown in 5.15.

the animal's nitrogen economy is even greater, 573 mg kg$^{-0.75}$ d^{-1}, equivalent to twice the maintenance requirement. In other words, in the absence of caecotrophy, the amount of nitrogen required daily to maintain the animal in nitrogen balance, would not be the 290 mg kg$^{-0.75}$ measured by Chilcott & Hume (1984b), but 620 mg kg$^{-0.75}$. This would be close to the maintenance requirement of the non-coprophagic greater glider (560 mg kg$^{-0.75}$ d^{-1}) on *E. radiata* foliage. It is the combination of selective retention of solutes and small particles of digesta in the caecum with caecotrophy that allows the common ringtail to specialise on *Eucalyptus* foliage despite its small body size and thus relatively high mass-specific metabolic requirements and small gut capacity.

Another factor contributing to the common ringtail's low maintenance nitrogen requirement, albeit in a small way, is the extensive recycling of endogenous urea to the digestive tract. Chilcott & Hume (1984b) found that, in common ringtails maintained on *E. andrewsii* foliage, 96% of urea synthesised in the liver was recycled to the gut rather than being excreted in the urine; this was equivalent to 11% of the animals' intake of truly digestible nitrogen.

As mentioned above in Section 5.4, protein digestibility can be adversely affected by the presence of condensed tannins. The deleterious effects of tannin–protein interactions can be reversed experimentally with polyethylene glycol (Jones & Mangan 1977; Foley & Hume 1987c). In a similar experiment to that described below with common brushtail possums (Section 5.5.5), McArthur & Sanson (1991) fed common ringtails either *E. ovata* (swamp gum) leaves or leaves dipped in polyethylene glycol (PEG) (molecular weight 20 000). Although leaf intake was unaffected by the PEG treatment, PEG reduced rather than increased digestibility of dry matter and gross energy, while nitrogen digestibility remained high (92–95%). McArthur & Sanson (1991) concluded that natural *E. ovata* condensed tannins did not affect protein availability to common ringtails, and that the tannins themselves were highly digestible. The tannin breakdown products were probably absorbed and excreted in the urine rather than in the faeces. In this way the faecal loss of nitrogen as tannin–protein complexes would be minimised. Dissociation of the complexes could well occur in the caecum, where selective retention of small particles would prolong exposure to microbial attack. O'Brien, Lomdahl & Sanson (1986) had seen massive populations of bacteria attached to digestion-resistant plant fragments in the common ringtail caecum. Many of these 'microbial rafts' are ingested as caecotrophes and the bacteria are largely digested in the small intestine.

However, when a condensed tannin (quebracho) was added at 4.6% to a non-eucalypt basal diet, intakes of digestible dry matter, digestible energy and metabolisable energy by common ringtails were significantly reduced, as also was intake of apparently digestible nitrogen and nitrogen balance (McArthur & Sanson 1993b). In this case much of the tannin was recovered in the faeces, indicating limited digestion and absorption of this condensed tannin.

Like koalas, ringtails appear to use a range of strategies to meet the additional energetic demands of reproduction. Munks & Green (1995) studied the energetics of lactation in free-living common ringtails feeding predominantly on the foliage of coast teatree (*Leptospermum laevigatum*) in Tasmania. The FMR of non-lactating adult ringtails was 2.2 times BMR, similar to greater gliders in Queensland. However, in contrast to Krockenberger's (1993) koalas, FMR increased during lactation, to 2.9 times BMR (Table 1.4). Female ringtails met the increased energy requirements of reproduction by four means. First, they increased food intake by 33% when they were suckling young out of the pouch (stage 3 of lactation). Second, they probably utilised body tissue accumulated during earlier stages of lactation, as body tissue mass declined by 32% between stage 2b of lactation (when they were suckling pouch young not permanently attached to the teat) and stage 3. Third, they expended less energy on thermoregulation in stage 3, which coincides with the warmer spring months. Fourth, the relatively long period of lactation in common ringtails means that the energy costs of reproduction are spread over time, which reduces the peak daily food requirement. The milk is also relatively dilute (Munks *et al.* 1991), and does not increase in lipid content at the time of pouch exit (Fig. 5.9c). All of these features help to explain how the common ringtail possum, the smallest of the four eucalypt specialist folivores, meets the energy costs of free existence, including reproduction, from foliage diets.

5.5.4 **Other ringtail possums**

Little information is available on either nutrient requirements or digestive tract function of other ringtail possums. From their analysis of preserved museum specimens, Crowe & Hume (1997) concluded that all the Australian Pseudocheiridae probably selectively retain solutes and small digesta particles in their caecum. However, several differences were observed in digestive tract characteristics within the family. The six genera of Australian pseudocheirids can be grouped into three pairs on either phylogenetic or morphological grounds: *Pseudocheirops* (green ringtail) with *Petropseudes* (rock ringtail); *Hemibelideus* (lemuroid ringtail) with *Petauroides* (greater glider); and *Pseudochirulus* (Herbert River ringtail) with *Pseudocheirus* (common and western ringtails) (Kingdon 1990). The first mentioned genus of each pair is a tropical rainforest inhabitant while the second is a drier forest inhabitant.

Despite their different habitats, each pair shares several common features. For instance, Crowe and Hume (1997) found generally similar digestive tract morphology and patterns of digesta nitrogen concentrations in the common and Herbert River ringtails. Kingdon (1990) calculated nominal surface areas of the intestine of several north Queensland possums. Although nominal surface areas do not account for the considerable increases in absorptive capacity due to elaborations of the mucosa into villi and microvilli (Snipes, Snipes & Carrick 1993), they nevertheless provide a means of comparing the relative sizes of different regions of the intestine. Thus nominal surface areas of the small intestine, caecum and colon of the common and Herbert River ringtails are quite similar (33%, 40% and 27% of total intestinal surface area respectively in the common ringtail, and 34%, 40% and 26% in the Herbert River ringtail). Yet their diets are remarkably different, as described earlier (Section 5.3).

In both the lemuroid ringtail and greater glider the caecum is heavier than that of the common ringtail (Crowe & Hume 1997). Likewise, Kingdon (1990) calculated nominal surface areas of 34%, 45% and 21% of the total intestine for the small intestine, caecum and colon of the lemuroid ringtail, and 31%, 46% and 23% for the greater glider. Both sets of data suggest a larger caecum and a higher degree of folivory in this pair than in the common ringtail. The lemuroid dentition is also closely similar to that of the greater glider in having a greater number of cutting edges on the upper molars compared with the common ringtail. Finally, the basic skeletal structure of the lemuroid ringtail and greater glider is very similar, even though the greater glider is longer limbed and more lightly built, both adaptations for gliding. The beginnings of that adaptation are seen in the lemuroid ringtail in its readiness to drop 2–3 m from one stratum to another in its rainforest habitat (Kingdon 1990).

The green ringtail–rock ringtail pair differs from the others in numerous ways. In both species the small intestine is short, the caecum is either small (in the green ringtail) or simple (in the rock ringtail), and the colon is long, particularly in the green ringtail (Crowe & Hume 1997). As a proportion of the green ringtail's total intestine, nominal surface areas calculated by Kingdon (1990) were 31%, 22% and 47% for the small intestine, caecum and colon respectively. The simple caecum of the rock ringtail may be related to the relatively low degree of folivory exhibited by this species. It is the least folivorous member of the Pseudocheiridae, its diet containing a large proportion of the fruits and flowers of a variety of trees and shrubs, including figs, found in and around its rocky habitat (Kingdon 1990; Kerle & Winter 1995). This is reflected in the low nitrogen levels found throughout its digestive tract (Crowe & Hume 1997).

Six species of New Guinean ringtails examined by Hume et al. (1993) were also similar in having a large, haustrated caecum that contained 43% of total tract digesta. All were primarily folivorous, based on qualitative analysis of stomach contents, even though the two smallest species also included tissues of mosses, ferns, lichens and fungi.

5.5.5 Common brushtail possum

Because it is the least specialised of the four eucalypt foliage feeders, the common brushtail possum has been used in numerous comparisons of digestive tract function and metabolism with other, more highly specialised species, namely the greater glider and the common ringtail possum.

The less folivorous and more frugivorous nature of the common brushtail's diet compared with those of most Australian pseudocheirids is reflected in its dentition. Compared with pseudocheirids, common brushtails have larger incisors, suggesting more incisal biting, and rather simple four-cusped upper and lower molars that are sublophodont and very similar to those of colobine monkeys (Kay & Hylander 1978). Only one upper and one lower premolar occlude with the cheek tooth row. This dentition emphasises tearing and crushing actions rather than cutting. Consequently, stomach contents of the common brushtail are less finely comminuted than those of the greater glider and common ringtail, and contain long strings of fibrous sclerenchyma tissue not present in the pseudocheirid stomach (Gipps 1980).

The digestive system of the common brushtail has been described by Lönnberg (1902) and the stomach by Oppel (1896). Simple in form, the stomach is lined mostly by fundic glandular mucosa, with a limited area of pyloric glandular mucosa on and caudal to the deep indenture on the lesser curvature, and a smaller area still of squamous epithelium near the cardia. There is no cardiac glandular mucosa (Fig. 5.17).

In the small intestine there is a dense concentration of Brunner's glands immediately distal to the pyloric sphincter (Fig. 5.17), with individual glands scattered along approximately 6 cm, but the glands are not organised into large lobes (Krause 1972). This is similar to the other arboreal folivores. The wide range of helminths found in the common brushtail (Table 5.8) reflects this marsupial's diet which includes a significant proportion of herbage from the ground layer. Many of the helminths were presumably introduced to Australia with domestic stock, and acquired by common brushtails grazing in paddocks containing infected ruminants (Beveridge & Spratt 1996). Most are associated with the small intestine, with only one, an oxyuroid nematode, found in the hindgut.

Compared with the stomach, the caecum and proximal colon of the common brushtail are well developed (Fig. 5.18), but the tissue mass of the caecum is only half that of the greater glider (Crowe & Hume 1997), and its

Figure 5.17 The stomach of the common brushtail possum (*Trichosurus vulpecula*), longitudinal section. After Oppel (1896).

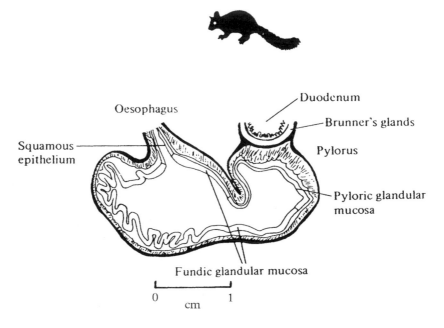

Table 5.8. *Helminth parasites of the common brushtail possum* (Trichosurus vulpecula)

	Site
Trematoda	
Fasciola hepatica	Bile duct
Cestoda	
Bertiella trichosuri	Small intestine
Nematoda	
Rhabditoidea	
Parastrongyloides trichosuri	Small intestine
Trichostrongyloidea	
Profilarinema hemsleyi	Stomach
Cooperia curticei	Small intestine
Trichostrongylus axei	Stomach
T. colubriformis	Small intestine
T. retortaeformis	Small intestine
T. nigatus	Small intestine
T. vitrinus	Small intestine
Paraustrostrongylus trichosuri	Small intestine
Metastrongyloidea	
Filostrongylus trideraiticus	Lung
Marsupostrongylus longilarvatus	Lung
Oxyuroidea	
Adelonema trichosuri	Hindgut
Spiruroidea	
Mastophorus muris	Stomach
Filarioidea	
Breinlia trichosuri	Abdominal cavity
Sprattia venacavincola	Vena cava

Note: Only adult forms of naturally occurring fully identified species are included.
Source: After Beveridge & Spratt 1996.

nominal surface area is only 20% of the total intestine versus 46% in the greater glider (Kingdon 1990). These data suggest more emphasis on enzymatic digestion of cell contents in the small intestine and less on microbial digestion in the caecum in the common brushtail. This is true, but unlike that of most pseudocheirids, the phalangerid caecum is augmented by the proximal colon. Nevertheless, the common brushtail caecum and proximal colon together are still only 70% of the greater glider caecum alone in tissue mass.

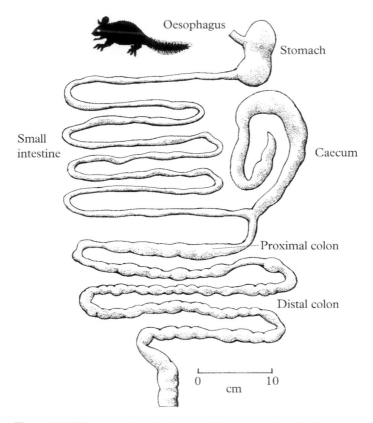

Figure 5.18 The gastrointestinal tract of the common brushtail possum (*Trichosurus vulpecula*).

DIGESTIVE FUNCTION

Early nutritional studies with the common brushtail possum were based on animals fed fruit-based diets. Honigmann (1941) recorded high crude fibre digestibilities (79–82%) on such diets, which he attributed to a slow passage of food through the gut measured with wheat grains stained with Loeffler's methylene blue. Long passage times were confirmed by Gilmore (1970).

A more detailed study of digestive tract function by Wellard & Hume (1981a) was based on animals fed semi-purified diets of honey, wheat bran, ground oat hulls, purified wood cellulose, casein, salt and minerals and vitamins. The dietary ingredients were manipulated to provide diets of low fibre (17% cell walls) and high fibre content (41% cell walls). The former approximated the fibre level in fruit, while the latter was similar to that of mature leaves of *Eucalyptus viminalis* (ribbon gum), a species preferred by common brushtails.

Increasing the fibre level from 17% to 41% cell walls (neutral-detergent fibre) had no effect on food intake but depressed digestibility of dry matter, fibre and energy. Defaecation patterns were erratic on the low-fibre diet despite regular food intake, reflecting its highly digestible nature (92% on a dry-matter basis), and probably insufficient bulk to maintain normal motility patterns in the hindgut. Consequently, on the high-fibre diet (but not the low-fibre diet), digestibilities of dry matter and fibre were negatively related to level of food intake (Fig. 5.19).

Erratic defaecation also meant that mean retention times (MRTs) of the two markers used (^{51}Cr-EDTA for solutes and ^{103}Ru-Phenanthroline for particles) could not be calculated for the low-fibre diet. On the high-fibre diet, MRTs were 64 hours for the solute marker and 71 hours for the particulate marker. These values confirmed earlier indications that passage through the common brushtail gut was slow, and also suggested that there was no selective retention of either phase of digesta in the gut. The latter feature can be seen clearly when compared with the koala (Fig. 5.8), and was corroborated on a semi-purified diet by Sakaguchi & Hume (1990).

MRTs in common brushtails on a natural foliage diet of *Eucalyptus melliodora* (yellow box) were a little lower (51 hours, and 49 hours for ^{51}Cr-EDTA and ^{103}Ru-Phenanthroline respectively) than on the semi-purified diet used by Wellard & Hume (1981a), but the lack of any selective passage remained (Foley & Hume 1987a). Cell-wall digestibility was much lower (27%) on *E. melliodora* foliage than on the semi-purified diet (58%), and the digestible energy content of the leaves was low (45%), reflecting the highly lignified nature of eucalypt cell walls (Fig. 5.2). Foley & Hume (1987c) found few bacteria attached to lignified tissues in the hindgut.

Figure 5.19 Relationships between dry matter intake and digestibility of (a) dry matter, and (b) neutral-detergent fibre (NDF) in common brushtail possums (*Trichosurus vulpecula*) fed high-fibre diets (●). Data from low-fibre diets (○) included for comparison. From Wellard & Hume (1981a).

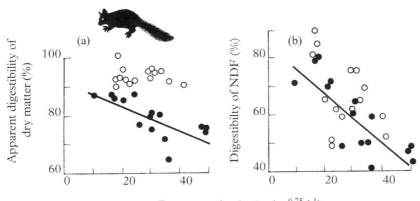

Dry matter intake (g kg$^{-0.75}$d^{-1})

Table 5.9. *Inhibition of the effects of* Eucalyptus melliodora *leaf tannins on intake and digestion in common brushtail possums* (Trichosurus vulpecula) *with polyethylene glycol 4000 (PEG). Intake of chemical constituents in g* $kg^{-0.75}$ d^{-1}, *and of energy in MJ* $kg^{-0.75}$ d^{-1}

Parameter	Without PEG	With PEG
Dry matter intake	36.8	48.7
Dry matter digestibility (%)	50.0	49.6
Cell wall digestibility (%)	27.1	48.1
Cell contents digestibility (%)	59.1	50.2
Metabolic faecal nitrogen	0.25	0.31
Digestible energy intake	0.35	0.54
Metabolisable energy intake	0.27	0.45

Source: After Foley & Hume 1987c.

The other factor inhibiting fibre digestion was probably the high tannin content (20–29% total phenolics) of *E. melliodora* leaves. This was thought responsible for the low ratio of metabolisable to digestible energy in the leaves (0.77), probably because of the urinary excretion of leaf phenolics (the essential oil content of *E. melliodora* foliage is only 1% of dry matter). When the inhibitory effect of leaf tannins on fibre digestion was blocked by supplementing the animals with polyethylene glycol 4000 (PEG), there were dramatic increases in digestive performance; digestibility of cell walls increased by 77%, and consequently intake of metabolisable energy increased by 67% (Table 5.9). These effects were attributed to the reversal by PEG of tannin-microbial enzyme complexes in the hindgut. The inhibition of cellulases by tannins is well documented (Mandels & Reese 1965).

In contrast, digestibility of cell contents decreased on the PEG treatment. This was explained by a combination of an increased microbial biomass in the hindgut from the extra energy from the fermentation of cell walls, and the lack of a mechanism in the common brushtail's hindgut to selectively retain solutes and small particles (including bacteria) in the caecum. The increased microbial biomass was therefore largely lost in the faeces (metabolic faecal nitrogen increased by 24% on the PEG treatment). The net effect of the decreased cell contents digestibility and the enhanced cell wall digestibility was that dry matter digestibility remained unchanged, yet dry matter intake increased by 32%. This brought the level of dry matter intake of *Eucalyptus* foliage by common brushtails up to a level similar to those recorded for common ringtails, greater gliders and koalas. One reason why the common brushtail feeds to only a limited extent on *Eucalyptus* foliage, with its highly lignified tissues, may therefore be its lack of a mechanism to facilitate the passage of large particles, so that food intake is normally inhibited by the gut-filling effect of a mass of indigestible fibre in the hindgut (Foley & Hume 1987c).

Fermentation rates measured in the caecum and proximal colon of captive common brushtails *in vitro* were similar (18 and 20 mmol short-chain fatty acids (SCFA) L^{-1} h^{-1}, respectively), and similar to those in the caecum of captive greater gliders (19 mmol L^{-1} h^{-1}) (Foley *et al.* 1989). However, because of the lower food intakes of the common brushtails, SCFA production contributed more significantly to their digestible energy intake (15%) than in the greater glider (7%).

Field metabolic rates have not been measured in common brushtail possums, so it is not clear how brushtails meet the additional energy demands of reproduction. This is an area that needs research. The pattern of changes in the macronutrient composition of milk of brushtails is similar to that of common ringtail possums (Fig. 5.9). Compared to eutherians, copper and iron concentrations in brushtail milk analysed by Jolly *et al.* (1996) in New Zealand were high, as in other marsupials; zinc levels were exceptionally high.

NITROGEN METABOLISM

The profound effects that eucalypt plant secondary metabolites can have on metabolism of arboreal folivores are seen in the comparison between common brushtails fed either semi-purified or eucalypt foliage diets (Table 1.8). Wellard & Hume (1981b) reported that the maintenance nitrogen requirement on their semi-purified diets was 203 mg$kg^{-0.75}$ d^{-1} on a dietary basis, and 189 on a truly digestible basis. Although fibre content of the diet had no effect on the estimates, metabolic faecal nitrogen was very sensitive to dietary fibre level, increasing from 78 mg $100g^{-1}$ dry matter intake on the 10% cell-wall diet to 184 on the 17% cell-wall diet and to 336 on the 41% cell-wall diet. This sensitivity is probably a further result of the absence of a separation mechanism for the selective retention of bacteria and other small particles in the common brushtail's hindgut.

When Foley & Hume (1987b) measured the maintenance nitrogen requirement of common brushtails on *E. melliodora* foliage, values were much higher than those on the semi-purified diets. On a dietary basis the estimate was 560 mg$kg^{-0.75}$ d^{-1}, and on a truly digestible basis it was 420. The principal reason for the more than doubling of the requirements is the large faecal loss of nitrogen on the foliage diet, probably because of the high phenolics content of yellow box leaves. Apparent digestibility of nitrogen was only 33%, in contrast to 70% on a semi-purified diet of similar nitrogen and fibre content (Wellard & Hume 1981b). Loss of metabolic faecal nitrogen was 740 mg 100 g^{-1} dry matter intake, more than double the semi-purified diet value. This difference is probably due to the low lignin content and finely ground form of the fibre sources used in the semi-purified diets.

WATER METABOLISM

The widespread occurrence of common brushtails throughout many parts of semi-arid Australia seems to be explained in part by its relatively low water requirements (Table 1.7), but its frugal water economy cannot be explained by an outstanding urine concentrating ability. Deprivation of food and water for three days increased urine osmolality from 184 to 1503 mOsm kg^{-1} water (Reid 1977), 4.8 times the plasma concentration but slightly less than would be expected for a mammal of this body size (Beuchat 1990a,b).

In common with carnivorous and omnivorous marsupials coexisting with species of *Gastrolobium* and *Oxylobium* in south-western Australia, common brushtails have evolved a high degree of tolerance to the toxic compound fluoroacetate (Twigg & King 1991). Why they appear not to have evolved similar tolerance to the diformylphloroglucinol compounds (DFPCs) of *Eucalyptus* is a question waiting to be answered.

5.5.6 Other brushtail possums

As with ringtail possums, the bulk of our knowledge on the nutrition of the Phalangeridae comes from one species, in this case *T. vulpecula*. From their analysis of preserved museum specimens, Crowe & Hume (1997) concluded that, in contrast to the numerous differences found among pseudocheirids, the Australian Phalangeridae were quite uniform in digestive tract morphology. As a proportion of total gut tissue mass, the small intestine of six phalangerid species was 37% and the caecum was 21%; in contrast, the small intestine of seven pseudocherid species was 30% and the caecum was 32%. Also, there was no significant change in nitrogen concentration in digesta along the hindgut of any of the six phalangerids, suggesting no selective retention of solutes and bacteria. Thus Crowe & Hume (1997) concluded that the digestive strategy of the common brushtail is probably representative of other Australian phalangerids as well.

Similar patterns of digesta nitrogen concentration were seen in the four New Guinean phalangerid marsupials examined by Hume *et al.* (1993). One of these species, *Phalanger gymnotis* (ground cuscus), is reported to be the most frugivorous of the eight New Guinean phalangerid species, and is perhaps the most frugivorous of all phalangerids. Hume, Runcie & Caton (1997) examined the gastrointestinal tract morphology and measured digesta passage rates in six captive ground cuscus maintained on a mixture of fruit and tree leaves. The animals selected a diet of more than 90% fruit. The longest and heaviest region of the tract was the small intestine (Fig. 5.20). Total nitrogen concentration of digesta was low in the stomach and small intestine, but increased four-fold in the hindgut, due to microbial activity, but there were no differences along the hindgut, suggesting no selective

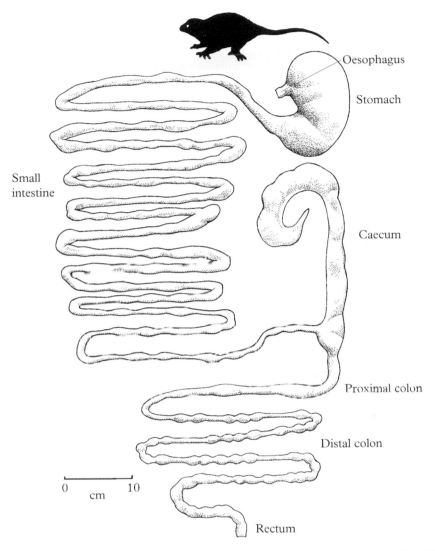

Figure 5.20 The gastrointestinal tract of the ground cuscus (*Phalanger gymnotis*) from New Guinea.

retention of bacteria in the caecum. This was confirmed by measuring the rate of passage of fluid and large particle markers (Co-EDTA and Cr-mordanted cell walls respectively); the mean retention times of the two markers were similar. Although fermentation rates in the caecum and proximal colon were similar to those in common brushtails on a foliage diet, fluid volumes were less than one-third those of brushtails and consequently calculated daily production rates of SCFA were less than half that in brushtails, and contributed only 5% of digestible energy intake (versus 15% in brushtails). These results seem consistent with reports that the natural diet of the ground cuscus is based largely on fruit rather than foliage.

5.6 SUMMARY AND CONCLUSIONS

Comparative studies of the digestive physiology and metabolism of the four common marsupial folivores that feed on *Eucalyptus* foliage have yielded much valuable information on how these mammals have evolved to exploit food resources not available to most vertebrates. The value of eucalypt leaves as food is limited by their low concentration of nutrients and their often high content of plant secondary metabolites, mainly phenolics and terpenes, including (DFPCs). Much remains to be discovered about how the plant secondary metabolites of eucalypt foliage interact with nutrients within the foliage and with the metabolism of koalas, greater gliders, common ringtail possums and common brushtail possums. This is an important area of research because it will help us to better understand the effects of different forestry management practices on the well-being of these marsupials, and to better predict the likely consequences for them of climate change.

Three of the four common folivorous marsupials have been shown to selectively retain fluid, solutes and small particles (including bacteria) in their caecum or, in the case of the koala, the caecum and proximal colon. However, only the smallest species, the common ringtail possum, is caecotrophic like the rabbit. This digestive strategy probably explains why such a small mammal can maintain itself on a sole diet of *Eucalyptus* foliage. Indications are that caecotrophy is a common feature of the digestive strategies of most if not all ringtail possums, both Australian and New Guinean, but further work will be necessary to confirm this.

The fourth species, the common brushtail possum, has no separation mechanism for particles of different sizes in its hindgut and consequently has only a limited ability to use *Eucalyptus* foliage as a sole diet. However, other aspects of its metabolism, especially its tolerance of plant secondary metabolites, also play a major role in determining its diet in the wild. Other phalangerid marsupials appear to have similar digestive strategies to that of the common brushtail, but as with the ringtails, further work is needed to confirm this.

The alternative herbivore digestive strategy to hindgut fermentation is foregut fermentation, which in the Marsupialia is seen in the kangaroos, wallabies and rat-kangaroos. These groups are discussed in Chapters 6, 7 and 8.

6 Foregut fermenters – kangaroos and wallabies

6.1 CONCEPTS

The kangaroos and wallabies (family Macropodidae), together with the rat-kangaroos (family Potoroidae), make up the group of herbivorous marsupials in which the primary site of microbial fermentation is the forestomach, formed by expansion of the cardiac region of the stomach. In the Macropodidae the expansion is grossly tubiform in nature, while in the Potoroidae it is sacciform. This basic difference in gastric morphology has important nutritional consequences, and it is mainly for this reason that the rat-kangaroos are dealt with separately in Chapter 8. Macropodid and potoroid marsupials are two of several groups of mammalian foregut fermenters. Others include ruminants, camelids, peccaries, hippos, sloths and colobine monkeys (Langer 1988).

6.1.1 Foregut fermentation versus hindgut fermentation

Because foregut fermentation precedes enzymatic digestion in the hind-stomach and small intestine it has several advantages over hindgut fermentation, but also some drawbacks. Perhaps the most important advantage is that microbial cells synthesised in the forestomach are digested in the small intestine, whereas microbial cells synthesised in the hindgut are lost in the faeces unless the faeces are ingested (coprophagy); microbial cells contain high-quality protein and B vitamins. This makes foregut fermenters far less dependent on the quality of dietary nitrogen; any form of nitrogen that can be degraded to ammonia in the forestomach can be used, because most of the microbial protein is synthesised *de novo* from ammonia.

Another advantage of foregut fermentation is that degradation of plant cell wall polysaccharides in the forestomach can not only provide a significant proportion of the energy absorbed from the digestive tract, but also can facilitate access by enzymes to cell contents. This is of particular importance with poor-quality forages high in lignified cell walls. However, on high-quality forages high in cell contents and low in lignified cell walls, foregut fermentation becomes a disadvantage. This is because fermentation of the readily digestible components of the cytoplasm (proteins, sugars, storage polysaccharides and organic acids) is energetically inefficient; energy is lost as heat and gases (hydrogen, methane). The animal would be better off if these plant constituents were digested enzymatically in the small intestine and the products directly absorbed into the blood stream.

Degradation of plant toxins by forestomach bacteria is often cited as being a major advantage of foregut over hindgut fermentation systems. This is true for certain classes of allelochemicals, such as alkaloids, but not for phenolics or terpenes; both of these are bacteriocides. Only hindgut fermenters are able to utilise *Eucalyptus* foliage, which contains both phenolics and terpenes (Chapter 5). The only foregut fermenting arboreal folivores are the tree-kangaroos, none of which feeds on eucalypts.

Because most of the advantages of foregut fermentation are likely to be important when the herbivore is consuming food of high fibre content, Hume & Warner (1980) considered that foregut fermenters probably evolved in regions where forage of often poor nutritive value was available. The rapid radiation of the Ruminantia among the Artiodactyla (Janis 1976) and of the Macropodidae among the Marsupialia in the Miocene and Pliocene concurrent with the spread of grasslands throughout the world (Hume 1978) is consistent with this view.

6.1.2 Classification of herbivore diets – ruminant and macropod

The diets of present-day wild ruminants vary widely with body size of the species. This was recognised by Hofmann (1973), who studied East African ruminants that differed in body size from the 4 kg dik-dik to the 1000 kg eland. He classified these herbivores into three grades: bulk and roughage eaters (most of which are large), concentrate selectors (most of which are small), and an intermediate group of mixed feeders (species that change their diet from grasses when they are available to browse in poorer times). A similar classification of extant members of the marsupial family Macropodidae has been constructed by Sanson (1978): large kangaroos such as the eastern grey kangaroo (*Macropus giganteus*) (Fig. 6.1) tend to be bulk and roughage feeders (or grazers), the small wallabies are concentrate selectors (or browsers), and the rock-wallabies (*Petrogale* spp.) are mixed feeders.

6.2 THE MACROPODID DIGESTIVE TRACT

Variations in macropodid dentition are dealt with in Chapter 7 together with diet. Briefly, the molars of grazers are characterised by strong links between the lophs, which reduce the surface area of contact during occlusion, resulting in a cutting action, ideal for comminuting higher-fibre material (Sanson 1980). In contrast, the weak links of browsers allow for a large surface area of contact during occlusion, which results in a crushing action that is best suited to lower-fibre plant material.

Figure 6.1 Female eastern grey kangaroo (*Macropus giganteus*) with large pouch young. (Pavel German)

Many of the early descriptions of the macropodid digestive tract by European anatomists in the nineteenth century referred to the apparent extensive analogy between the foregut fermentation of the kangaroos and that of the eutherian ruminants. However, with the benefit of several detailed physiological as well as anatomical studies over the last 20 years, it is clear that the macropodid forestomach has more in common with the equine colon than with the ruminant forestomach.

The digestive tract of the eastern grey kangaroo is shown in Fig. 6.2. This preparation has been cleared of most mesenteric attachments and the stomach has been partially uncoiled to show the main external features. The stomach is viewed from the left side. Fig. 6.3 shows the stomachs of the eastern grey kangaroo (*Macropus giganteus*), tammar wallaby (*M. eugenii*) and red-necked pademelon (*Thylogale thetis*) as examples to illustrate some differences that are found in the relative proportions of the sacciform and tubiform regions of the forestomach.

6.2.1 Oesophagus

The macropodid oesophagus is relatively long, with a substantial part extending beyond the diaphragm into the abdominal cavity (Owen 1868; Mackenzie 1918). Although the oesophagus of all species is lined with a

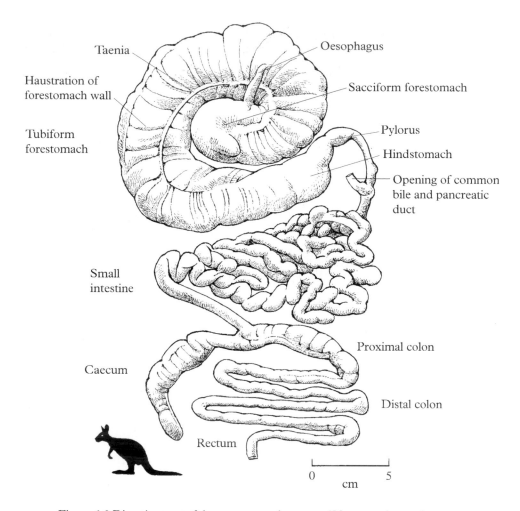

Figure 6.2 Digestive tract of the eastern grey kangaroo (*Macropus giganteus*).

stratified squamous epithelium containing no glands, Obendorf (1984a) recognised four distinct morphological types of oesophageal lining that could be related to differences in dietary preferences and habitat. Type I species have an oesophagus with a smooth lining that is relatively unspecialised. The group includes the rock-wallabies (*Petrogale* and *Peradorcus*), tree-kangaroos (*Dendrolagus*), pademelons (*Thylogale*), nailtail wallabies (*Onychogalea*), hare-wallabies (*Lagorchestes*), dorcopsis wallabies (*Dorcopsis* and *Dorcopsulus*) and the quokka (*Setonix*). All are browsers or mixed feeders.

Type II species have an oesophagus with a smooth lining with large, irregular, pleated longitudinal folds. All members are large grazing kangaroos of the genus *Macropus*. In Type III species the entire oesophagus is lined with finger-like papillae. The papillae are predominantly keratinised projections derived from a highly thickened stratified epithelium with little

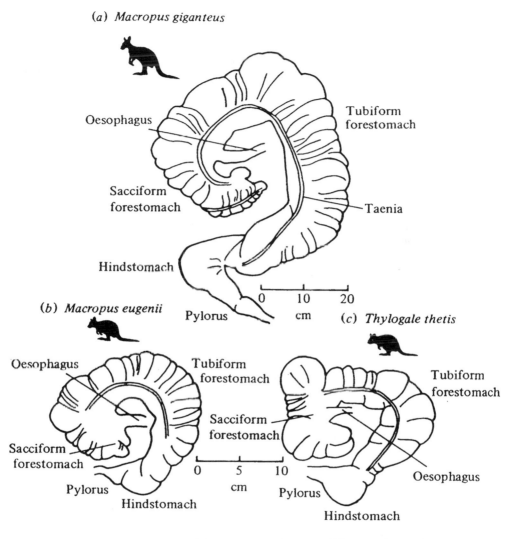

(a) *Macropus giganteus*

Oesophagus

Tubiform
forestomach

Sacciform
forestomach

Taenia

Hindstomach

0   10   20
cm

Pylorus

(b) *Macropus eugenii*

Oesophagus

Tubiform
forestomach

Sacciform
forestomach

Sacciform
forestomach

Pylorus

Hindstomach

0   5   10
cm

(c) *Thylogale thetis*

Tubiform
forestomach

Oesophagus

Pylorus

Hindstomach

Figure 6.3 The stomachs of the eastern grey kangaroo (*Macropus giganteus*), tammar
wallaby (*Macropus eugenii*) and red-necked pademelon (*Thylogale thetis*) cleared of
mesentery and partially uncoiled to illustrate some differences in the proportions of the
sacciform and tubiform regions of the forestomach. From Hume (1982).

connective tissue extending into them. Type III species include the smaller
brush wallabies of the genus *Macropus* , together with the swamp wallaby
(*Wallabia*). Type IV species, the red-necked and Bennett's wallabies (*Macropus rufogriseus*) have an oesophagus which is lined with longitudinal folds
proximal to the diaphragm but with papillae distal to the diaphragm.

The oesophagus of all macropodids except the tree-kangaroos also has a
well-formed keratinised layer. The other marsupial grazers, the wombats,
also have well keratinised oesophageal linings. Conversely, all the arboreal

folivores discussed in Chapter 5 have poorly keratinised oesophageal epithelia, similar to that of the tree-kangaroos (Obendorf 1984a).

6.2.2 **Stomach**

The macropodid stomach is a long, tubular organ that has a gross morphology not unlike that of the equine ventral colon (Stevens & Hume 1995). It consists of two functionally distinct regions, the enlarged forestomach where microbial fermentation of ingested food occurs, and the hindstomach, which secretes hydrochloric acid and pepsinogen. The forestomach can be divided into the sacciform forestomach and the tubiform forestomach on the basis of their gross morphology. The two regions are delimited *in situ* by a perpendicular plane running from a permanent ventral fold, adjacent to the cardia, to the dorsal wall. The ventral fold can be seen clearly in the photographs by Langer, Dellow & Hume (1980) in animals which, after death, were preserved while suspended in a standing position prior to dissection (Fig. 6.4). Fig. 6.4b shows the fold in the forestomach of *Macropus eugenii* during dissection. It is more difficult to see in eviscerated specimens, which are invariably dissected free from mesenteric attachments and partially uncoiled during preparation.

Division of the macropodid forestomach into sacciform and tubiform regions is not only a morphological convenience; it has a functional basis as well, as we shall see presently. Other terminologies can be found in the literature. The sacciform region has been referred to as the 'left cul-de-sac' by Owen (1868), the 'left end of the stomach' or 'cardiac fundus' by Schäfer & Williams (1876), the 'blind sacs of the stomach' by Wilckens (1872), the 'left cul-de-sac' or 'conical fundus' by Flower (1872), and 'cul-de-sac' by Moir, Somers & Waring (1956). Hume (1978) and Kennedy & Hume (1978) referred to it simply as the 'forestomach'.

The tubiform region of the forestomach (the main tubular body of the organ) was called the 'midstomach' by Hume (1978) and Kennedy & Hume (1978). It corresponds to the 'middle stomach compartment' of Owen (1868), and the 'intestinal section of the stomach (der Darmtheil des Magens)' of Wilckens (1872). Other authors have not distinguished between the sacciform and tubiform regions as such. Griffiths & Barton's (1966) first gastric region of *Macropus rufus* includes the sacciform region with the cranial part of the tubiform region to the caudal end of the gastric sulcus (spiral groove). Their second gastric region consists of the rest of the tubiform region. Moir *et al.* (1956) included the sacciform region (cul-de-sac) of the stomach of *Setonix brachyurus* with the cranial section of the tubiform region, calling the combined region the forestomach, region I, or 'rumen'. The caudal section of the tubiform region was termed the non-sacculated area or region II. More recently, Richardson (1980) divided the stomach of *Macropus eugenii* into proximal, middle and distal compartments

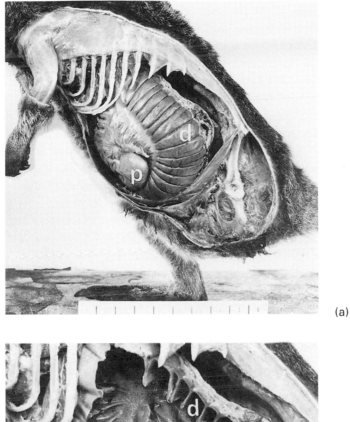

(a)

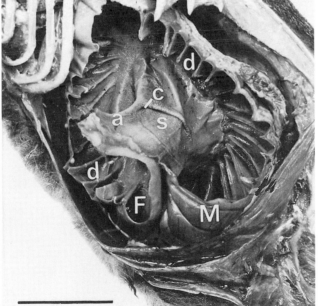

(b)

Figure 6.4 Stages in the dissection of a tammar wallaby (*Macropus eugenii*) from the left side. In (a) the stomach is shown *in situ*. p, parietal blind sac of the sacciform forestomach; d, haustrations of the tubiform forestomach. The scale line is 20 cm. In (b) the permanent ventral fold (a) separating sacciform (F) from tubiform (M) regions of the forestomach can be seen in relation to the cardia (opening of the oesophagus) (c) and the gastric sulcus (s). (d) semi-lunar folds on the inner surface of the sacciform and tubiform regions of the forestomach. From Langer, Dellow & Hume (1980).

on the basis of the position of two of the three major gastric flexures, the caudal gastric flexure and the pyloric flexure. Unfortunately, the caudal gastric flexure is not always readily found in eviscerated specimens, and its functional significance is not clear.

Schäfer & Williams (1876), Oppel (1896) and Gemmell & Engelhardt (1977) divided the macropodid forestomach on the basis of histology rather than morphology. Thus region A of Schäfer & Williams (1876) and Oppel (1896) corresponds to the oesophageal region of Gemmel & Engelhardt (1977) in that it is lined with squamous epithelium. Region B (Schäfer & Williams 1876; Oppel 1896) is termed the cardiac region by Gemmell & Engelhardt (1977) because it is lined with cardiac glandular mucosa. The problem with this system is that the proportion of the forestomach wall lined with cardiac glands varies enormously among macropodid species, as can be seen in the drawings of the stomach of the grey dorcopsis (*Dorcopsis luctuosa*) and of *Macropus giganteus* from Schäfer & Williams (1876) (Fig. 6.5).

Figure 6.5 The stomachs of (a) the grey dorcopsis wallaby (*Dorcopsis luctuosa*) from New Guinea, and (b) the eastern grey kangaroo (*Macropus giganteus*) to show variations between species in distribution of the epithelial lining. After Schafer & Williams (1876). Note that in *M. giganteus* the cardiac glandular mucosa of the sacciform forestomach is continuous with that of the tubiform forestomach. It is separated by squamous epithelium in the same species dissected by Langer, Dellow & Hume (1980) (Fig. 6.10).

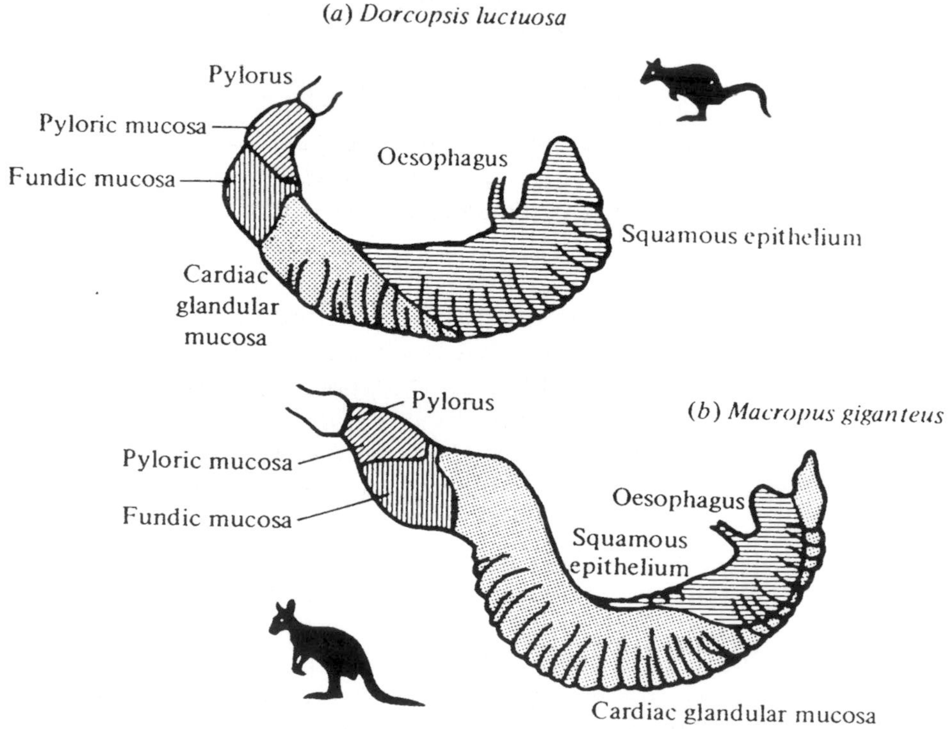

(*a*) *Dorcopsis luctuosa*

(*b*) *Macropus giganteus*

The hindstomach as used by Hume (1978), Kennedy & Hume (1978) and Langer *et al.* (1980) consists of the gastric pouch (the site of hydrochloric acid secretion) and the pyloric region. It corresponds to the glandular pouch of Flower (1872), the pyloric fundus (region C) and that part of region B caudal to this (Schäfer & Williams 1876), the fundic and pyloric regions of Gemmell & Engelhardt (1977), and the distal compartment of Richardson (1980). In all macropodid species the hindstomach is the smallest gastric region (Fig. 6.3).

In most species the tubiform forestomach is the largest gastric region. In *M. eugenii* it contains about 58% of the total dry matter in the stomach, in *M. giganteus* about 72% (Langer *et al.* 1980), in *L. hirsutus* 71–74% (Bridie, Hume & Hill 1994) and in the two Australian tree-kangaroos, *Dendrolagus lumholtzi* and *D. bennettianus*, 73–76% (Hume & Flannery unpubl.). However, in *Thylogale thetis* (red-necked pademelon) and *T. stigmatica* (red-legged pademelon) the largest stomach region is the sacciform forestomach, as can be seen for *T. thetis* in Fig. 6.3. It constitutes 52% of total stomach capacity in *T. thetis* and probably a similar value in *T. stigmatica* (Langer 1979). In contrast, *T. billardierii* (Tasmanian pademelon) lacks the dorsal pouch which is so prominent in the sacciform forestomach of the other *Thylogale* species above, and so this gastric region is not nearly so voluminous in *T. billardierii* (Dellow 1979). Thus generalisations are often difficult, even within the one genus.

The other major external features of the macropodid stomach are the taeniae and associated haustrations. The longitudinal muscles of the forestomach (but not the hindstomach) are organised into three bands, the taeniae, one on the left side, one on the right side, and the third under the line of attachment of the greater omentum on the greater curvature. Contractions of the circular muscles form non-permanent semi-lunar folds between the taeniae on the internal surface, creating the external haustrations which give the macropodid stomach its 'colon-like' appearance (Owen 1868). Differences in the patterns of haustrations are particularly evident in the two Australian tree-kangaroos (Fig. 6.6).

The contractions are of two main types (Dellow 1979). The first consists of localised contractions that involve each haustration and occur over a 4–6 second cycle. They are independent of, but associated with, sequential contractions of two or more adjacent haustrations. Their function appears to be local mixing of digesta. The second form of contraction is a stronger sequential wave of contractions that travels a short distance along the greater curvature wall. Each is a caudal displacement of a semi-lunar fold along the stomach wall, associated with relaxation of each successive caudal contraction and formation of another fold cranial to the wave of contraction. This type of contraction has more of a propulsive function.

Motility patterns of the stomach of *Macropus eugenii* and *Setonix brachyurus* (quokka) have been examined by Richardson & Wyburn (1983;

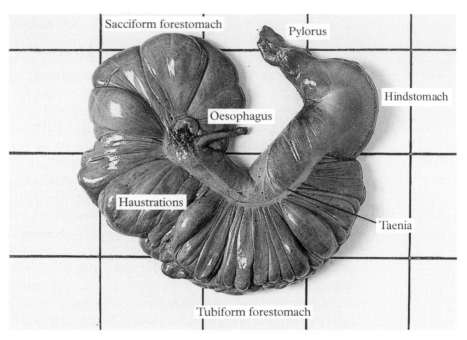

(a)

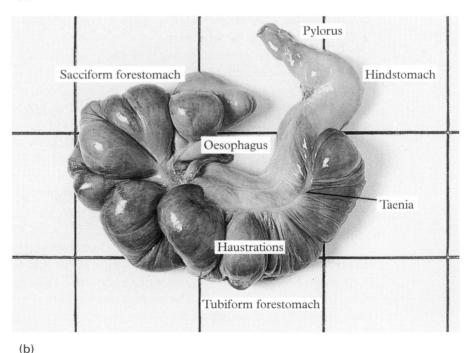

(b)

Figure 6.6 The stomachs of the two Australian tree-kangaroos (a) *Dendrolagus lumholtzi* (Lumholtz's tree-kangaroo) and (b) *Dendrolagus bennettianus* (Bennett's tree-kangaroo) to show differences in the degree and pattern of haustration between species. Note the short but wide tubiform forestomach in each case. Scale: 10 cm × 10 cm grid in background. Hume & Flannery, unpublished.

1988) and Wyburn & Richardson (1989), with some surprising results. Slow waves (slow changes in electrical potential) were recorded from the tubiform forestomach and the hindstomach of both species; the sacciform region was not examined. They originated from the vicinity of the cardia (opening of the oesophagus) and were propagated caudally along the tubiform forestomach towards the pylorus at about 3 mm per second. Radiographic observation showed haustral contractions to occur at the same frequency of the slow wave. These are the sequential waves of contraction observed earlier by Dellow (1979). This is very different from other mammals, from simple-stomached species to ruminants, in which slow waves are found only in the pyloric glandular region, and are not associated with mechanical activity. Peristaltic contractions of the stomach in these mammals, including the forestomach of ruminants, are always associated with a series of action potentials or spiking activity. Action potentials were recorded from the quokka tubiform forestomach but not from the tammar forestomach. These differences between two macropodid species and between macropods and other mammals are worthy of further investigation.

The peristaltic contractions of other mammals are controlled by higher centres via the vagus nerve; their main function is to push digesta towards the pylorus, and thus to control the rate of gastric emptying. Richardson & Wyburn's (1988) results suggest that there is probably little higher neural control over gastric emptying via the vagus nerve.

THE GASTRIC SULCUS

Internally, there are differences among macropodid species in the presence and degree of development of the gastric sulcus, the position of the cardia (opening of the oesophagus) in relation to the permanent ventral fold between sacciform and tubiform forestomach regions, and the relative distributions of squamous epithelium and cardiac glandular mucosa.

In the ruminant forestomach a reticular groove connects the cardia with the reticulo-omasal orifice. Contraction of its muscular walls forms a tube which allows milk in suckled young to bypass the reticulum and rumen, where it would be fermented, and to pass directly into the omasum. Here it runs along the short omasal canal into the abomasum, the site of peptic digestion. It is an ingenious device to allow for efficient utilisation of milk. The stimulus to closure of the sulcus is the sucking reflex. It does not appear to operate in adult ruminants unless they have been suckled throughout their development.

In most macropods a gastric sulcus extends from the cardia caudally along the lesser curvature of the tubiform forestomach. It is prominent in such species as *Macropus eugenii*, *M. robustus*, *M. rufogriseus*, *Wallabia bicolor* (swamp wallaby) (Dellow 1979) and *Lagorchestes hirsutus* (rufous hare-wallaby) (Bridie, Hume & Hill 1994). In *M. parma* and *M. rufus* only the right lip is well developed. In *M. giganteus* the sulcus is well developed in pouch young, but in adults it is relatively much smaller, and the lips are poorly defined. This is an example of reduction during ontogenetic development.

A gastric sulcus is absent in the pademelons *Thylogale thetis* and *T. stigmatica*, in both adult and pouch young animals. However, it is present in *T. billardierii* (Dellow 1979). A sulcus is also absent in *Peradorcus concinna* (narbalek or little rock-wallaby), at least in adult animals; the situation in pouch young has not been described.

It has been assumed that the gastric sulcus functions in macropodid pouch young in a fashion similar to that described for ruminant sucklings (Langer *et al.* 1980). However, evidence from Griffiths & Barton (1966) in the red kangaroo does not support this assumption. These workers found proteolytic activity to be distributed throughout the stomach of pouch young until 200 days of age. Only then was it restricted to the gastric pouch. The young left the pouch permanently at about 236 days. Histological observations supported the enzymatic findings. Columnar epithelium lined the interior of the whole stomach at birth. At 200 days of age there was a transition in histology between pouch young and adult. Gastric glands began to differentiate in the region destined to become the gastric pouch, and the gastric sulcus began to generate squamous epithelium. At 236 days of age true gastric tissue was found in the gastric pouch, and cranially the forestomach was lined with cardiac glands.

Thus it appears that the gastric sulcus, if it does function to channel ingested milk directly to the hindstomach, may be important in this regard only after the young begins to eat grass. The 'young at foot' red kangaroo will continue to suck from outside the pouch for another 120 days (Tyndale-Biscoe 1973).

There may, however, be a role for the gastric sulcus in adult macropods. Dellow (1979) was able to relate the pattern of initial distribution and subsequent dispersion of barium sulphate meals in the stomach of *Macropus eugenii*, *M. giganteus* and *Thylogale thetis* to the relative sizes of the sacciform and tubiform regions of the forestomach, the presence or absence of a gastric sulcus, and the position of the cardia relative to the permanent ventral fold between the sacciform and tubiform forestomach regions. His results, compiled from single radiographs and video-tape data, are shown in Fig. 6.7.

In *Macropus eugenii* the cardia opens on the caudal side of the sacciform–tubiform dividing fold. Consequently most of the contrast medium entering the stomach was directed into the tubiform forestomach. Here it moved caudally along the floor and immediate vicinity of the gastric sulcus, and mixed with digesta in the cranial and central regions of the tubiform forestomach. Contrast medium which entered the sacciform forestomach mixed with digesta close to the cardia.

In *M. giganteus* most of the barium sulphate entered the sacciform

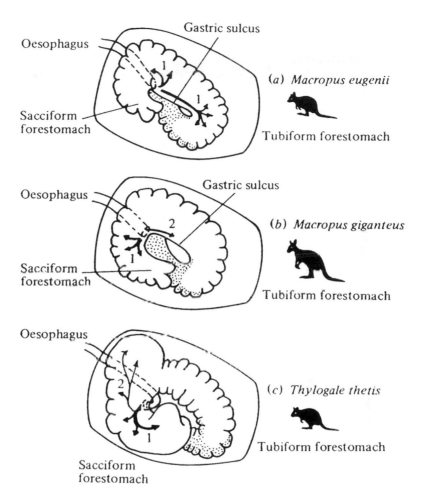

Figure 6.7 Diagrams of the stomachs of (a) the tammar wallaby (*Macropus eugenii*), (b) eastern grey kangaroo (*Macropus giganteus*) and (c) red-necked pademelon (*Thylogale thetis*) *in situ* to show the dispersion of barium sulphate into the sacciform and/or tubiform forestomach regions, depending on the position of the cardia and the presence in *Macropus* of a gastric sulcus. 1, direction of primary dispersion; 2, direction of secondary dispersion. Stippled areas indicate residual barium sulphate 25 hours later. After Dellow (1979).

forestomach; in this species the cardia opens onto the sacciform–tubiform dividing fold. The contrast medium which did enter the tubiform forestomach was directed along the lesser curvature of its cranial region.

In *Thylogale thetis* the cardia opens into the sacciform forestomach. Consequently all of the barium sulphate entered this gastric region, first mostly into the ventral area, later into the dorsal pouch. Up to an hour elapsed before any contrast medium appeared in the tubiform forestomach.

In all three species contrast medium was then gradually transported caudally along the tubiform forestomach and, 8 hours after administration, the whole of this region was outlined and contrast medium was seen in the hindstomach. Food ingested 6 hours after barium sulphate administration did not mix with the contrast medium, reflecting the tubular nature of digesta flow, and the presence of sequential mixing pools in the kangaroo stomach. These are characteristics of a modified plug-flow reactor system, which operates as a number of small stirred-tank reactors arranged in series (see Chapter 4).

Twenty-four hours after dosing, Dellow (1979) found that any contrast medium remaining in the stomach was confined almost entirely to the hindstomach. Contrast medium injected into the hindstomach outlined only this region of the stomach, and was transferred quite rapidly to the duodenum.

Thus it appears that the gastric sulcus plays a role in the nutrition of adult macropods by assisting the caudal movement of liquid digesta to the more distal parts of the tubiform forestomach, a function first suggested by Owen (1868). This may be important in maintaining the fermentation rate in this part of the stomach.

GASTRIC HISTOLOGY
As suggested by Fig. 6.5, the distribution of epithelial types in the macropodid stomach shows considerable variation among species.

Squamous epithelium
The cellular structure of the squamous epithelium of the macropodid stomach is similar to that described for other mammals. Squamous epithelium from the sacciform forestomach of *Thylogale thetis* and *Macropus giganteus* is shown in Fig. 6.8. The basal cells are compact and cuboidal; they have few mitochondria and are separated by intercellular spaces. Towards the lumen of the stomach the cells are compacted and fibre-filled. The squamous epithelium of both species in Fig. 6.8 is cornified, and interdigitates with the underlying lamina propria mucosae. Interdigitation tends to be most pronounced over permanent folds and in the gastric sulcus, presumably for maximum strength. Gemmell & Engelhardt (1977) described the squamous epithelium from *M. eugenii* as being cornified also, but in neither their illustration nor that of Langer *et al.* (1980) is there any evidence of cornification in *M. eugenii*. Whether these differences in extent of cornification among species has a dietary basis is not known.

Cardiac glandular mucosa
The cardiac mucosa is several-fold thicker than the squamous epithelium. Sections of cardiac mucosa from *T. thetis* and *M. giganteus* (Fig. 6.9) show the tubular mucin-secreting glands of this tissue. Griffiths & Barton (1966) found that water extracts of cardiac mucosa from *M. rufus* exhibited weak amylase activity, but considered that this was probably of microbial origin. No other enzymatic activity in adult cardiac mucosa has been reported.

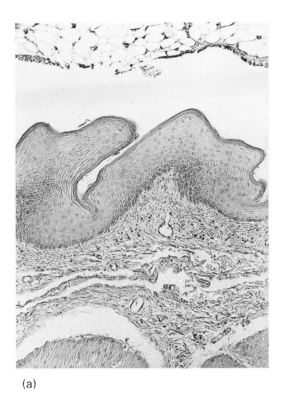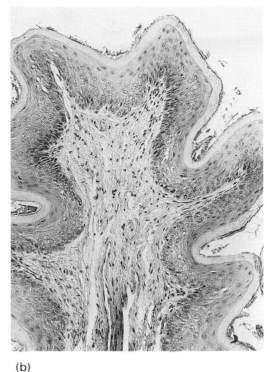

(a)   (b)

Figure 6.8 Squamous epithelium from the sacciform region of the forestomach of (a) *Thylogale thetis* (red-necked pademelon) (× 100) and (b) *Macropus giganteus* (eastern grey kangaroo) (× 200). From Dellow (1979).

Gastric glands

The epithelium of the gastric pouch is distinctly red in colour and strongly rugose. In section these folds are seen to be clothed with a deep epithelium of typical gastric tissue. Ultrastructural observations by Gemmell & Engelhardt (1977) confirmed the presence of all four cell types normally associated with mammalian fundic mucosa: surface epithelial cells, mucous neck cells, chief cells (pepsinigenic cells) and parietal (hydrochloric acid secreting) cells.

Pyloric glands

These are mucus-secreting glands of the same type as found in the cardiac glandular mucosa.

Distribution of epithelial types

Differences among three macropodid species in the distribution of stratified squamous epithelium and cardiac glandular mucosa are illustrated in Fig. 6.10. Squamous epithelium is most extensive in *Macropus giganteus*, lining most of the sacciform region, one-third of the tubiform region and

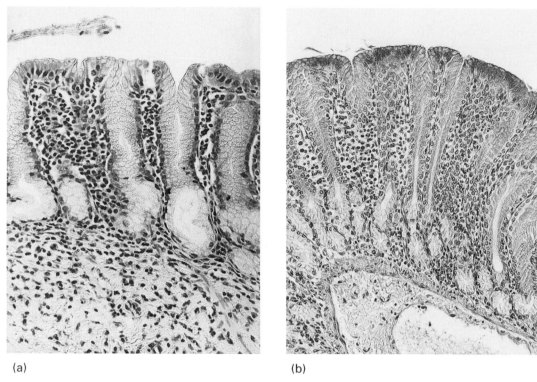

(a) (b)

Figure 6.9 Cardiac glandular mucosa from the tubiform region of the forestomach of (a) *Thylogale thetis* (red-necked pademelon) (× 200) and (b) *Macropus giganteus* (eastern grey kangaroo) (× 200). From Dellow (1979).

the length of the gastric sulcus. In contrast, in the two Australian tree-kangaroos, sqamous epithelium is restricted to the gastric sulcus and a small area close to the cardia (Hume & Flannery, unpubl. 1998). Intraspecific differences also occur, as demonstrated in *Macropus giganteus* in Figs. 6.4 and 6.9; Schäfer & Williams (1876) found that the cardiac glandular mucosa in the sacciform forestomach was continuous with that in the tubiform region, whereas Langer *et al.* (1980) showed it as a patch isolated by squamous epithelium. Both forms have been seen by D. L. Obendorf (pers. comm. 1981) in *M. giganteus* in Victoria. The nutritional significance of these differences within and among species remains obscure.

In several species, particularly *M. giganteus* and *M. rufus*, pyloric glandular mucosa extends beyond the pyloric sphincter into the proximal duodenum (Krause 1972).

6.2.3 Intestine

Both the small and large intestine of macropods are relatively short for herbivores (Stevens & Hume 1995). Most of the small intestine lies loosely coiled caudal to the sacciform forestomach. In *Macropus robustus robustus*

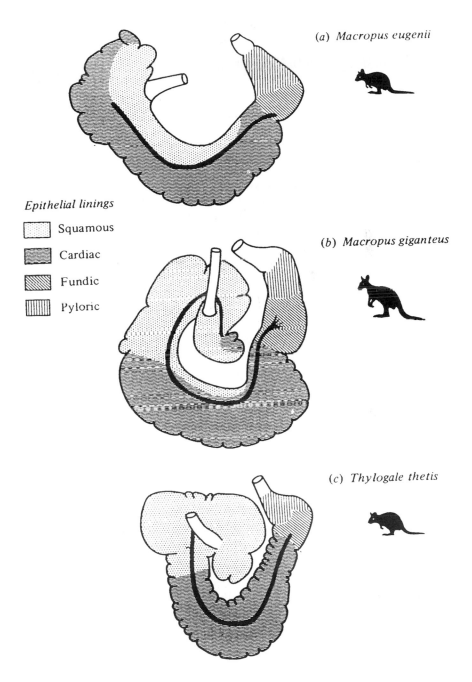

Figure 6.10 Diagrams of the right-hand aspects of the stomachs of (a) *Macropus eugenii* (tammar wallaby), (b) *Macropus giganteus* (eastern grey kangaroo) and (c) *Thylogale thetis* (red-necked pademelon). After Langer, Dellow & Hume (1980).

(eastern wallaroo) and *M. r. erubescens* (euro) the small intestine contains 9% and 10% of total gastrointestinal tract contents respectively (Freudenberger 1992). Brunner's glands of the Macropodidae are remarkably similar to those of the Dasyuridae (Chapter 2), despite obvious dietary differences. Those of *M. giganteus* are organised into very large lobes reminiscent of small salivary glands (Krause 1972). They empty onto the pyloric glandular mucosa which extends into the duodenum in this species, and in *M. rufus*. In *M. parma* (parma wallaby), *Petrogale penicillata* (brush-tailed rock-wallaby) and *Lagorchestes conspicillatus* (spectacled hare-wallaby) they empty onto both pyloric and intestinal epithelia. In *Setonix brachyurus* (quokka) they empty largely onto intestinal epithelium (Krause 1972).

The entire inner surface of the small intestine is lined with villi, as it is in other mammals. The villi of *Macropus agilis* (agile wallaby), *M. rufus* and *Thylogale thetis* are elongated and tapered, and are radially arranged to the long axis of the small intestine and with their bases almost parallel to each other (Osawa & Woodall 1992). Villus size and morphology were related to diet quality. On a low-fibre diet, captive *M. agilis* and *M. giganteus* and wild *Wallabia bicolor* (swamp wallabies) in the wet season had significantly longer villi than animals on a high-fibre diet (wild *M. agilis* and *M. giganteus*, and wild *W. bicolor* in the dry season) (Osawa & Woodall 1990). Longer villi mean a greater absorptive surface area of the small intestine when nutrient supply is likely to be greatest. The increase in villus length observed by Osawa (1987) in *M. agilis* during lactation, when food intake increases, can be interpreted in the same way. The effect of diet quality on villus length in *M. agilis* is shown in Fig. 6.11.

As in the stomach, Richardson and Wyburn (1983;1988) found differences in motility patterns of the small intestine between macropods and other mammals. Migrating myoelectrical complexes were recorded in the small intestine of both the tammar and quokka. They consisted of three phases: (a) a quiescent phase during which there were only slow waves; (b) a phase of irregular spiking activity during which there are action potentials associated with only some of the slow waves; and (c) a phase of regular spiking activity when there are action potentials on every slow wave. This pattern is typical of all mammals so far examined. The slow waves act as a pacesetter for contractions of the small intestinal wall and, in all mammals except the two macropods examined by Richardson & Wyburn (1988) and the guinea pig (Calligan, Costa & Furness 1985), their frequency declines from duodenum to ileum in a series of gradient plateaux. This decreasing frequency gradient has been thought to be essential for the caudal movement of digesta along the small intestine, even though there is no mechanical activity associated with the slow wave.

In the tammar, quokka and guinea pig there is no gradient in frequency of slow waves along the small intestine. This suggests that either there is only one gradient plateau extending the entire length of the small intestine in

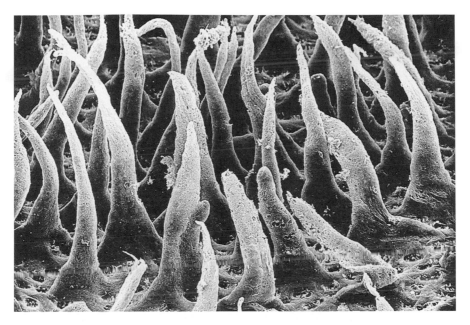

(a)

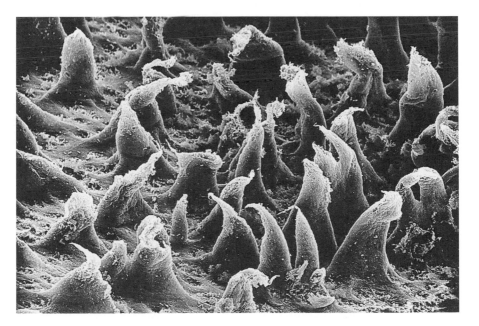

(b)

Figure 6.11 The effect of diet quality on the length of villi in the small intestine of the agile wallaby (*Macropus agilis*). Scanning electron micrographs from (a) a captive animal fed high-quality lucerne (alfalfa) hay (magnification × 85), and (b) a free-living animals (magnification × 85). Stomach contents from the captive wallabies contained 40% more crude protein and 36% less fibre than the wild wallabies. From Osawa (1987).

these three species, or that there is no plateau and the frequency of the slow waves is an intrinsic property of the smooth muscle of the intestinal wall (Richardson & Wyburn 1988).

There is relatively little information available on the activities of hydrolytic enzymes along the small intestine of macropods. Adult eastern grey kangaroos have low levels of disaccharidase activity (Table 2.3), as would be expected in a foregut fermenter; most non-structural carbohydrates would be fermented before they reached the small intestine. However, in the suckling pouch young, lactase activity is high (Table 2.3). Messer, Crisp & Czolij (1989) showed that these lactase activities were due entirely to an acid β-galactosidase that is located within the lysosomes of the enterocytes of the brush border. This is in distinct contrast to eutherians, in which lactose is digested extracellularly by a neutral β-galactosidase located on the microvillous membrane. In macropods the lactose appears to be taken up into the enterocytes by pinocytosis. This is a relatively slow process, but normally adequate because macropod milk contains little lactose. However, it may limit the rate at which lactose-containing milk can be fed to orphaned pouch young without causing severe diarrhoea, as often happens.

The large intestine or hindgut of the Macropodidae consists of a simple caecum with a mobile body and apex (Richardson & Wyburn 1980), and a colon that forms loose coils until it straightens out where the descending colon terminates in a small rectum which opens at the cloaca. There are no villi in any part of the hindgut. The colon can be divided into proximal and distal regions (Fig. 6.2). The lumen of the proximal colon is wider than that of the distal colon, and is confluent with that of the caecum. Taeniae are present on the caecal wall and extend along the wall of the proximal colon. However, they are not well developed, and the associated haustrations are poorly defined. A gastrocolic ligament attaches the proximal colon to the tubiform forestomach. Together the caecum and proximal colon function as a secondary site of microbial fermentation. The distal colon is distinguished by both its smaller calibre and the appearance of faecal pellets as water is absorbed from the digesta throughout its length.

The lengths of both the caecum and colon relative to body size are greater in grazers such as *M. giganteus* and *M. rufus* than in browsers such as *W. bicolor* and *T. thetis* (Osawa & Woodall 1992), no doubt because of the often higher fibre content of grasses and sedges compared with browse. A longer colon in *M. rufus* than in *M. giganteus* (Osawa 1987) can also be related to environmental effects; the range of *M. rufus* extends further into the arid zone, and there is thus a greater need to conserve water by absorption in the hindgut (Hume & Dellow 1980; Woodall & Skinner 1993). Similarly, Freudenberger & Hume (1993) reported that the colon of the arid-zone euro (*M. robustus erubescens*) was 37% longer than that of the more mesic but closely related eastern wallaroo (*M. r. robustus*). This correlated with drier faeces in the euro (54% water versus 59% in the eastern wallaroo) when water intake was restricted, and a lower water intake by the euro when water was available *ad libitum*.

Habitat aridity is also reflected in the relative lengths of proximal and distal regions of the colon, as seen in Chapter 4 with the wombats; the distal colon of the arid-zone *Lasiorhinus* is 59% of total colon length, versus 32% in the more mesic *Vombatus*). Most net uptake of water occurs in the distal colon (Stevens & Hume 1995). In the arid-zone rufous hare-wallaby (*Lagorchestes hirsutus*), the distal colon is 93% of total colon length (Fig. 6.12). This can be compared with approximately 70% in the eastern grey kangaroo (Fig. 6.2).

Figure 6.12 The digestive tract of the rufous hare-wallaby (*Lagorchestes hirsutus*). From Bridie, Hume & Hill (1993). Note the very long distal colon, a feature of many arid-zone species.

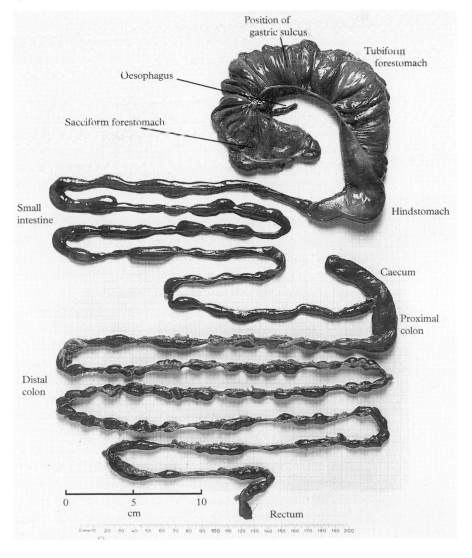

6.2.4 **Regurgitation**

The strong desire by the early anatomists to compare macropodid digestion and metabolism with that of ruminants led them to refer to food regurgitation in the macropods as 'rumination'. Rumination is only found in ruminants and their close relatives, the camelids. It consists of controlled and rhythmic regurgitation, remastication, reinsalivation and swallowing of the chewed bolus. The complete cycle of rumination is closely integrated with cyclic contractions of the ruminant and camelid forestomach. It allows these animals to ingest in haste and masticate at leisure and in greater safety (Stevens & Hume 1995). It also allows for thorough mastication of plant material that has already been softened and partly digested in the fermentation fluid (Hume & Warner 1980).

Home (1814) quoted Banks' observations on 'rumination' in the kangaroo; the kangaroos kept by Banks ruminated 'when fed on hard food', but he conceded that 'it is not however their constant practice, since those kept in Exeter Change have not been detected in that act'. Owen (1834) stated that he had

more than once observed the act of rumination in the kangaroos. It does not take place while they are upon the tripod of their hinder legs and tail. The abdominal muscles are in violent action for a few minutes; the head is a little depressed; and then the cud is chewed by a quick rotatory motion of the jaws. This act was more commonly noticed after physic had been given to the animals, which we may suppose to have interrupted the healthy digestive processes; it by no means takes place with the same frequency as in the true Ruminants.

Wood Jones (1924) also reported on regurgitation in kangaroos, wallabies and bandicoots: 'The animal, after a meal, makes a vigorous heaving movement of its chest and abdomen, and the stomach contents, which are forced up into the mouth appear to be re-swallowed without any further chewing.' He did not refer to the act as rumination. More recently, regurgitation has been reported in the quokka (*Setonix brachyurus*) by Calaby (1958) and Moir *et al.* (1956), and in the red-necked wallaby (*Macropus rufogriseus*) and Tasmanian pademelon (*Thylogale billardierii*) by Mollison (1960). Calaby's (1958) caged quokkas ejected food boluses at irregular intervals which fell through the mesh floor of the cages and could not be recovered by the animals. Thus there are differences in detail, sometimes considerable, in the various descriptions of regurgitation by marsupials, but Barker, Brown & Calaby (1963) concluded that 'it is not analogous to rumination in ruminants, and the term "rumination" should not be used in connection with kangaroos'. They proposed the term 'merycism' (from a Greek word meaning 'chewing the cud'). Merycism is often applied to regurgitation, remastication and reswallowing in humans and animals generally, and does not have the specialised meaning of rumination as applied to ruminants.

Indeed, macropodid marsupials would appear to have no real requirement for rumination, as they eat more slowly and masticate their food more thoroughly than do ruminants. However, the observation by Dellow (1979) that the frequency of occurrence of merycism can be increased by the addition of crushed wheat grain to a chopped lucerne hay diet suggests that merycism may aid digestion in kangaroos and wallabies by stimulating saliva flow. A faster fermentation resulting from starch ingestion would tend to lower digesta pH, and greater saliva flow and buffering of acids in the forestomach would be advantageous under these conditions.

A second type of jaw movement has been observed by a number of workers (Moir *et al.* 1956). It occurs more frequently than merycism, and usually only while the animal is resting, sometimes several hours after feeding. This process does not involve regurgitation of a food bolus, but rhythmic jaw movements can continue for up to 30 minutes, leading Dellow (1979) to suggest that this also has the effect of stimulating salivary secretion.

6.2.5 Salivary glands

The salivary glands of red kangaroos (*Macropus rufus*), eastern grey kangaroos (*M. giganteus*) and western greys (*M. fuliginosus*) were examined by Forbes & Tribe (1969). The positions of the three main types of salivary

Figure 6.13 Diagram of the location of the main paired salivary glands of the red kangaroo (*Macropus rufus*). After Forbes & Tribe (1969).

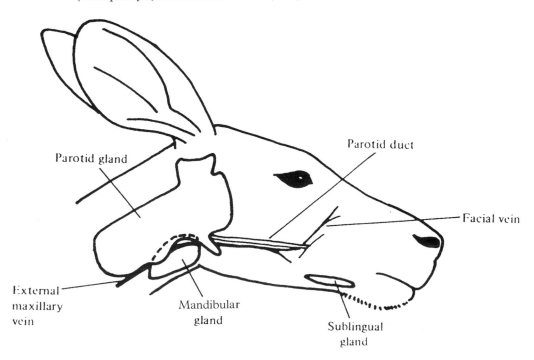

glands in macropods are shown in Fig. 6.13. As in many other foregut fermenters, the macropodid parotids are the largest of the three types, and are histologically similar to those of sheep, but twice as heavy in relation to body mass. On the same basis the mandibular glands appear to be of similar size in kangaroos and sheep (Tribe & Peel 1963). The sublinguals are the smallest glands in both kangaroos and sheep.

Although Forbes & Tribe (1969) reported on salivary flow and composition in the red kangaroo, their data were based on limited samples from animals of unknown sodium status. Detailed studies by Beal (1984, 1986, 1989) in red and eastern grey kangaroos and euros (*M. robustus erubescens*) of known sodium status provide a much better data base. Parotid saliva is relatively high in sodium, calcium, phosphate and bicarbonate ions. Eastern grey kangaroos had higher maximal flow rates of parotid saliva than red kangaroos and euros, indicative of a larger gland relative to body mass. At equivalent flow rates, eastern greys still had higher rates of parotid urea clearance and higher rates of phosphate secretion. Bicarbonate and phosphate ions are important in buffering the SCFAs produced in the forestomach fermentation. Higher digestive efficiencies reported in western grey kangaroos (*M. fuliginosus*) by Prince (1976) could thus be at least partly explained by differences in salivary secretion of phosphate. Inorganic phosphorus concentrations in the forestomach fluid of the animals studied by Prince (1976) were higher in western greys and euros (10–11 mmol L^{-1}) than in red kangaroos (4–7 mmol L^{-1}).

Mandibular glands of all three species had similar rates of secretion of nitrogen (as protein plus urea), and low rates of secretion of amylase activity, bicarbonate and phosphate, and therefore mandibular saliva does not have the high buffering capacity of parotid saliva. However, another function of saliva in kangaroos is in thermoregulation; they spread it on their forelimbs during heat stress. Water from water-secreting glands in the nose is also used. The region of the forearm used has a dense superficial network of fine blood vessels. In hot conditions there is a marked increase in blood flow through these vessels, so significant heat loss is possible (Dawson 1995). The mandibular gland is the major contributor to the saliva secreted during thermoregulatory salivation. Despite their smaller size, maximal flow rates (mL min^{-1}) stimulated by acetylcholine were similar to those of the parotids under equivalent stimulation (Beal 1984). Mandibular saliva is ideally suited as an external coolant because it is essentially a hypotonic solution of sodium chloride and potassium chloride (Beal 1986), which minimises electrolyte loss. Its role in evaporative cooling of koalas was mentioned in Chapter 5 (Section 5.5.1).

In addition to its buffering function, phosphate in the saliva of both macropods and ruminants is an important endogenous source of phosphorus for the symbiotic bacteria in the forestomach. Because of their high growth rates, bacteria contain a high ratio of ribonucleic acid-nitrogen to

Table 6.1. *Concentrations of ribonuclease in the pancreas of herbivores*

	Ribonuclease content (μg g^{-1} wet weight pancreas)
Forestomach fermenters	
Red kangaroo (*Macropus rufus*)	600
Eastern grey kangaroo (*M. giganteus*)	530
Tammar wallaby (*M. eugenii*)	515
Sheep (*Ovis aries*)	1080
Bison (*Bison bison*)	1180
Cow (*Bos taurus*)	1200
Elk (*Cervus canadensis*)	550
Uganda kob (*Kobus kob*)	270
Hippopotamus (*Hippopotamus amphibius*)	62
Hindgut fermenters with limited forestomach fermentation	
Rat (*Rattus*)	260
Guinea-pig (*Cavia*)	240
Golden hamster (*Mesocricetus auratus*)	260
Mouse (*Mus*)	395
Rabbit (*Oryctolagus cuniculus*)	0.5
Hindgut fermenters	
Virginia opossum (*Didelphis virginiana*)	20
Horse (*Equus caballus*)	25
Pig (*Sus*)	80
Elephant (*Loxodonta africana*)	0.7

Source: After Barnard 1969.

total nitrogen. This means that they have a substantial requirement for phosphate to synthesise the ribose-phosphate chain of the RNA molecule. Ruminants salvage some of this phosphate when bacteria are digested in the abomasum and small intestine. The enzyme ribonuclease occurs at far greater concentrations in the pancreas of ruminants than of non-ruminant eutherians. A close parallel exists among the Marsupialia, at least in those species so far examined. As can be seen in Table 6.1, the concentrations of ribonuclease in the pancreas of three macropodid species are much higher than that in the omnivorous Virginia opossum (*Didelphis virginiana*).

A similar parallel is likely (but not yet adequately tested) in the form of high activities of the enzyme lysozyme in the mucosa of the macropodid hindstomach and the ruminant abomasum. Lysozyme hydrolyses the β 1–4 linkages in the polysaccharide chain of bacterial cell walls, and hence may play an important role in foregut fermenters in making available to the animal the contents, including RNA, of bacterial cells flowing out of the forestomach. Dobson, Prager & Wilson (1984) found much higher activities of lysozyme c in the mucosa from the fundic region of the abomasum of cows than in the forestomach or the pyloric region of the abomasum. High lysozyme c activity was also detected in the hindstomach of other ruminants (roe and black-tailed deer), and other foregut fermenters (camel and langur monkey), but not in the stomach of non-foregut fermenters (pig and pig-tailed macaque monkey). Preliminary data (F. Clark & Hume, unpubl., 1982) indicate lysozyme activity in the gastric pouch of the hindstomach of eastern grey kangaroos at 33% of that in the abomasum of sheep, but no activity in the sacciform or tubiform regions of the forestomach or in the pyloric gland region of the hindstomach.

6.2.6 The helminth fauna

The helminth fauna of the Macropodidae is dominated by the anoplocephalid cestodes and by strongyloid, trichostrongyloid, metastrongyloid, oxyuroid and filarioid nematodes (Beveridge & Spratt 1996). By far the most spectacular radiation within the Macropodidae has been that of the strongyloid nematodes, with 40 genera and 171 species described, and many species still undescribed. Large populations are often reported in the macropodid stomach. For instance, M. J. Wolin & T. L. Miller (pers. comm. 1997) suggested that in red kangaroos, western grey kangaroos (*Macropus fuliginosus*) and euros (*M. robustus erubescens*), the biomass of nematodes appeared to exceed that of the microbes. Whether these nematodes are parasites or commensal symbionts has not been investigated. Helminths commonly ferment carbohydrates to short-chain fatty acids (Saz 1981; Köhler 1985). This is an aspect of macropodid digestion that deserves study.

Macropus giganteus in south-eastern Australia are commonly infected with the introduced liver fluke (*Fasciola hepatica*), and are potential contaminators of pasture for ruminants, in contrast to common wombats, which appear to be resistant to infection.

The large populations of nematodes found by Obendorf (1984c) in the oesophagus of several macropodid species all had specialised adaptations of their outer cuticle and adopted body conformations that helped to hold them in close association with the oesophageal lining, where they appeared to feed on the adherent bacterial plaque.

6.2.7 Microbiology

Moir, Somers & Waring (1956) provided the first preliminary description of the microbiology of the macropodid stomach in their study of the quokka (*Setonix brachyurus*). They reported a 'dense bacterial population strikingly similar to that of the sheep's rumen under similar conditions. This population consisted mainly of Gram-negative rods and cocci, with a few spiral forms. Gram-positive rods were also present and these dominated the population where the pH was below 5.5.' Although only about 15 types of bacteria were discernible, compared with over 30 for the sheep, the total density of the population (10^{10} mL^{-1}) and the proportion of cellulolytic bacteria are similar (R. E. Hungate, pers. comm. to Moir, 1968). Later, Dellow et al. (1988) recorded direct counts of bacteria in forestomach digesta ranging from 17 to 76×10^{10} g^{-1} in eastern grey kangaroos and 21 to 52×10^{10} g^{-1} in red-necked pademelons, all shot while foraging in the wild. In red-necked pademelons and tammar wallabies maintained in captivity on chopped lucerne hay the direct counts ranged between 8 and 51×10^{10} g^{-1}; in sheep on the same diet the direct count in the rumen was 44×10^{10} g^{-1}.

The other region of the macropodid digestive system that has been examined microbiologically is the oesophagus. Obendorf (1984b) found that the oesophageal lining of the 13 macropodid species he investigated was colonised by a large and diverse population of bacterial forms. Many had extracellular coats and capsules, apparently for attaching to the surface epithelium. Only the superficial layers of cells were colonised, and cells with ruptured cell membranes were invaded by bacteria. Sloughing of damaged cells provided a new surface for bacterial attachment. Obendorf (1984b) proposed that swallowing of sloughed cells with bacteria attached provided a continuous source of inoculum to the macropodid forestomach fermentation system. How important this is in maintaining the fermentation is so far untested.

Although Moir et al. (1956) were unable to detect any protozoa in the quokka stomach, they remarked that this could possibly have been the result of the diet (commercial sheep pellets and chopped oaten hay). Subsequent work with foraging animals on Rottnest Island confirmed the presence of three unidentified ciliates at total concentrations of from 0.5 to 15×10^6 g^{-1} (Moir 1965). Yadav, Stanley & Waring (1972) first detected ciliate protozoa in pouch-young quokkas when they began to ingest plant material. Ciliates were then found in the stomach of tammar wallabies (Lintern 1970; Obendorf 1984d) and adult red kangaroos (Harrop & Barker 1972; Obendorf 1984d). Obendorf (1984d) also recorded ciliates from the stomach of eight other macropodid species.

Dellow et al. (1988) found ciliate protozoa in the forestomach of the majority of free-living animals of four macropodid species, as well as captive tammar wallabies. Total numbers were variable, ranging from 0.3 to 15×10^4 g^{-1}. The only free-living species in which no protozoa were detected was the red-necked pademelon. The highest counts were always recorded in samples from the sacciform forestomach, and total numbers decreased progressively along the length of the tubiform forestomach. A similar decrease in protozoal numbers had been earlier noted by Obendorf (1984d). This consistent pattern could be due to the rapid disappearance of soluble carbohydrates in the cranial regions of the forestomach (Dellow &

Hume 1982c), to rapid dilution rates in the tubiform forestomach (Dellow 1982), or to both of these factors. Patterns of digesta flow and fermentation in the macropodid stomach are discussed below (Sections 6.3 and 6.4).

In most cases the ciliates present in the animals examined by Dellow *et al.* (1988) were holotrichs (i.e. with an even covering of cilia over their surface) (Fig. 6.14). Small numbers of spirotrichs were present in most samples. A. Jankowski (pers. comm. 1982) found the protozoa of the macropodid forestomach to have no taxonomic similarities with the ciliate protozoa of sheep, cattle or horses. Thus Dehority (1996) proposed a new family of entodiniomorph protozoa from the macropodid forestomach (the Macropodiniidae), on the basis of at least three distinct features which are incompatible with any of the eight established families in the order Entodiniomorphida.

A biological basis for the host specificity of the macropodid forestomach protozoa has not been established, but may well be related to the relatively short digesta retention times in the forestomach of kangaroos and wallabies discussed below. When S. K. Baker *et al.* (1995) inoculated ten defaunated sheep with forestomach contents from kangaroos the ciliates disappeared from the rumen within 14 days. Also, no rumen protozoa were observed in the kangaroo forestomach, nor were kangaroo ciliates found in rumen contents from sheep grazing the same pasture.

No protozoa have been found in the hindgut of macropods (Dellow *et al.* 1988). The only other site in the macropodid digestive system that is inhabited by protozoa is the oesophagus. Obendorf (1984d) found protozoa in the oesophagus as well as the forestomach of all ten species of free-living kangaroos and wallabies that he examined. Within any one animal, the ciliates in the oesophagus were similar to those in the sacciform forestomach. In the large grazing kangaroos the protozoa were located between the longitudinal mucosal folds of the oesophageal lining (see Section 6.2.1 above). In the brush wallabies they were commonly found between the oesophageal papillae. In both macropodid groups the ciliates were in the lumen of the oesophagus and present in low numbers, suggesting that they had been regurgitated during merycism. However, in one *Macropus dorsalis* (black-striped wallaby) large numbers of ciliates were found closely associated with the epithelial lining. The protozoa were so closely packed that they assumed unconventional elongated shapes, and completely displaced the bacterial plaque normally adherent to the oesophageal lining.

The other microbes found in the macropodid forestomach are anaerobic fungi. In ruminants fungi play an important role in plant fibre breakdown (Gordon & Phillips 1993). Dellow *et al.* (1988) reported fungal sporangia, similar to those found in ruminants, in forestomach samples from all free-living animals they examined except *T. thetis*; nor were they present in samples from captive *T. thetis* or *M. eugenii* fed chopped lucerne hay. The fungi populated the plant fragments of the digesta (Fig. 6.15). In ruminants

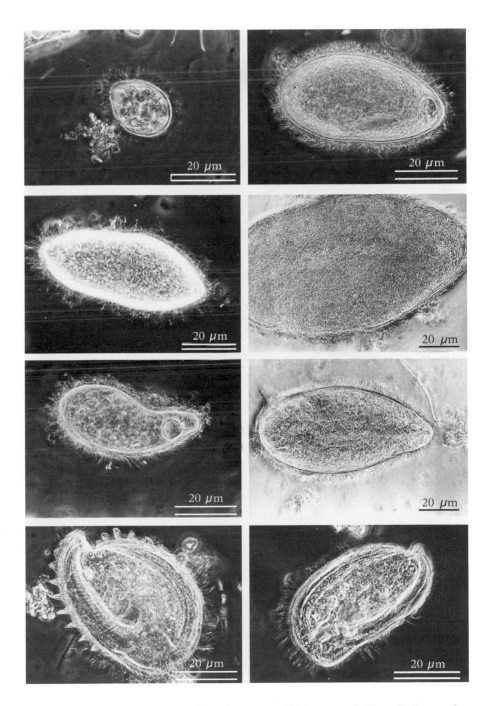

Figure 6.14 Examples of protozoa from the macropodid forestomach. From Dellow *et al.* (1988).

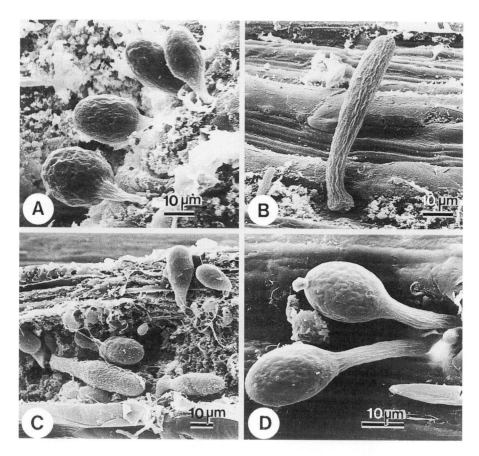

Figure 6.15 Examples of fungal sporangia on plant fragments from the forestomach of the eastern grey kangaroo (*Macropus giganteus*) (A, B) and swamp wallaby (*Wallabia bicolor*) (C, D). From Dellow *et al.* (1988).

Bauchop (1979) found that fungi were attached to the more slowly digested fibrous material, and were virtually absent in animals grazing soft, leafy pasture. The absence of fungi in *T. thetis* may be explained on a similar basis, as the digesta were composed mainly of fleshy leaves low in fibre content.

6.3 PASSAGE OF DIGESTA THROUGH THE GASTROINTESTINAL TRACT

Early studies of digesta passage in macropodids were based on the use of stained hay particles. Results are summarised in Table 6.2. They indicate that, even on a lower voluntary food intake, excretion times in the quokka (*Setonix*), eastern grey kangaroo (*Macropus giganteus*) and red kangaroo (*M. rufus*) are all shorter than in the sheep.

If the results are plotted on a cumulative basis (Fig. 6.16), differences in

Table 6.2. *Dry matter intake and times for excretion of 50% and 90% of stained hay particles in macropods and sheep fed chopped lucerne (alfalfa) hay* ad libitum

Species	Dry matter intake (g kg$^{-0.75}$ d^{-1})	Excretion time (h) 50%	90%	Reference
Setonix brachyurus	47	–	38	Calaby 1958
Macropus giganteus	49	39	50	Forbes & Tribe 1970
Macropus rufus	58	35	45	Foot & Romberg 1965
Macropus rufus	38	41	58	McIntosh 1966
Macropus rufus	63	28	39	Forbes & Tribe 1970
Macropod mean	52	36	48	
Ovis aries	64	38	–	McIntosh 1966
Ovis aries	72	41	67	Foot & Romberg 1965
Ovis aries	67	52	89	Forbes & Tribe 1970
Sheep mean	67	44	75	

Figure 6.16 Cumulative appearance of stained hay particles in the faeces of red kangaroos (*Macropus rufus*), eastern grey kangaroos (*Macropus giganteus*) and sheep (*Ovis aries*) fed chopped oaten straw *ad libitum*. The four curves shown are for the kangaroos (●) and the sheep (○) with the fastest and slowest rates. After Forbes & Tribe (1970).

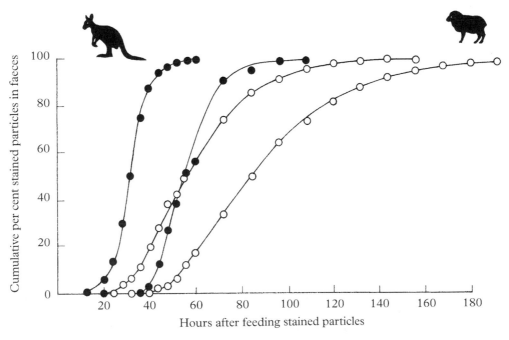

the pattern of excretion as well as in excretion times become apparent. For example, the time difference between 50% and 90% excretion times is considerably shorter in the macropods (10–12 hours) than in the sheep (26–37 hours); this is reflected in the steeper cumulative excretion curves for the kangaroos. Mean retention times (MRTs) calculated from these curves are 42.9 ± 5.5 (SEM) hours for *M. rufus*, 44.5 ± 6.5 hours for *M. giganteus* and 70.1 ± 10.1 hours for the sheep.

Dellow (1982) has made a much more detailed study of digesta passage in macropods using the two markers ^{51}Cr-EDTA (for solutes) and ^{103}Ru-Phenanthroline (which associates with particles). Animals were offered additional food (chopped lucerne hay) four-hourly to encourage steady-state conditions in the gut. The patterns of appearance of the two markers in the faeces of sheep, the eastern grey kangaroo, red-necked pademelon (*Thylogale thetis*) and tammar wallaby (*Macropus eugenii*) are shown in Fig. 6.17. Cumulative percentage recovery of the markers is shown in Fig. 6.18; marker MRTs calculated from the curves are also shown. Similar patterns were obtained in the tammar by Warner (1981b).

The most striking difference is between the sheep and the eastern grey kangaroo, especially in the pattern of appearance of the two markers. Separation of the markers was minimal during their passage through the digestive tract of the sheep. This is partly because the flow characteristics in the reticulo-rumen are those of a stirred-tank reactor (Chapter 2); there is one primary mixing pool of digesta. The other reason is that during its retention in the reticulo-rumen there is substantial exchange of the ^{103}Ru-Phenanthroline between particles, and migration from larger to smaller particles as digestion proceeds (Faichney & Griffiths 1978). Smaller particles leave the rumen faster than larger ones, and so excretion times of large particles are underestimated by this particle-associated marker (Faichney 1993). The differential between the MRTs of solutes and particles in the reticulo-rumen of sheep measured with ^{51}Cr-EDTA and indigestible acid-detergent lignin (an internal marker which does not migrate between particles) is usually between 2 and 3, at least on forage-based diets. This is much greater than the value of 1.2 measured with ^{51}Cr-EDTA and ^{103}Ru-Phenanthroline by Dellow (1982).

In contrast to the sheep, there was a distinct separation of the markers (a differential of 1.5–2.0 in the MRTs of the two markers) in the eastern grey kangaroo, the solute marker passing through the gut much more rapidly than the particle marker (Table 6.3). This difference can be seen quite clearly in the cumulative excretion curves in Fig. 6.18, which reflect the tubiform morphology of the macropodid stomach and the tubular nature of digesta flow through the organ. Flow through the macropodid stomach is modelled best as a modified plug-flow reactor (PFR); that is, as a number of small stirred-tank reactors in series (Chapter 4). Migration of ^{103}Ru-Phenanthroline between particles in this system is not such a problem

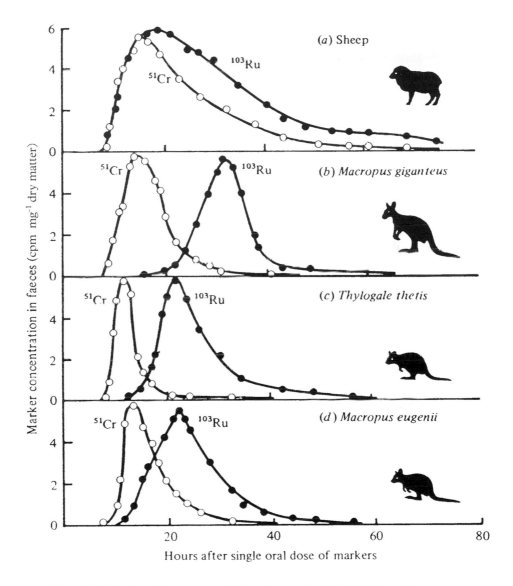

Figure 6.17 Patterns of appearance of the solute marker ^{51}Cr-EDTA and the particle-associated marker ^{103}Ru-Phenanthroline in the faeces after a pulse oral dose in (a) sheep; (b) eastern grey kangaroos (*Macropus giganteus*); (c) red-necked pademelons (*Thylogale thetis*) and (d) tammar wallabies (*Macropus eugenii*) fed chopped lucerne (alfalfa) hay *ad libitum*. After Dellow (1982).

because of shorter retention times in each mixing pool, so that the movement of large particles is more nearly approximated by this marker in the macropods than it is in sheep.

When the markers were injected through catheters into the hindstomach of the pademelon and tammar the patterns of appearance in the faeces were identical for the two markers (Fig. 6.19), showing that there was no

Table 6.3. *Kinetics of a solute and a particle-associated marker in the digestive tract of sheep and three macropodid marsupials fed chopped lucerne (alfalfa) hay* ad libitum

| Species | Body mass (kg) | Mean retention time (h) | | DRR[a] |
		Solute marker (^{51}Cr-EDTA)	Particle marker (^{103}Ru-Phenanthroline)	
Sheep (*Ovis aries*)	49.5	22.7	27.2	1.2
Eastern grey kangaroo (*M. giganteus*)	20.8	16.1	31.5	2.0
Red-necked pademelon (*Thylogale thetis*)	4.9	12.5	25.1	2.0
Tammar wallaby (*Macropus eugenii*)	4.8	16.7	25.8	1.5

Note: [a]DRR (Differential retention ratio) = $\dfrac{\text{MRT particle marker}}{\text{MRT solute marker}}$

Particle marker MRTs in the sheep are underestimates for reasons given in the text (p. 236).

Figure 6.18 Cumulative appearance of the solute marker ^{51}Cr-EDTA and the particle-associated marker ^{103}Ru-Phenanthroline derived from the curves in Fig. 6.17. The upper curve in each case is for the solute marker. Values are the mean retention times (h) for each marker. After Dellow (1982).

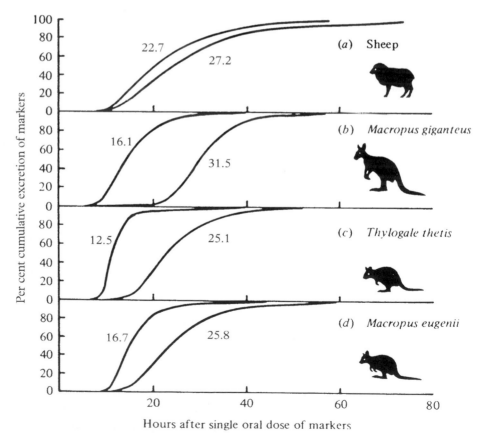

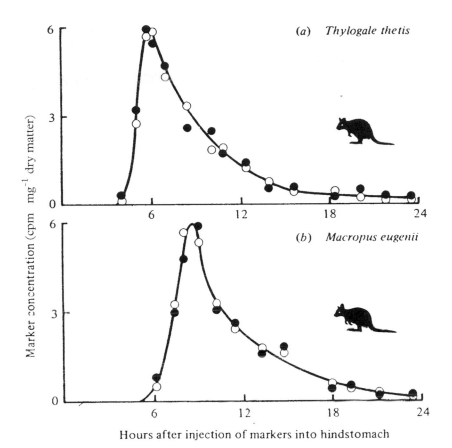

Figure 6.19 The pattern of appearance of the solute marker ^{51}Cr-EDTA and the particle-associated marker ^{103}Ru-Phenanthroline in the faeces of (a) red-necked pademelons (*Thylogale thetis*) and (b) tammar wallabies (*Macropus eugenii*) after a single injection of both markers into the hindstomach. Note the absence of any separation of the markers, and their short elimination times. After Dellow (1982).

differential digesta flow through the intestine. MRTs for the solute and particle-associated markers in the hindstomach and intestine were, respectively, 9.8 and 11.9 hours. The forestomach accounted for only 23–29% of total tract MRT for the solute marker, but 54–61% for the particle marker. Similarly, Warner (1981b) concluded that the stomach accounted for about two-thirds of the total retention time for particles, and was the major site of separation of the solute and particle markers; there was no evidence for any other major mixing compartments.

The conclusions from these results are that:
(a) Retention of particulate digesta in the stomach of macropods is, like ruminants, much longer than in any other part of the digestive tract.
(b) Separation of the fluid and particulate phases of digesta in the macropodid gut occurs only in the stomach.

(c) In contrast to ruminants, mixing of digesta in the macropodid for-estomach does not occur within a single large pool but in a series of smaller pools.

(d) Total mean retention times (MRTs) of solute markers in the macro-podid gut are probably similar to those in the gut of ruminants of equivalent body size.

(e) Direct comparisons between macropods and ruminants for the par-ticle-associated marker [103]Ru-Phenanthroline are not possible be-cause of the problem of exchange of this marker between particles and the different extents to which this results in net migration from larger to smaller particles in the two digestive systems. However, based on the results of Forbes & Tribe (1970), it appears that the MRTs of large particles are shorter in the macropodid gut, especially on poor quality (high-fibre) forages.

The nutritional consequences of the morphology of and digesta flow through the macropodid digestive system are explored below.

6.4 DIGESTION IN THE MACROPODID FORESTOMACH

6.4.1 Fermentation products

SHORT-CHAIN FATTY ACIDS

Moir *et al.* (1956) confirmed that the forestomach of the quokka (*Setonix brachyurus*) was a fermentation organ when they reported the presence of short-chain fatty acids (SCFA). The level was clearly related to time after feeding, rising from 23 mmol L^{-1} in animals fasted for 22 hours to 105 mmol L^{-1} in those recently fed. There was also a close relationship between SCFA concentration and pH, values of which were as low as 5.0 in recently fed animals. The highest pH of 8.0, in a fasted animal, was close to the pH of saliva (8.5). The very low level of SCFA in the hindstomach (3 to 11 mmol L^{-1}) strongly suggested that the SCFA were absorbed directly from the forestomach as in the ruminant. This suggestion was supported by an increase in SCFA levels in portal blood after feeding, from 8 to 26 mg dL^{-1}. Later work by J. M. Barker (1961) confirmed the absorption of SCFA from the quokka stomach.

The proportions in which the individual SCFA are found in the macro-podid forestomach are similar to those in the rumen (Table 6.4). In both herbivores the hindgut contains SCFA at lower total concentrations, and higher molar proportions of acetate and lower proportions of propionate. This is because the substrate entering the hindgut is lower in readily fermentable carbohydrates and higher in structural carbohydrates than is the ingested food . Similar molar proportions of SCFA in the forestomach of the red kangaroo, as well as similar differences between the forestomach and caecum, were reported by Henning & Hird (1970).

Table 6.4. *Concentration, molar proportions and rate of production* in vitro *of short-chain fatty acids (SCFA) in the forestomach and hindgut (caecum plus proximal colon) of red-necked pademelons (*Thylogale thetis*), red-necked wallabies (*Macropus rufogriseus*) and sheep fed chopped lucerne (alfalfa) hay ad libitum*

	Thylogale thetis	Macropus rufogriseus	Ovis aries
Body mass (kg)	5.5	11.3	37.8
Forestomach SCFA:			
Total concentration (mmol L^{-1})	120	129	99
Molar proportion (%)			
Acetic	68.0	65.1	69.5
Propionic	19.0	20.9	18.3
i-Butyric	1.0	1.0	1.5
n-Butyric	9.0	8.2	8.0
i-Valeric	0.9	1.2	1.5
n-Valeric	2.2	3.6	1.2
Production			
mmol L^{-1} h^{-1}	39.4	51.9	22.9
mmol d^{-1}	416	1085	2418
% of DE^a intake	20.5	42.0	29.0
Hindgut SCFA:			
Total concentration (mmol L^{-1})	54	66	50
Molar proportion (%)			
Acetic	71.6	70.3	73.5
Propionic	15.1	16.0	17.3
i-Butyric	1.3	2.0	1.4
n-Butyric	9.0	7.8	4.8
i-Valeric	1.2	2.2	1.7
n-Valeric	1.7	1.7	1.5
Production			
mmol L^{-1} h^{-1}	28.7	26.7	16.0
mmol d^{-1}	38	32	596
% of DE intake	1.9	1.3	6.9

[a] DE = digestible energy
Source: From Hume 1977a.

SCFA concentrations are usually highest in the sacciform forestomach and decrease along the length of the tubiform forestomach. At the same time, molar proportions of acetate tend to increase and those of propionate to decrease caudally from the sacciform forestomach. These patterns, demonstrated in three captive macropodids (Dellow & Hume 1982c) and five free-living species (Dellow *et al.* 1988), can be explained by the progressive disappearance of the more readily fermentable components of the food along the length of the forestomach, and an increase in absorption rate as carbon-chain length of the SCFA increases (Stevens & Hume 1995).

AMMONIA

Ammonia is produced in the herbivore forestomach by microbial degradation of dietary and microbial protein, and of many forms of non-protein nitrogen, including urea of both dietary and endogenous origin. This ammonia can be the major source of nitrogen for bacterial protein synthesis in the forestomach. Lintern-Moore (1973a) demonstrated the incorporation of 64–85% of plant nitrogen into microbial protein in the forestomach of the tammar wallaby by following changes in the concentration of various nitrogenous fractions in the forestomach with time after feeding. Her results are shown in Fig. 6.20. The decline in plant and soluble nitrogen was quite rapid, and there were concomitant increases in both bacterial and protozoal nitrogen. Bacteria constituted 85–94% of total microbial nitrogen. Thus protozoa made only a small contribution to the total microbial biomass. This is explained by the fall in protozoal numbers along the length of the macropodid forestomach (Section 6.2.7), and by the pooling of sacciform and tubiform forestomach digesta by Lintern-Moore (1973a) for analysis.

There was no discernible change in ammonia concentration in Lintern-Moore's (1973a) study. However, the findings of Brown (1969) and Lintern-Moore (1973b) that nitrogen from dietary urea was retained just as efficiently as nitrogen from casein, a readily degradable protein, in euros and tammars respectively, confirms that ammonia must be a key intermediate in microbial protein synthesis in the macropodid forestomach, just as it is in the rumen of sheep.

GASES

Considerable quantities of gas are produced in the forestomach fermentation of macropods. In the quokka the gas consists of 65–75% carbon dioxide, as well as hydrogen and methane (Moir 1968). Kempton, Murray & Leng (1976) could not detect any methane in respired gas collected from eastern grey kangaroos fed chopped lucerne hay, nor in the gas produced from *in vitro* incubation of forestomach contents from other eastern greys fed the same diet. However, Engelhardt *et al.* (1978) found that tammar wallabies fed lucerne hay produced 7–11 mL methane kg^{-1} body mass h^{-1}, equivalent to 1–2% of their digestible energy intake. Although less than production in sheep, in which methane can account for about 10% of digestible energy intake, it is still significant.

The only field data available are those of Dellow *et al.* (1988). They found 5–10% methane in stomach gas collected from red-necked wallabies and swamp wallabies which were actively grazing or browsing when they were shot. Lower concentrations (1–2%) were found in eastern wallaroos

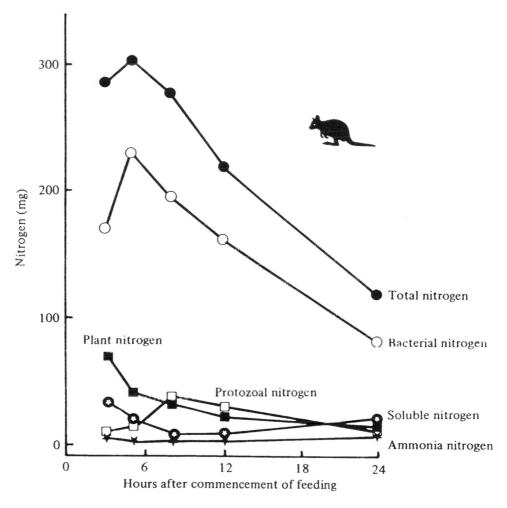

Figure 6.20 Distribution of nitrogen in the forestomach digesta of tammar wallabies (*Macropus eugenii*) at intervals after the commencement of feeding. After Lintern-Moore (1973a).

and eastern grey kangaroos, but these animals were resting. Thus, as would be expected, methane production seemed to be associated with feeding. Hydrogen concentrations were relatively high (1–3%) in the red-necked wallabies, eastern wallaroos and eastern greys, and even higher (10–11%) in the swamp wallabies. In all species oxygen was always less than 0.1%, nitrogen less than 0.2%, and carbon dioxide greater than 70% of sampled gas.

In ruminants, methanogenesis from CO_2 serves as an efficient trap for excess reducing equivalents generated in the forestomach fermentation. Reduction of CO_2 to acetic acid has been found to be an alternative to methanogenesis in the hindgut of humans, rodents and termites (Wolin &

Miller 1994). However, M. J. Wolin & T. L. Miller (pers. comm. 1997) found that microbes that reduce CO_2 to either methane or acetate were not major contributors to the forestomach fermentation in red kangaroos, western grey kangaroos or euros. When present, they probably produce acetate and not methane.

Why methanogens are present in such low numbers is not clear, but their virtual absence from the tubiform forestomach may be related to the short retention time of the fluid phase of digesta in the macropodid forestomach and the slow growth rates of methanogenic archaebacteria. This may limit them to the sacciform forestomach.

6.4.2 Patterns of fermentation and digestion

IN VITRO INCUBATION STUDIES

Rates of SCFA production measured *in vitro* appear to be more rapid in macropods than in the sheep, both in the forestomach and the hindgut (Table 6.4). As a proportion of the animal's intake of digestible energy (i.e. energy absorbed), the contribution from forestomach SCFA in the sheep ($29 \pm 3\%$) in Hume's (1977a) study fell in between the values for the red-necked pademelon ($21 \pm 3\%$) and the red-necked wallaby ($42 \pm 5\%$). The hindgut contribution was smaller in all three species, but greater in the sheep ($7 \pm 1\%$) than in the wallabies (1–2%) because of the relatively larger caecum and proximal colon of the sheep.

The faster rate of fermentation in the macropodid forestomach is associated with the faster digesta flow in that organ (a modified plug-flow reactor) compared with the ovine rumen (a stirred-tank reactor) (see Chapter 4), and hence faster turnover of the microbial population. Presumably faster turnover of the microbes maintains a higher concentration of microbial enzymes in the forestomach contents. In the case of Hume's (1977a) study, it is also associated with the smaller body sizes of the wallabies (6 and 11 kg) compared with the sheep (38 kg). Among ruminants, rapid fermentation rates are associated with small body size (Hoppe 1977) and high rumen turnover rates (Hungate *et al.* 1959) and their selection of more fermentable food. The same probably holds for macropods; *in vitro* SCFA production in the forestomach of eastern grey kangaroos (mean body mass 19 kg) fed chopped lucerne hay was 34 mmol L^{-1} h^{-1} (Hume & Dellow 1980), slower than in the smaller wallabies. However, because of the larger absolute capacity of the eastern grey's forestomach, SCFA produced there contributed 28% of digestible energy intake, a similar value to that of Hume's (1977a) sheep on the same diet.

The fastest rates of SCFA production recorded in macropodids have been in red-necked pademelons shot while grazing on improved pasture adjacent to their rainforest refuge. On this occasion sacciform and tubiform forestomach digesta were incubated separately. In two animals, SCFA production was 102 and 99 mmol L^{-1} h^{-1} in the sacciform forestomach, and 60 and 65 mmol L^{-1} h^{-1} in the tubiform forestomach (Hume 1982). No doubt the higher concentration of rapidly fermentable carbohydrate in the fresh grass pasture compared with the lucerne hay explains much of the difference in fermentation rates between laboratory and field.

The limitations of *in vitro* estimations of fermentation rate were discussed in Chapter 4. There is no doubt that *in vitro* procedures underestimate SCFA production, because of the time lag between removal of contents from the animal and commencement of the incubation, during which many soluble constituents of the digesta may disappear. Thus Faichney (1968), using an *in vitro* technique, estimated that SCFA production in the reticulo-rumen of sheep fed chopped lucerne hay accounted for 34% of digestible energy intake. In contrast, estimates made *in vivo* by an isotope dilution technique in sheep fed similar diets yielded values ranging from 53% (Leng, Corbett & Brett 1968) to 62% (Bergman *et al.* 1965). Nevertheless, there is little reason to doubt the validity of comparisons made *in vitro* between species or between different regions of the gut, provided the animals are receiving similar diets.

MEASUREMENT OF SCFA PRODUCTION *IN VIVO*

Isotope dilution techniques can only be used satisfactorily when it can be assumed that mixing of the injected or infused isotopically labelled SCFA with a single large pool of SCFA is instantaneous. This requirement is only met in ideal continuous-flow, stirred-tank reactors (CSTRs). It is closely approached in the reticulo-rumen, which is best modelled as a stirred-tank reactor. It is not applicable to the macropodid forestomach because of its 'colon-like' morphology, in which digesta flow is best modelled as a modified plug-flow reactor (a series of small stirred-tank reactors). Hence most estimates of fermentation rate in macropodids are based on *in vitro* incubation of contents.

The problem of incomplete mixing of infused isotope due to the presence of more than one mixing pool of digesta in the macropodid stomach was successfully solved by Dellow, Nolan & Hume (1983) by treating the forestomach of pademelons and tammars as a series of four mixing pools; one in the sacciform region and three in the tubiform region, with no retrograde flow to a preceding pool (i.e. a modified plug-flow reactor). The ^{14}C-acetate was infused for 48 hours into the sacciform forestomach, the primary pool. Rate of infusion into each of the other pools was then the rate of flow of labelled acetate from the preceding pool. Flow of acetate was estimated with reference to the fluid marker ^{51}Cr-EDTA, which was infused into the sacciform region along with the labelled acetate. Total net production of acetate was then equal to the sum of net production in each pool. From the molar proportions of the individual acids in the digesta, net production of total SCFA was calculated (Fig. 6.21).

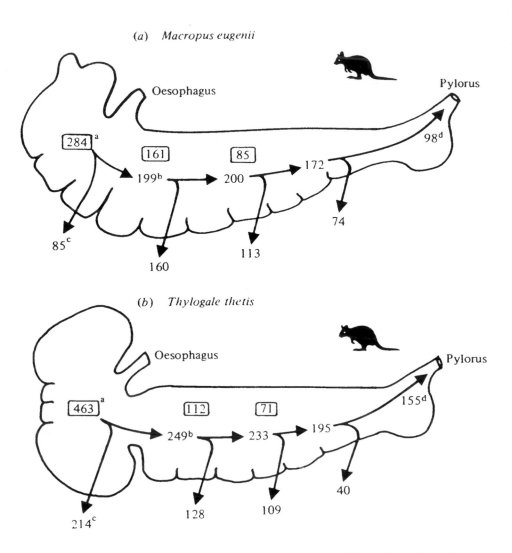

Figure 6.21 Diagram showing the production, absorption and flow of short-chain fatty acids (SCFA) (mmol d^{-1}) in the stomach of (a) tammar wallabies (*Macropus eugenii*) and (b) red-necked pademelons (*Thylogale thetis*). [a] (in boxes) net production; [b] flow along the stomach; [c] net absorption; [d] flow through the hindstomach. From the *in vivo* experiments of Dellow, Nolan & Hume (1983).

In line with data obtained *in vitro*, net SCFA production and absorption was highest in the sacciform region and decreased along the length of the forestomach of both wallabies. The ratio of SCFA produced (mmol d^{-1}) to digestible organic matter intake (g per day) was similar in the two species (9–10 mmol g^{-1}). Both are substantially higher than *in vitro* estimates (2–3), indicating again that *in vitro* procedures underestimate true production. The *in vivo* results suggest that microbial digestion in the macropodid forestomach is as efficient as it is in the reticulo-rumen; for instance,

Czerkawski (1978) reported a ratio of SCFA produced to digestible organic matter intake of 8.3 mmol g^{-1}. The lower fibre digestibility often reported in macropods compared with sheep must therefore be a function of the faster passage through the forestomach described above in Section 6.3.

MICROBIAL PROTEIN SYNTHESIS

We have already seen that ammonia is a key intermediate in the synthesis of microbial protein in both the ruminant and macropodid forestomachs. By infusing ^{15}N-ammonium chloride into the sacciform forestomach along with the ^{14}C-acetate and ^{51}Cr-EDTA, Dellow et al. (1983) estimated net microbial protein synthesis in both the red-necked pademelon and tammar wallaby concurrently with their measurements of SCFA production. After 48 hours of infusion, considered to be more than sufficient for steady-state labelling of microbial protein with ^{15}N, the animals were killed and the flow of bacterial nitrogen that arose from ammonia and passed through the hindstomach was calculated from the relationship:

$$\text{Bacterial N flow (g d}^{-1}) = \text{non-ammonia N flow (g d}^{-1}) \times \frac{\text{non-ammonia N enrichment in HS}}{\text{bacterial N enrichment in TFS}}$$

where HS = hindstomach and TFS = tubiform forestomach.

This procedure assumes that enrichment of bacterial nitrogen with ^{15}N does not change significantly between the central region of the tubiform forestomach and the hindstomach.

In the sacciform forestomach, 44% of bacterial nitrogen was derived from ammonia in pademelons and 40% in tammars. These values rose to 74% and 84% in the tubiform forestomach of the two species respectively. The lower values in the sacciform region are indicative of more direct incorporation of peptides and/or amino acids of either dietary or en-dogenous origin, or both.

Dellow et al.'s (1983) estimates of net microbial protein synthesis (25–27 g N kg^{-1} organic matter apparently fermented) in the forestomach of the two wallabies are within the range of values from ruminants (Czerkawski 1978). Thus, as we have seen with SCFA production, microbial protein production in the macropodid forestomach is at least as efficient as in the ruminant forestomach.

PATTERNS OF DIGESTION

The patterns of decreasing fermentation rates and changing molar propor-tions of acetate and propionate along the length of the macropodid stomach are reflected in the patterns of disappearance of various dietary com-ponents (Fig. 6.22). Note particularly the very rapid disappearance of total soluble sugars, with 95% digested in the sacciform and cranial tubiform

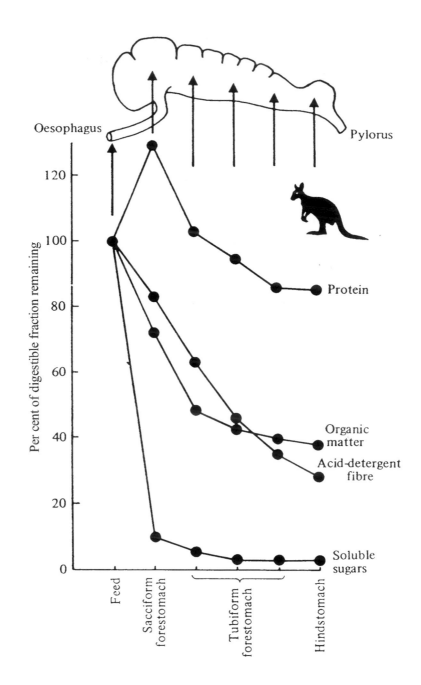

Figure 6.22 The disappearance of the digestible components of chopped lucerne (alfalfa) hay along the stomach of the eastern grey kangaroo (*Macropus giganteus*). Note the rapid disappearance of soluble sugars compared with fibre. The increase in crude protein in the sacciform forestomach is due to the incorporation of ammonia from endogenous sources of nitrogen into microbial protein. From Dellow & Hume (1982c).

Table 6.5. *Intake and digestion of two forage diets and their components by red kangaroos* (Macropus rufus), *euros* (M. robustus erubescens) *and sheep*

	Red kangaroo	Euro	Sheep
Body mass (kg)	27–33	23–31	33–39
Dry matter intake (g kg$^{-0.75}$ d^{-1})			
Lucerne hay	53	53	92
Wheat straw	18	35	49
Apparent digestibility of dry matter (%)			
Lucerne hay	55	62	61
Wheat straw	36	38	44
Digestibility of non-structural carbohydrates (%)			
Lucerne hay	88	88	87
Wheat straw	86	87	88
Digestibility of cell-wall constituents (%)			
Lucerne hay	35	44	44
Wheat straw	32	32	40
Digestibility of acid-detergent fibre (%)			
Lucerne hay	36	46	44
Wheat straw	33	28	39

Source: From Hume 1974.

forestomach regions. Organic matter digestion is also more rapid in the sacciform region, whereas acid-detergent fibre (ADF), which consists of cellulose and lignin, was digested more slowly along the length of the forestomach. The initial rise in crude protein is because of recycling of endogenous nitrogen, probably mainly via saliva, to the forestomach and its incorporation into microbial protein. The remainder of the curve for crude protein reflects the balance between continued incorporation of endogenous nitrogen into microbial protein and digestion of this and dietary protein.

6.5 FOOD INTAKE AND DIGESTION

There are two principal nutritional consequences of the tubular nature of digesta flow through the macropodid stomach and the resultant shorter whole-tract mean retention times of digesta. The first, alluded to above, is that digestibility is often lower in kangaroos than in ruminants of equivalent size. This is especially so for the structural polysaccharides of plant cell walls (Table 6.5).

The second consequence is that kangaroos are much less affected by increasing dietary fibre levels than are sheep. This is demonstrated in Fig. 6.23. The solid curve in this figure is based upon a compilation of studies of

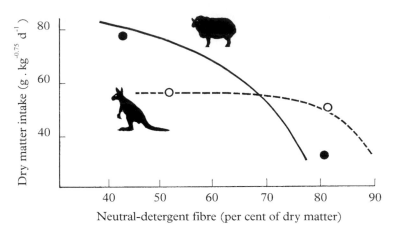

Figure 6.23 Relationship between dry matter intake by domestic ruminants and eastern wallaroos (*Macropus robustus robustus*) and neutral-detergent fibre (cell wall) content of forage. Ruminant line from Van Soest (1965). Sheep (dark) and wallaroo (light) data from Hollis (1984).

sheep and cattle fed a range of forages, both grass and legume, in North America, by Van Soest (1965). This curve shows that, above a certain fibre level, food intake by ruminants declines as a result of increasing levels of ruminal fill as forage fibre content increases. This is because particles have to be reduced to a certain critical size (approximately 1 mm for sheep and 1.3 mm for cattle) before they have a high probability of leaving the reticulo-rumen (Faichney 1993). The higher the forage fibre content the more slowly are the particles degraded, the higher the level of rumen fill, and the more food intake is inhibited.

However, this model for food intake control does not apply to kangaroos, because particle flow out of the macropodid forestomach is relatively unaffected by particle size. This is partly because of the modified plug-flow reactor characteristics of the kangaroo stomach versus the stirred-tank reactor characteristics of the ruminant forestomach, but also because the particle sorting characteristics present in the latter (Faichney 1993) are not present in the kangaroo. Fig. 6.23 includes data points for sheep and eastern wallaroos (*Macropus robustus robustus*) fed either chopped lucerne hay of relatively low fibre content or chopped oaten straw of high fibre content (Hollis 1984). The two data points for the sheep fell close to the general ruminant curve of Van Soest (1965). On the low-fibre forage the wallaroos ate about 30% less than did the sheep, a reflection of their lower maintenance energy requirements (see Sections 1.5 and 1.6). However, on the high-fibre forage, dry matter intake fell by only 17%, in contrast to the 57% fall in the sheep. Consequently, the wallaroos ate more dry matter than the sheep on the high-fibre diet. Although the broken line in Fig. 6.23 is drawn through only two data points, the model of food intake control in

Table 6.6. *Intake and digestion of chopped lucerne (alfalfa) hay and chopped oaten straw by young red kangaroos* (Macropus rufus) *and sheep*

	Kangaroo	Sheep	Kangaroo Sheep
Body mass (kg)	11.4	38.8	
Dry matter intake (g kg$^{-0.75}$ d^{-1})			
Lucerne hay	58	72	0.81
Oaten straw	40	29	1.40
Apparent digestibility of dry matter(%)			
Lucerne hay	54	63	0.86
Oaten straw	36	39	0.91
Digestible dry matter intake (g kg$^{-0.75}$ d^{-1})			
Lucerne hay	31	45	0.70
Oaten straw	15	11	1.27

Source. From Foot & Romberg 1965.

kangaroos described by this line is consistent with the colon-like morphology of the kangaroo forestomach.

Supporting evidence for the kangaroo model comes from two studies. The first is that of Foot & Romberg (1965). When young red kangaroos (*Macropus rufus*) and sheep were fed diets similar to those used by Hollis (1984), the macropodids ate 19% less of the low-fibre forage than did the sheep (Table 6.6). Dry matter intake by the kangaroos was 31% lower on the high-fibre diet, in contrast to the 60% lower intake by the sheep. Consequently, as in the Hollis (1984) study, the kangaroos ate more dry matter than the sheep on the high-fibre diet. Dry matter digestibility was lower in the kangaroos on both diets, so in terms of digestible dry matter intake the differences between herbivores was a little less.

The second study is that of Freudenberger & Hume (1992). By offering eastern wallaroos, euros (*M. r. erubescens*) and goats isonitrogenous diets containing 40%, 60% or 80% milled barley straw, they found that the macropodids compensated for decreasing dry matter digestibility by actually increasing intakes of the higher-fibre diets. In contrast, the goats were unable to fully compensate for falling digestibility. Consequently, intakes of digestible dry matter by the goats fell by 20% while those of the macropodids fell by only 11%. Freudenberger & Hume (1992) were able to conclude that macropodids can maintain relatively greater intakes of increasingly fibrous diets if the constraint of mastication is removed by milling and/or pelleting the food on offer.

The advantage that the large kangaroos have over ruminants of similar size in utilising high-fibre forages is shared by colon fermenters, including the marsupial wombats (Chapter 4) and equids (horses, zebras and donkeys) (Hume 1989). This is understandable on the basis of the similar tubiform morphology of their main fermentation organ, the proximal colon, and the performance characteristics of modified PFRs. The advantage is not shared by the smaller wallabies or the rat-kangaroos. Smaller body sizes and/or more sacciform forestomachs mean that high-fibre forages cannot be processed fast enough by these marsupials. Thus Hollis (1984) found that food intake by 6 kg tammar wallabies fell by 50% (almost as much as in sheep) when changed from the low-fibre diet to the high-fibre diet in her study.

The ecological consequences of the differences in digestive strategy between large kangaroos and ruminants are explored in Chapter 7.

6.6 SCFA AND CARBOHYDRATE METABOLISM

There is considerable conversion of acetic acid to butyric acid in the forestomach of both macropods and ruminants. Dellow's (1979) estimate of 24–41% in the stomach of *Thylogale thetis* and *Macropus eugenii* is similar to some estimates in sheep (Leng & Leonard 1965), although conversion of 61% of acetic acid to butyrate was reported by Bergman *et al.* (1965).

The wall of the fermentation chamber uses SCFA as its energy source and in the process modifies the proportions of SCFA entering the portal blood system. Henning & Hird (1970) showed that the forestomach mucosa of *Macropus rufus* and *M. giganteus* metabolises butyric acid to a much greater extent than it does acetate or propionate. Most of the butyrate is converted into ketone bodies, principally acetoacetate. In this the macropodid forestomach wall is similar to that of the ruminant forestomach and the guinea pig caecum. The partial oxidation of butyrate could satisfy the energy needs of the epithelial tissue and at the same time provide a substrate, ketone bodies, to other tissues for further oxidation. Another consequence of ketogenesis is the regeneration of coenzyme A, a shortage of which could limit the rate of oxidation in cells heavily loaded with fatty acids (Henning & Hird 1970).

Henning & Hird's (1970) study suggests that in the macropodid forestomach ketogenic activity is restricted almost entirely to cardiac glandular mucosa, with negligible activity in the squamous epithelium. This contrasts with the ruminant forestomach, which is lined entirely with squamous epithelium, but which exhibits ample ketogenic activity throughout. Differences in absorptive and metabolic activities between the different epithelial types found in the macropodid stomach deserve further investigation.

Carbohydrate metabolism in macropods appears to be similar to that in other foregut fermenters, including the ruminants and the camelids. For instance, levels of activity of various disaccharidases in homogenates of small intestinal mucosa from *Macropus giganteus* (Table 2.3) are low, indicating that little digestible carbohydrate normally reaches the small intestine. This agrees with the rapid disappearance of total soluble sugars in the cranial forestomach in Fig. 6.22. It also complements the observations of Moir *et al.* (1956) that, like ruminants, quokkas have lower blood glucose levels than do simple-stomached mammals; of Ballard, Hanson & Kronfeld (1969) that the rate of glucose incorporation into glycogen in the liver of quokkas is less than 4% of the rate in rat liver (Table 6.7); and of J. M. Barker (1961) that quokkas apparently release glucose into the blood at all times, as do sheep. These observations are in contrast to those in the fed dog, which show that glucose is taken up by the liver (Fig. 6.24).

That the livers of quokkas and sheep do not take up glucose from the blood is further reflected in the absence of glucokinase activity (Table 6.7). In the rat liver, hexokinase is saturated with respect to substrate at low glucose concentration, while glucokinase has a maximum activity at much higher glucose concentrations. During long-term fasting of rats, glucokinase activity is lost, while hexokinase activity remains unchanged (Ballard *et al.* 1969). Since glucokinase is both the adaptive and the more active glucose phosphotransferase in rat liver, macropods and other foregut fermenters have apparently adapted to the low glucose absorption from the gut by the loss of a major part of the glucose phosphorylating activity.

Table 6.7. *Hexokinase and glucokinase activities, and glucose incorporation into glycogen in liver slices*

	Hexokinase (units[a] g^{-1})	Glucokinase (units g^{-1})	Glucose incorporation (μmol g^{-1} h^{-1})
Omnivores			
Rat	0.66	3.83	25.4
Mouse	0.70	3.08	7.4
Pig	0.64	3.17	5.8
Herbivores – hindgut fermenters			
Guinea pig	0.39	1.25	6.5
Rabbit	0.28	1.32	11.8
Herbivores – foregut fermenters			
Sheep	0.13	< 0.03	0.3
Cow	0.39	< 0.03	0.2
Quokka (*Setonix brachyurus*)	0.39	< 0.03	1.0
Tammar (*Macropus eugenii*)	0.21	< 0.03	0.6

Note: [a] One unit of enzyme activity hydrolyses 1 μmol substrate min^{-1}.
Source: After Ballard, Hanson & Kronfeld 1969.

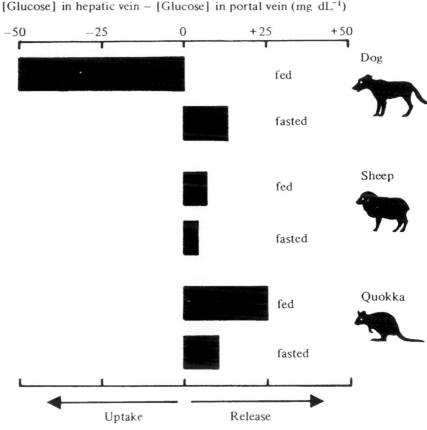

[Glucose] in hepatic vein – [Glucose] in portal vein (mg dL^{-1})

Figure 6.24 Differences between glucose concentration in the hepatic vein and the portal vein in fed and fasted dogs, sheep and quokkas (*Setonix brachyurus*). Both the sheep and quokka release glucose synthesised by gluconeogenesis in the liver into the blood at all times. After Ballard, Hanson & Kronfeld (1969).

Quokkas (J. M. Barker 1961) and red kangaroos (Griffiths, McIntosh & Leckie 1969) show considerable tolerance to large intravenous injections of insulin, even though their blood glucose levels are substantially reduced (Table 6.8). They are much less tolerant of hyperglycaemia; red kangaroos made diabetic by destroying the Islets of Langerhans in the endocrine pancreas with injections of alloxan, develop a marked hyperglycaemia and die unless treated with insulin (Griffiths *et al.* 1969). This is similar to the responses of sheep, but contrasts with the much greater tolerance of simple-stomached mammals, both marsupial and eutherian, to hyperglycaemia (Table 6.8).

J. M. Barker (1961) demonstrated that propionate is a gluconeogenic precursor in macropods. She found no increment in blood glucose level following intravenous acetate injections, but a noticeable rise when

Table 6.8. *Responses to changes in blood glucose levels in marsupial and eutherian herbivores*

	Normal plasma glucose (mg dL^{-1})	Response to	
		Hypoglycaemia (Insulin)	Hyperglycaemia (Alloxan)
Foregut fermenters			
Quokka	78	Coma at 20	–
(*Setonix brachyurus*)			
Red kangaroo	73–85	No effect at 50	Coma at 300
(*Macropus rufus*)			
Sheep			
(*Ovis aries*)	65	No effect	Coma
Hindgut fermenters			
Common brushtail possum	120	Coma at 40	No effect below 400
(*Trichosurus vulpecula*)			
Rabbit	120	Coma at 50	No effect
(*Oryctolagus cuniculus*)			

Source: After Tyndale-Biscoe 1973.

propionate was injected. This is similar to the responses of both sheep and rabbits, and reflects the marked glucogenicity of propionate.

Freudenberger & Nolan (1993a) measured glucose entry rate (an estimate of gluconeogenic activity in foregut fermenters) in euros (*Macropus robustus erubescens*) and goats. Gluconeogenic substrates were not identified, but would have included propionate along with glucogenic amino acids. Glucose entry rates were remarkably similar in the two herbivores (2.7 and 2.8 mg min^{-1} kg$^{0.75}$), despite the greater energy intakes of the goats; in ruminants glucose entry rate is positively correlated with intake of digestible energy.

6.7 LIPID METABOLISM

The depot fats of *Macropus rufus* and *M. giganteus* contain *trans* acids (Hartman, Shorland & McDonald 1955), an indication that microbial modification of food constituents in the macropodid forestomach is significant. In ruminants, depot fats are highly saturated, the degree of saturation being altered significantly only if unsaturated dietary lipid is protected from microbial hydrogenation in the forestomach in some way (Cook *et al.* 1970). However, the depot fat of *M. giganteus*, like that of the horse (a hindgut fermenter), is less saturated than that of ruminants (Redgrave & Vickery 1973), indicating that although microbial modification of dietary

Table 6.9. *Lipid content and fatty acid composition of muscle from ruminants and marsupials*

| Species | Lipid content % | % total fatty acids[a] | | | |
		Monosaturated	Polyunsaturated	unsaturated
Ruminant (lean)				
Sheep (*Ovis aries*)	3.4	44	46	10
Cow (*Bos taurus*)	1.9	46	44	10
Marsupial (foregut fermenters)				
Red kangaroo (*Macropus rufus*)	–	35	32	33
Eastern grey kangaroo (*M. giganteus*)	–	32	29	39
Antilopine wallaroo (*M. antilopinus*)	1.5	30	27	43
Eastern wallaroo (*M. robustus robustus*)	–	28	24	48
Black-footed rock-wallaby (*Petrogale lateralis*)	2.0	31	38	31
Swamp wallaby (*Wallabia bicolor*)	–	25	15	60
Long-nosed potoroo (*Potorous tridactylus*)	–	31	26	43
Marsupial (hindgut fermenters)				
Common wombat (*Vombatus ursinus*)	–	37	13	50
Koala (*Phascolarctos cinereus*)	0.8	31	12	57
Common brushtail possum (*Trichosurus vulpecula*)	1.1	36	15	49
Common ringtail possum (*Pseudocheirus peregrinus*)	–	30	35	35

Note: [a]Saturated fatty acids are mainly 14:0, 16:0 and 18:0; monounsaturated fatty acids are mainly 16:1 and 18:1; polyunsaturated fatty acids are mainly 18:2, 18:3, 20:3, 20:4, 20:5, 22:4, 22:5 and 22:6.

Source: Compiled from Naughton, O'Dea & Sinclair 1986 and Mann *et al.* 1995.

fats is substantial, it is less than in ruminants. Intramuscular fat is also less saturated in macropods than in ruminants (Table 6.9). These differences are consistent with the shorter retention times of food constituents in the macropodid forestomach noted in Section 6.3. Variations among macropod species presumably reflect differences in diet (see Section 7.3.2). For instance, the high polysaturated content of the lipids of the swamp wallaby, the only browsing macropod listed, is more similar to that of the koala, also a browser, than to any of the other macropods.

6.8 NITROGEN METABOLISM AND UREA RECYCLING

Ammonia is a key intermediate in microbial protein synthesis in the forestomach of herbivores. About 63% of bacterial nitrogen in the tubiform forestomach of *Macropus eugenii* is derived from ammonia (Kennedy & Hume 1978). This is within the range reported in the ruminant forestomach (50–80%). On low-protein diets a large part of this ammonia may be derived from endogenous urea recycled to the gut. When Lintern (1970) injected ^{15}N-urea intravenously into tammar wallabies fed either low (0.34%) – or high (2.60%) – nitrogen diets, much less of the injected dose was excreted in the urine on the low-nitrogen diet (Table 6.10). Instead, at slaughter, more labelled nitrogen was found in the forestomach and the caecum. Incorporation of labelled nitrogen into the bacterial fraction of forestomach digesta was also higher in the tammars fed the low-protein diet. Incorporation into the protozoal fraction was insignificant, in line with Dellow *et al.*'s (1988) observations that protozoa are only found in high numbers in the sacciform forestomach.

Using intravenous infusions of urea labelled with either ^{15}N or ^{14}C, Kennedy & Hume (1978) also showed that urea was transferred from the blood to the gut of tammars. The proportion of urea synthesised in the liver which was transferred to the gut was similar and high (74–86%) on both a high-nitrogen chopped lucerne hay and a low-nitrogen chopped oaten hay diet. However, incorporation of nitrogen from endogenous urea into microbial protein in the gut was equivalent to only 34–53% of nitrogen intake on the high-protein diet, but to 103–112% of nitrogen intake on the low-protein diet. This latter result indicates that urea recycling in macropods may be of sufficient magnitude to sustain microbial function in the gut during periods of nitrogen shortage, in the same way as has been demonstrated in ruminants (Kennedy & Milligan 1980).

That the macropodid kidney concentrates urine on low-nitrogen diets was simply demonstrated by Lintern & Barker (1969). They fed two groups of tammars either a high (1.2%) or low (0.4%) nitrogen diet for 28 days. In the high-nitrogen group, which remained in positive nitrogen balance throughout, plasma urea concentration remained unchanged, and the ratio of concentrations of urea-nitrogen to total nitrogen in the urine, the UR

Table 6.10. *Fate of ^{15}N-urea injected intravenously into tammar wallabies* (Macropus eugenii) *fed either low or high nitrogen diets*

Animal	Diet	Urea (mg) injected	Injected urea (mg) appearing after 3 h in		
			Urine	Forestomach	Caecum
1	Low	463	2	62	1.5
2	nitrogen	471	2	63	1.9
3	(0.34%)	604	2	59	1.6
4	High	471	214	16	0.6
5	nitrogen (2.60%)	532	245	19	0.6

Source: After Lintern 1970

ratio (Kinnear & Main 1975), was consistently 0.85. The low-nitrogen group dropped into negative nitrogen balance, plasma urea concentrations fell to half their initial value, and the UR ratio fell to about 0.35, indicating that proportionately more urea was recycled to the digestive tract.

Measurements of rates of whole-body protein turnover in marsupials have been based on the fraction of ^{15}N injected as ^{15}N-glycine which is recovered in urea in the urine over a certain period. White, Hume & Nolan (1988) found that ^{15}N enrichment of urinary urea increased rapidly after injection to reach a peak value 4–12 hours after intramuscular injection (Fig. 6.25). Thereafter there were at least two clearly discernible decay components. The fastest component probably represents rapidly turning over nitrogen pools such as hepatic and digestive tract enzymes. The slowest component represents muscle protein turnover, together with re-cycling of the label through urea synthesis in the liver, its recycling to the gut and incorporation into microbial protein, and subsequent return to the liver after digestion of labelled microbial protein. The recycling component leads to an underestimate of protein turnover rate. Thus selection of the most appropriate time period over which to collect urinary urea is problematic. White *et al.* (1988) considered 45 hours to be long enough to allow all the injected ^{15}N (except that recycled from body protein) to have cleared from the metabolic pool, yet short enough to minimise the amount of recycled ^{15}N appearing in urinary urea.

The low protein turnover rate in the tammar wallaby (Table 6.11) is consistent with this species' low maintenance nitrogen requirement (Table 1.8). The higher protein turnover rates and higher maintenance nitrogen requirements of *Thylogale thetis* and *Macropus parma* are probably explained in terms of their more mesic distributions.

Based on a 43-hour urinary urea collection, Freudenberger & Nolan (1993b) found a rate of whole-body protein turnover in the eastern wal-

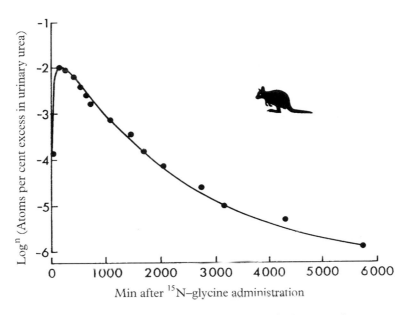

Figure 6.25 Build-up and decay curve of enrichment of urinary urea in tammar wallabies (*Macropus eugenii*) after intramuscular injection of ^{15}N–glycine. From White, Hume & Nolan (1988).

Table 6.11. *Protein turnover, synthesis and catabolism in five macropodid marsupials and the goat. All values in g crude protein per $kg^{-0.75}$ d^{-1}. (Means $\pm$ standard errors)*

Species	Protein turnover	Protein synthesis	Protein catabolism
Red-necked pademelon (*Thylogale thetis*)	10.4 ± 0.5	5.4 ± 0.8	4.7 ± 0.7
Parma wallaby (*Macropus parma*)	10.8 ± 0.6	5.6 ± 0.6	4.3 ± 0.6
Tammar wallaby (*Macropus eugenii*)	7.3 ± 0.8	5.1 ± 0.7	3.7 ± 0.7
Eastern wallaroo (*M. robustus robustus*)	7.5 ± 1.2	2.9 ± 1.2	0.8 ± 1.4
Euro (*M. robustus erubescens*)	10.2 ± 0.1	6.1 ± 0.1	4.3 ± 0.3
Goat (*Capra hircus*)	10.3 ± 0.3	5.2 ± 0.7	3.0 ± 0.5

Source: From White, Hume & Nolan 1988 and Freudenberger & Nolan 1993b.

laroo (*Macropus robustus robustus*) that was similar to that of the tammar. The rate in the other subspecies, the euro (*M. r. erubescens*), was higher and similar to rates in *Thylogale thetis* and *M. parma*, and in the goat. These differences do not correlate with differences in maintenance nitrogen requirements, and it is clear that more work is needed to refine and standardise the measurement of whole-body protein turnover in marsupials.

6.9 **CONCLUSIONS**

The family Macropodidae (kangaroos and wallabies) is the most extensively studied group of marsupials with respect to their digestive physiology and metabolism. Comparisons with ruminants have unfortunately masked some of the significant differences between these two types of foregut fermenters. The basic difference is the type of fermentation system; while the ruminant forestomach can best be modelled as a continuous-flow, stirred-tank reactor (CSTR), the macropodid forestomach functions as a series of small CSTRs or, in other words, a modified plug-flow reactor. Although the fermentation pathways in the two systems are no doubt similar, patterns of digesta flow, and thus the composition of the populations of fermentative microorganisms and their end-products differ, as do patterns of digestion and absorption.

Within the Macropodidae there are differences in the relative capacities of the sacciform and tubiform regions of the forestomach; the sacciform region tends to be relatively larger in smaller species. However, there are exceptions, notably the hare-wallabies which, at 1–2 kg, have a basically tubiform forestomach (Fig. 6.12), as we see in the large kangaroos. Not enough is known about the implications of these gastric variations in terms of the intake and digestion of natural forages. This is an area that needs further research. Relationships among dentition, gastric morphology and natural diets of kangaroos and wallabies are explored in Chapter 7.

7 Nutritional ecology of kangaroos and wallabies

This chapter provides an opportunity to discuss some aspects of the ecology of kangaroos and wallabies in relation to their nutrient requirements, principally energy, protein and water (Chapter 1), and their digestive and metabolic adaptations to herbivorous diets (Chapter 6). The classification of macropodid diets and the relationship between dentition and diets of macropods were briefly introduced in Chapter 6.

7.1 CONCEPTS – DENTITION AND DIET

Teeth are involved in the mechanical preparation of food in a number of ways, by cutting, shearing, crushing and grinding. Cutting is a process in which a sharp edge is pressed against another surface (Rensberger 1973). Shearing is a type of cutting in which two sharp edges appress food, narrowly missing each other. In crushing, blunt surfaces oppose each other with no translational motion. Grinding is crushing with a translational force (Sanson 1989).

Shearing is effective in breaking ductile or elastic materials which are crack resistant. Grasses, with high levels of fibre, are composite materials of cellulose microfibrils running through a hemicellulose matrix. The cellulose microfibrils are oriented in a fashion that provides for maximum strength. As such they are quite resistant to breakage. Consequently, they require high shearing forces to break them down into small particles. Also, the vascular bundles of grass leaves are evenly distributed and relatively narrow. Separation into smaller particles is more economically achieved by shearing rather than crushing or grinding. Therefore grinding surfaces are not useful to grazers, which means that more small ridges, or cutting edges, can be packed into a given tooth to emphasise the shearing action.

In contrast to grasses, which are reasonably homogeneous structurally, browse is a heterogeneous material with different thicknesses and levels of resistance to breakage. Consequently, mastication of browse requires teeth to perform several different actions. Stems are often highly lignified and so require large or coarse shearing blades to break them. Leaves contain vascular bundles that may be secondarily thickened or even lignified, and usually are concentrated into a reticulated system of veins. Consequently, leaves require fine shearing. A grinding action is also advantageous, as it catches tissue, including turgid mesophyll cells, between tooth faces, causing them to burst, exposing both the contents and walls to enzyme attack.

Tooth wear differs between grazers and browsers. Teeth adapted to masticating grass must cope with two problems. The first is the mechanical treatment of the plant tissue. This is achieved by increasing the sharpness of edges, and reducing the number of edges in contact. Both of these features maximise shearing force. The second problem is tooth wear. Tooth wear is generated by contact of occluding teeth (usually the molars), and by abrasive materials in the diet. Grasses tend to have high levels of abrasive silica. Silica is abrasive because it is harder that tooth enamel, and consequently wears down the enamel edges. Wear may initially increase the masticatory capacity of the tooth because it erodes the dentine faster than enamel, leaving enamel ridges exposed. Eventually however, these enamel ridges wear to such an extent that surface area of contact increases (the tooth becomes blunt) and shearing forces are reduced. This has been demonstrated by McArthur & Sanson (1988) in the eastern grey kangaroo (*Macropus giganteus*), in much the same way as Lanyon & Sanson (1986b) showed in koalas (see Chapter 5).

One response to the high abrasive levels in grasses is an increase in hardness of the enamel. There is some evidence that the enamel of *Macropus giganteus*, a grazer, is harder than that of the swamp wallaby (*Wallabia bicolor*), a browser (Palamara *et al*. 1984). Another response seen in grazing macropods is molar progression, the sequential anterior movement of molars through the dental mill, after which they are lost. This is not a device to counteract wear in grazers, because a grazer still wears out only four molars in its life, the same as a browser. The grazer simply wears them out two at a time in a curved tooth row rather than all at once in a flat tooth row (Sanson 1980). The curved tooth row of grazers means that shearing forces are concentrated on a smaller area of occlusion, which maximises masticatory effectiveness. Only *Peradorcus* (nabarlek or little rock-wallaby) can use molar progression as an anti-wear device, and that is because it has extra molars (Sanson 1989).

7.2 DENTITION OF MACROPODID MARSUPIALS

Although it is convenient to divide macropodid dietary types into a browsing grade, a grazing grade and an intermediate grade of browser/grazers (Sanson 1978, 1980), the situation is in reality more complex. When the information now available is examined (see below), it is clear that there is a spectrum of dietary preferences ranging from strict browsing to almost exclusive grazing. Nevertheless, the recognition of grades is still useful in organising the variety of dental morphologies and dietary preferences for description (Sanson 1989). The living genera of the Macropodidae can be assigned to the three grades identified on their character states (Table 7.1)

Table 7.1. *Organisation of extant genera of the Macropodidae into three grades on the basis of dental morphology and diet; (a) – (d) indicate increasing levels of grass in the diet within a grade*

Grade	Diet	Genera
Browser	Low-fibre browse	(a) *Dorcopsis, Dorcopsulus, Dendrolagus, Thylogale, Setonix, Wallabia*
		(b) *Petrogale*
Intermediate browser/grazer	Mixed browse and grass	(a) *Petrogale xanthopus*
		(b) *Peradorcus*
		(c) *Macropus, Lagorchestes*
		(d) *Lagostrophus*
Grazer	Grass	*Macropus, Onychogalea*

Source: Modified from Sanson 1989.

Teeth are conventionally divided into incisors, canines, premolars and molars. Canines are rarely found in the Macropodidae, and are not important in food preparation. Incisors are involved in food prehension and ingestion. Molars are involved in mastication. Premolars aid both the incisors and the molars in their respective functions. The two basic types of macropodid molar morphology are shown in Fig. 7.1, first showing the initial contact between opposing lophs, and again at maximum interdigitation. The weak longitudinal ridges (links) between the transverse lophs of Type B molars result in a crushing action; there is a large surface area of contact between the upper and lower teeth. This is characteristic of the browser grade. In contrast, Type G molars are characterised by strong links between the lophs. When occluded, the well-developed links are in contact with opposing lophs, and the surface area of contact is reduced. This results in more of a shearing action as the upper and lower molars move past each other. This is characteristic of the grazer grade.

The changing relationships between the incisors, premolars and molars with diet in browsers, intermediate browser/grazers and grazers among the Macropodidae are summarised in Fig. 7.2.

7.2.1 **Browsers**

As browsers are highly selective feeders, their incisors need to be precise at selecting particular plant species or plant parts in an array of vegetation. The first upper incisor is larger than the second and third incisors. This minimises the amount of upper and lower incisor contact, which emphasises manipulation of food items by the incisors (Sanson 1989). The item is either plucked by holding it with the incisors and pulling, or by severing it directly. A large shearing premolar can be helpful in the latter process for

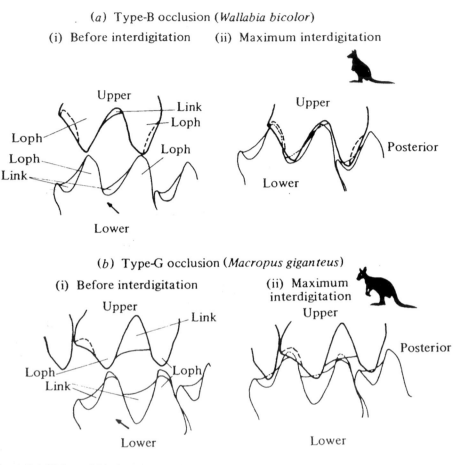

Figure 7.1 Molars of (a) the swamp wallaby (*Wallabia bicolor*), a browser, and (b) the eastern grey kangaroo (*Macropus giganteus*), a grazer, showing initial contact between opposing lophs (left) and maximal interdigitation (right). The arrows indicate the direction of movement of the lower jaw. Note the much larger surface area of contact in the swamp wallaby than in the eastern grey kangaroo. After Sanson (1980).

tough stems. Browsers have permanent premolars which exceed the length of the first molars. Large premolars prevent the forward drift of molars in the jaw (a natural tendency in macropods due to their posterio-anterior jaw movements during mastication). The most strictly browsing genera such as *Dendrolagus* (tree-kangaroos) have larger, more sectorial (cutting) premolars than those genera which generally include some grass in their diet, such as *Petrogale* (rock-wallabies). The molars are lophodont (i.e. have well-developed lophs), and have a predominantly crushing action, but with some fine shearing and grinding capacities as well, reflecting their heterogeneous diet. The tooth row is flat for precise molar occlusion, and all cheek teeth occlude at once.

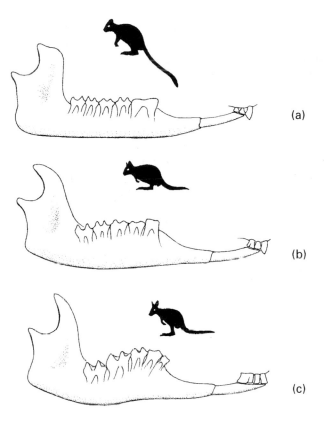

Figure 7.2 Lateral view of the lower jaw and upper incisors of (a) a browser; (b) an intermediate browser/grazer; and (c) a grazer among the family Macropodidae. In browsers the entire cheek tooth row occludes, and the large premolar blocks the anterior drift of the molars. In the intermediate grade the cheek tooth row is slightly curved, reducing the number of molars in occlusion, but anterior drift is allowed by the smaller premolar, so that all molars are used. In grazers the tooth row is very curved, which minimises the number of molars in occlusion, but the vestigial premolar allows molar progression, so all molars are used. After Sanson (1989).

7.2.2 Intermediate grade

In intermediate browser/grazers, grasses can constitute a major proportion of the diet at certain times of the year. Their dentition must therefore be able to cope with both browse and grasses. The first upper incisor is still larger than the other two, but the third incisor is broad. This increases the length of contact between the upper and lower incisors and increases the efficiency of cropping grasses, while still retaining some of the manipulative capabilities afforded by the large first incisor. The premolar is smaller and therefore less effective in shearing; it is also less effective at blocking the molars from moving forward in the jaw. Consequently there is molar progression in most species. The molars have both a shearing and grinding component, which makes them less specialised and therefore somewhat less efficient. However this is compensated by the cheek tooth row being slightly curved, which reduces the number of molars in occlusion and concentrates the force on the anterior molars, but all molars tend to be used because of molar progression.

As indicated by Table 7.1, there are several subgroups within the intermediate group, which reflect increasing levels of grass in the diet. *Peradorcus* is one of the most graminivorous of the rock-wallabies. It also feeds extensively on the fern *Marsilea*, which has an unusually high silica content (about 18% of dry matter) (Sanson, Nelson & Fell 1985). Its dentition reflects the highly abrasive nature of its diet. Its premolar is extremely reduced, and its tooth row markedly curved (Sanson 1989). The molars have the shearing and grinding capacity of a browser, but molar progression is well developed, and there is unlimited molar replacement; in this *Peradorcus* is unique among marsupials.

Lagostrophus (banded hare-wallaby) is unusual in having a reduced premolar but a flat tooth row. All molars occlude. There is molar progression, but the teeth do not fall out, and the whole tooth row drifts forward, reducing the length of the diastema between the incisors and premolar (Sanson 1989).

7.2.3 Grazers

In grazers, prehension and ingestion involve cropping by the incisors of many grass blades at once. This is facilitated by a reduction in the relative size of the first upper incisor, and a further increase in the width of the third incisor, to form a long occlusal contact between upper and lower incisors. The premolar is vestigial, which allows molar progression. The tooth row is very curved, and only two molars are in occlusion at any one time. The molars have an enhanced shearing capacity, and all four are used in the life of the animal. Shearing efficiency is increased by the addition of extra ridges and lateral movement of the mandible, the latter made possible by lack of impediment from the vestigial premolar.

7.3 STUDIES ON MACROPODID DIETS

7.3.1 The problem of analysis of herbivore diets

Analyses of the diets of herbivores suffer from a number of conceptual and methodological problems. These are detailed by Norbury & Sanson (1992) among others. There is a difference between describing foods that comprise the bulk of the diet and foods that are preferred. Measuring diet selection requires a comparison of the relative abundance of food items in the diet to the relative abundance of food items available. High relative abundance in the diet but low relative abundance of food items available means high selection for that food item.

Relative abundance of plant species available has been estimated by four main methods (Norbury & Sanson 1992). Each has its advantages and problems. *Frequency of occurrence* (presence or absence within a sampling unit) is rapid but does not account for differences in plant size, overemphasises rare species, and underemphasises common species. *Density* (the number of individual plants of each species within a sampling unit) is more accurate, but does not account for plant size. *Cover* (the proportion of ground covered by the perpendicular projection of plants onto the ground) takes into account plant size but does not account for differences in plant height. *Biomass* (the dry weight of each species per unit area) is potentially the most accurate method, but is labour intensive, and is destructive, meaning that repeated measures in the same sampling unit cannot usually be made.

Likewise, relative abundance of plant species and parts in the diet of wild herbivores has been estimated by several methods (Norbury & Sanson 1992), each with its strengths and weaknesses. Relative intake (usually estimated by the biomass of each species that regenerates when foraging is prevented) is potentially the most accurate measure but the most difficult. Feeding frequency measurement (direct observation of plant species and parts ingested) is potentially accurate but close observation is often impossible. Mouth contents analysis (identification of ingested plant material) is direct but sample sizes are small and many animals must be killed to maximise sample numbers. Stomach analysis (microscopic analysis of stomach contents) is potentially accurate but also requires killing animals and is thus limited to common species and has ethical constraints. Faecal analysis (microscopic analysis of undigested food residues) is widely used because faecal samples are easy to collect with minimal disturbance to the animals, but it has two major limitations. The first is the problem of differential digestion, already encountered with carnivores (Chapter 2) and omnivores (Chapter 3); some plant parts and species do not survive passage through the digestive tract. The second is the problem of identifying very small plant fragments in the faeces. A portion of the undigested material will always be unidentifiable microscopically, although this can be reduced substantially by using scanning electron microscopy (SEM) rather than light microscopy (Allen 1994). Norbury (1988) has proposed the use of conversion equations based on the known proportion of identifiable tissue for each plant species to improve the accuracy of light microscopic faecal analysis based on areas of the field of view covered by different plant fragment types.

Analysis of macropodid diets has relied on a range of the above techniques. This, and the relative advantages and disadvantages of each technique, must be considered in the interpretation of published findings.

7.3.2 **The major findings**

The macropodid species for which there is a reasonable amount of dietary information are organised in this section on the basis of Sanson's (1989) scheme (Table 7.1). An alternative organisation would have been on the basis of increasing body size. This is because body size is the single strongest determinant of dietary patterns of most mammalian herbivore groups (Demment & Van Soest 1985). Because of their relatively high mass-specific energy requirements (Chapter 1) small species must process food faster than larger species, and therefore would be expected to select highly digestible plant species and plant parts. This is why most small browsers are concentrate selectors and most grazers are large bulk and roughage feeders (Hofmann 1973; Hume 1989). However, as will be seen in this survey, differences in dental morphology are a strong modifying influence on this general pattern within the Macropodidae.

In his review of the general patterns of macropodid diets, Dawson (1989) divided macropodid diets into four major categories:

(a) Forbs – small herbaceous dicotyledonous plants (dicots), often annuals or ephemerals which are generally of good nutritional value.
(b) Shrubs – small bushes, often perennials such as salt bushes, with leaves of good nutritive value but with poor-quality twigs and stems.
(c) Browse – woody bushes and trees, with leaves that may be highly lignified and therefore of poor quality except when young.
(d) Grass – monocotyledonous grasses and sedges (monocots) that are of high quality when young but in the mature state are often of high fibre content and therefore of low nutritive value.

Jarman (1994) emphasised the importance of the eating of seeds and seedheads by a number of macropodid groups, and speculated on the role of seedhead eating in the evolution of grazing in the macropods.

THE BROWSER GRADE

The forest wallabies

There is little detailed information available on the diet of the New Guinea forest wallabies (*Dorcopsis* and *Dorcopsulus*). *Dorcopsis* wallabies are browsers and much use is made of their sectorial premolars when feeding (Menzies 1989). They pick up food with the incisors, then transfer it with the hands to the side of the mouth to be processed by the premolars. Sanson (1980) quoted two personal communications suggesting that *Dorcopsis* are browsers, taking soft vegetation, flowers and fruits. *Dorcopsulus vanheurni* (small dorcopsis) is the smallest extant macropodid marsupial known from New Guinea. Flannery (1995) thought that on the basis of its small size (1.5–2.5 kg), low-crowned molars and elongate premolars, it may fill the potoroid niche in New Guinea.

Tree-kangaroos

Both Australian tree-kangaroos inhabit complex notophyll vine forests in north Queensland. Although Proctor-Gray (1984) reported that *Dendrolagus lumholtzi* (Lumholtz's tree-kangaroo) was strictly folivorous, Martin (1992) found *D. bennettianus* (Bennett's tree-kangaroo) to be an opportunistic generalist folivore/frugivore. The list of food plants of Bennett's tree-kangaroo included 18 tree, 10 vine, 1 tree/epiphyte (*Schefflera actinophylla*) and one epiphytic fern species (*Platycerium hilli*) . The diet of Lumholtz's tree-kangaroo was similar in nitrogen content to that of sympatric green ringtail possums (*Pseudochirops archeri*), and both were higher than that of sympatric common brushtail possums (*Trichosurus vulpecula*) (Proctor-Gray 1984). Dietary fibre content was highest for the tree-kangaroo and lowest for the common brushtail. These differences are consistent with both the larger size and the foregut fermentation digestive strategy of tree-kangaroos compared with the hindgut fermentation strategy of the two possums, and the more folivorous nature of ringtail possums versus brushtail possums (Chapter 5).

Pademelons

Pademelons are inhabitants of rainforest and wet sclerophyll forest where they feed in forest openings and on adjacent pastures. Redenbach (1982) studied the diets of two pademelons, *Thylogale stigmatica* (red-legged pademelon) and *T. thetis* (red-necked pademelon). *T. stigmatica* took 88% dicot leaf and stem, 10% dicot fruits, 1% fern and 1% epigeous (aboveground) fungi; no grass fragments were present in the faeces. In contrast, sympatric *T. thetis* took mainly grasses (66% of faeces) and 31% dicot leaf and stem, with only 1% of each of dicot fruits, epigeous fungi and fern. Of twenty faecal pellets, 13 contained monocot seedhead epidermis (up to 7% of total epidermal area) and 7 contained dicot seed material (also up to 7% of total epidermal area). *T. billardierii* (Tasmanian pademelon) appears more similar to *T. thetis* than to *T. stigmatica* in its feeding habits, with a diet consisting mainly of short green grasses and herbs, with some browse from taller woody shrubs (Johnson & Rose 1995). No details are available on the diet of *T. brunii* (dusky pademelon) from New Guinea.

Quokka and swamp wallaby

The quokka (*Setonix brachyurus*) has been the subject of intense study on its island habitat of Rottnest Island, close to Perth in Western Australia. Less is known about the remnant mainland population in the south-west corner of the state. Storr (1964a) examined the diet of quokkas at three locations on Rottnest, and found that succulents dominated the diet in areas where surface water was never available (79% in spring, 97% in summer). The value was still 52–77% in an area where seepage water was available. The next major dietary component in summer was shrubs, up to 28% in one area. Forbs and grasses only assumed importance in winter and early spring; these were all annual species that respond to the predominantly winter rainfall (annual average 74 cm). In contrast to Rottnest Island, Storr (1964b) found that on the mainland south of Perth quokkas live in densely vegetated swamps. The quokkas feed on the vegetation of the swamps and swamp margins. Their diet thus contains no grass, but consists predominantly of sedges, perennial herbs, and leguminous and myrtaceous shrubs. Similarly, Algar (1986) found that quokkas maintained in a reserve near Perth were mainly browsers, with the leaves of larger shrubs accounting for 52–88% of the diet in all seasons. Sedge intake increased in late summer and autumn, during the annual summer drought. Little grass was eaten, but grass was not readily available in the reserve.

Two studies in north-eastern New South Wales (Harrington 1976; Hollis, Robertshaw & Harden 1986) and one in Victoria (Edwards & Ealey 1975) all suggest that *Wallabia bicolor* is a generalist feeder with a preference for forbs and shrubs of a wide range of species. However, significant amounts of grasses, ferns, fungi and seeds are also consumed at different times of the year. On North Stradbroke Island, Queensland, Osawa (1990) also found that swamp wallabies consumed a wide range of plant categories including shrubs, forbs, grasses, sedges and fungi. Monocots (sedges and introduced grasses) were actually the major component of stomach contents of road-killed animals in all seasons except summer. Most of these monocots were growing along the roadside, and had higher nitrogen contents compared with neighbouring shrubs, except during summer.

Rock-wallabies

Sanson (1989) divided the rock-wallabies between the browser and intermediate browser/grazer grades (Table 7.1). Of the species in the former, there is dietary information on only three species. *Petrogale assimilis* (allied rock-wallaby) is an animal of the wet-dry tropics. Horsup & Marsh (1992) reported that forbs were their major food item (50–65% of the identified epidermis in faeces), except at the height of a long dry season when they fell to only 22%. Browse was the next most important diet component (20–26% of identified epidermis), but this increased to a peak of 41% in the long dry season. Most of the increase was in the form of leaf fall from trees. Plants with stellate trichomes such as *Waltheria* were abundant but selected against, so that they made up only 6–18% of identified epidermis, except at the end of the long dry season when they accounted for 32%. Grasses comprised only a small portion of the diet (6–9%), except after drought-breaking rains when they peaked at 16%; most of the increase was in the form of new shoots. Fruits, flowers and seed heads were also seen to be eaten. Horsup & Marsh (1992) concluded that the flexible diet of *P. assimilis* was an essential requirement for a small sedentary herbivore in a seasonally dry habitat.

Petrogale brachyotis (short-eared rock-wallaby) of the Kimberleys and Arnhem Land in northern Australia also appears to be flexible in its diet. Browse, particularly *Terminalia latipes* leaf and seed, was prominent in stomach contents in both the wet and dry seasons, together with small amounts of other, unidentified browse, as well as grasses (mainly seeds), in the wet season. In the dry season some *Triodia* (spinifex) was taken as well (Sanson, Nelson & Fell 1985). Sanson (1989) observed *P. brachyotis* using its elongated premolars to crack *Terminalia* seeds before chewing the fragments.

Petrogale penicillata (brush-tailed rock-wallaby) has been studied in New South Wales and Victoria. Its diet at two locations in south-eastern New South Wales was dominated by grasses (35–50%) and forbs (25–40%), but the foliage of shrubs and trees was also significant (12–30%) (Short 1989). Its diet in the Grampians in western Victoria also contained shrubs (leaves, flowers and seeds) and forbs, but 90% of identified faecal epidermis was grass (Wakefield 1971). Seed capsules and flowers of rock orchids (*Dendrobium* spp.) were seen to be eaten by K. Joblin (pers. comm. in Jarman (1994)). A preliminary study by Blackbourn (1991) also found that seedheads formed a significant part of the diet of *P. lateralis* (black-footed rock-wallaby) in central Australia; up to a third of the monocot component.

THE INTERMEDIATE BROWSER/GRAZER GRADE

Rock-wallabies

Petrogale xanthopus (yellow-footed rock-wallaby) has been studied more intensively than other rock-wallabies because of concerns over its threatened status throughout its disjunct range through arid and semi-arid parts of South Australia, New South Wales and Queensland. Results from all three states support the idea of *P. xanthopus* as an intermediate browser/grazer (Dawson & Ellis 1979; Copley & Robinson 1983; Allen 1994). In good seasons forbs and grasses make up the bulk of the diet. As conditions dry, the leaves of woody shrubs and trees (the latter as leaf fall) become increasingly important, and in drought become the biggest single component of the diet.

This shift in the diet of yellow-footed rock-wallabies, and of feral goats, which are also intermediate browser/grazers, is illustrated in Fig. 7.3. In this three-and-a-half year study in south-western Queensland, conditions ranged from extremely wet in 1989 to extremely dry in 1992. Forbs, chenopods and grasses were abundant in 1989 and 1990 but had almost disappeared by the end of 1992. Dietary overlap between the goats and rock-wallabies was high throughout the study (Fig. 7.3a, b); the diet of both was dominated by forbs, chenopods and grasses early, but by the end of the study browse (mainly dry leaf litter from *Acacia* trees) was the major food of both herbivores. The implications of this pronounced dietary overlap for

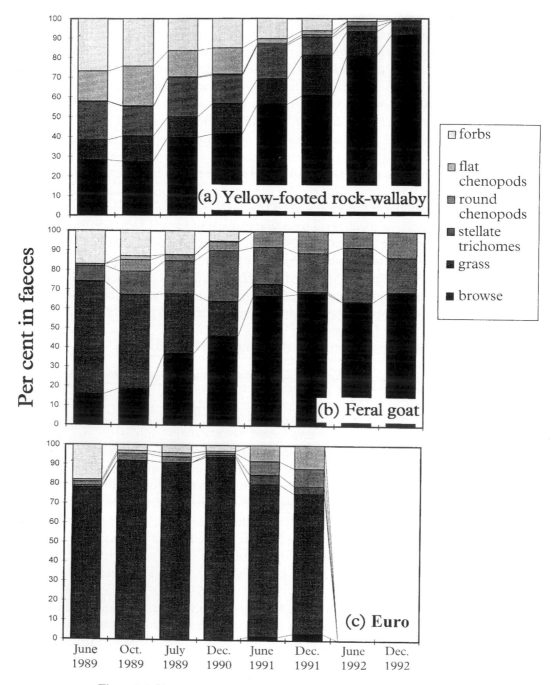

Figure 7.3 Changes in the diet of yellow-footed rock-wallabies (*Petrogale xanthopus*), euros (*Macropus robustus erubescens*) and feral goats in south-western Queensland from an extremely wet period to an extremely dry period. Both yellow-footed rock-wallabies and goats are intermediate browser/grazers, but the euro is a grazer. From Allen (1994).

the food supplies and future conservation of the rare yellow-footed rock-wallaby are of major concern. In contrast, euros in the same area consumed mainly grasses throughout the study (Fig. 7.3c), and so their diets overlapped that of the rock-wallabies only in times of abundant plant biomass. When the grasses disappeared in 1992 so did the euros, through starvation and migration.

Peradorcus concinna (little rock-wallaby or nabarlek) has been mentioned in Section 7.2.2 above in relation to its unique feature of apparently unlimited numbers of molars. Sanson *et al.* (1985) found that, in the wet season, *Peradorcus* feeds on the margins of flood plains which support dense lawns of grasses, which form the bulk of the diet. However, in the dry season the flood waters recede and the little rock-wallabies move out onto the flood plains to consume almost exclusively the fern *Marsilea crenata*. This fern is an unusual diet because it contains on average 18% silica. Silica is harder than tooth enamel, but the features of the *Peradorcus* dentition, including a sharply curved tooth row, molar progression and seemingly unlimited replacement of worn molars, allow it to utilise such an unpromising diet.

Brush wallabies

This group of seven small to medium-sized members of the genus *Macropus* fall into the subgenus *Notomacropus* (Flannery 1989). They typically inhabit shrub and forest communities.

Macropus irma (western brush wallaby), from the south west of Western Australia, is principally a browser (Algar 1986). Grasses and sedges made up 3–16% of the diet, forbs another 1–7% and browse 79–98% of the diet throughout the year. Most of the browse consumed consisted of the leaves of low shrubs.

Macropus dorsalis (black-striped wallaby), from south-eastern Queensland and north-eastern New South Wales, are largely grazers, although in dry seasons the proportion of browse in the diet can increase to 16% (Ellis, Tierney & Dawson 1992). Among the grasses they selectively feed on the leaves rather than stems of a wide range (often six or seven species) of mostly short, soft (low-fibre) grasses. Tougher wiregrasses (*Aristida* spp.) and bunched speargrass (*Heteropogon contortus*) were widely available but were not eaten. In New South Wales, Jarman, Phillips & Rabbidge (1991) also reported that *M. dorsalis* includes mainly monocot parts in their diet (74–93% at three different sites). At one of the sites, Wallaby Creek in the north east of the state, *M. dorsalis* took 21 different grass species, more than any of the other three sympatric macropods at Wallaby Creek.

Macropus rufogriseus (red-necked wallaby), *M. agilis* (agile wallaby), *M. parma* (parma wallaby) and *M. parryi* (whiptail wallaby) are generally regarded as primarily grazing species. *M. parryi* at Wallaby Creek includes 99% of monocot parts in its diet (Jarman & Phillips 1989). In savannah woodland in north Queensland, *M. parryi*'s staple diet is kangaroo grass (*Themeda australis*), although tips of bunch spear grass (*Heteropogon contortus*) are also eaten (Bell 1973). *M. parma* also uses *Themeda australis* as a preferred food (Maynes 1974). *M. agilis* appears to be more of a mixed feeder, favouring short green grass but also taking some long dry grass, as well as browsing on *Melaleuca* and other low shrubs (Bell 1973). *M. rufogriseus* also includes significant amounts of dicot leaf and stem (16%) in its mainly grass diet (Jarman & Phillips 1989).

Although *Macropus eugenii* (tammar wallaby) is the most commonly maintained macropod for experimental purposes, little detail is available on its natural diet. It appears to be principally a grazer, but Andrewartha & Barker (1969) noted that in the annual summer drought on Kangaroo Island tammars foraged beneath *Acacia retinoides* trees, suggesting that they may have been consuming the seeds of this legume, perhaps for protein.

Hare-wallabies

Lagostrophus fasciatus (banded hare-wallaby) is the sole survivor of the macropodid subfamily Sthenurinae, and is now found only on Bernier and Dorre Islands in Shark Bay, Western Australia. It is a mixed feeder, with grasses making up less than half the dietary intake, the remainder consisting of malvaceous and leguminous shrubs and other dicots (Prince 1995). Ride & Tyndale-Biscoe (1962) collected rectal pellets from seven animals from Dorre Island and ten from Bernier Island. Pellets from all seven Dorre animals contained grass fragments, the mean being 66% grass epidermis and the remainder unidentified plant tissue. In contrast, only three Bernier animals had grass in their faeces, two had *Carpobrotus* (a succulent) epidermis, and all ten had malvaceous hairs. The low incidence of grass in the Bernier animals, despite the abundance of grasses, particularly *Triodia* (spinifex), suggests that *Lagostrophus* is more of a browser than a grazer. Thomas (1887) also thought this on the basis of its dentition, which features a reduced premolar but a flat tooth row (see Section 7.2.2 above).

Lagorchestes hirsutus (rufous hare-wallaby) shares Bernier and Dorre Islands with *Lagostrophus*, but an isolated population also persists in the Tanami Desert in the Northern Territory, thanks to reintroductions from a captive breeding colony in Alice Springs. Lundie-Jenkins, Phillips & Jarman (1993) found that perennial grasses were the most consistent items in the diet of *L. hirsutus* in the Tanami. Grass seeds were seasonally important, as also were sedge seeds and bulbs. Dicots constituted only a minor part of the diet, but increased when the quality of other, more preferred plants declined. Insects also became noticeable diet components at these times, and may have been an important protein source in dry seasons. During a good season the summer diet contained no insects but instead the animals concentrated on the growing shoots of the grasses *Eragrostis falcata* and *Aristida browniana* (Pearson 1989). *Triodia pungens* (spinifex) was strongly avoided. Dicots made up 9–23% of identifiable plant fragments. Seeds of grasses and sedges were conspicuous in faecal samples, and by Bolton & Latz (1978) were observed to be eaten. Thus, like *Lagostrophus*, *L. hirsutus* is a mixed feeder.

Lagorchestes conspicillatus (spectacled hare-wallaby) is the only hare-wallaby that remains widespread. It occurs in tropical grasslands across the Northern Territory and north Queensland, and on Barrow Island off the north-western coast of Western Australia. In the dry season in the Northern Territory, *L. conspicillatus* preferred forbs and non-chenopod shrubs to grasses, although grasses always constituted at least 5% of the faeces (Ingleby & Westoby 1992). Seeds, of both dicot fruit and grass origin, were also identified in the faeces, from 3% to 54% of the faeces. Levels of browse in the faeces were generally low. On Barrow Island, Bakker & Bradshaw (1989) found that *L. conspicillatus* was a selective feeder, browsing mainly on colonising shrubs in areas disturbed by road construction but also eating the tips of *Triodia pungens* leaves in undisturbed and long-unburnt sites.

THE GRAZING GRADE

Nailtail wallabies

Onychogalea unguifera (northern nailtail wallaby) is reasonably secure in its habitat of open woodlands with a tussock grass understorey in the wet-dry tropics of Western Australia, the northern Territory and north Queensland (Ingleby 1991). Ingleby, Westoby & Latz (1989) found no evidence of *O. unguifera* browsing. Instead, they fed mainly on herbaceous (non-woody) dicots, including fruits and their seeds, as well as grass shoots when herbaceous dicots are less available. Rhizomes and buried stem bases of *Chrysopogon* tussocks were also consumed, but only rarely.

Onychogalea fraenata (bridled nailtail wallaby) (Fig. 7.4) is now one of Australia's rarest marsupials, persisting only in a small area in central Queensland. Its diet is diverse, consisting of herbaceous dicots (forbs, including the succulent *Portulaca*, and malvaceous species), grasses and browse (Ellis *et al.*1992). The diet is biased toward herbaceous dicots in good seasons. Grasses become increasingly important as the season worsens, but there are large differences in selection for different grass species. Browse is only important when forbs and grasses are scarce. Leaf fall from the tall shrub *Eremophila mitchelli* is a major component of the diet in the driest periods (Dawson 1989). Evans (1992b) observed *O. fraenata* selectively harvesting seedheads of grasses and forbs; in summer, seedheads formed 5–9% of available plant material but 21–24% in faecal material. Thus the nailtail wallabies appear to be highly flexible mixed feeders.

The large kangaroos

This is where most dietary information is available, because of the presumed competition between kangaroos and domestic stock. The diet of

Figure 7.4 A young bridled nailtail wallaby (*Onychogalea fraenata*) from central Queensland. *O. fraenata* is an endangered species. (Pavel German)

Macropus giganteus (eastern grey kangaroo) contains a large proportion of grass in all seasons (78–98%) (Taylor 1983). Forbs are generally avoided. Feeding is highly selective, with leaf preferred to other plant parts, although more stem and inflorescences are eaten in autumn and winter when leaf is less available. Seeds normally make up only a small proportion of the diet of *M. giganteus* (Jarman 1994), but Griffiths & Barker (1966) reported heavy seasonal use of the seedheads of *Triodia mitchellii* in south-west Queensland, the likely benefit being protein, since the total nitrogen content of the seeds (1.38%) was double that of the leaves (0.65%). Its highly selective feeding behaviour notwithstanding, *M. giganteus* is capable of surviving on poor quality plant material. Griffiths, Barker & MacLean (1974) found that during severe drought in south-west Queensland, *M. giganteus* ate the coarse grasses *Triodia mitchellii*, *Themeda australis* and *Aristida* spp.

Macropus fuliginosus (western grey kangaroo) also concentrates on monocots. Observations by Halford, Bell & Loneragan (1984) in a mixed pasture-woodland habitat, and by Algar (1986) on the coastal plain south of Perth, suggest a high intake of sedges. Halford *et al.* (1984) found that although pasture grasses made up more of the diet after the onset of winter rains, intake of sedges was maintained, as was a small proportion of forbs and shrubs. Similarly, Algar (1986) found that sedges made up 82% of the diet in all seasons except spring. This suggests that sedges were eaten because of their abundance, with grasses being preferred when available. In all seasons monocots formed more than 90% of the diet.

M. robustus robustus (eastern wallaroos) grazing sympatrically with *M. giganteus* on the New England Tablelands of New South Wales were also primarily grazers in all seasons (77–97% of the diet) (Taylor 1983), although there were some differences in the grass species selected. *M. robustus woodwardi* , the northern subspecies, is similarly chiefly a grazer (Croft 1987). The third subspecies, *M. robustus erubescens* (Euro) is primarily an arid-zone animal. Nevertheless, its diet is also primarily composed of grasses (Dawson *et al.* 1975). In dry seasons some shrubs are also eaten, but even in severe drought grasses made up more than 80% of its diet at Fowlers Gap in far western New South Wales (Dawson & Ellis 1996). Similar findings were reported by Allen (1994) in south-western Queensland (Fig. 7.3), and Lapidge (1999) in northern South Australia. *Macropus antilopinus* (antilopine wallaroo) is found in open *Eucalyptus* woodlands with a perennial grass understorey in monsoonal tropical Australia. It feeds largely if not entirely on grasses (Croft 1987; Calaby 1995).

Macropus rufus (red kangaroo), that magnificent animal of the semi-arid zone, also feeds mainly on grasses (Chippendale 1968; Low *et al.* 1973; Dawson *et al.* 1975), except in wet seasons when its diet width increases to include many forbs, up to 54% of identifiable particles in the faeces at Fowlers Gap (Dawson & Ellis 1994). In drier seasons shrubs are also eaten, but in severe drought Dawson & Ellis (1994) found that grass was 87–91% of identifiable particles. When the grasses disappear, red kangaroos shift to eating chenopods (Bailey, Martenesz & Barker 1971; Barker 1987).

7.3.3 Competition between kangaroos and domestic stock

Opportunities for competition for food would seem to be greatest between macropods in the grazing grade and sheep and cattle, which are also acknowledged to be primarily grazers. However, firm evidence for competition is scarce, even though it is often assumed to occur. This is because for competition to be established it has to be demonstrated that the presence of one species has an adverse effect on another, as Wilson (1991) did in the semi-arid woodlands of central-western New South Wales. This is a particular problem in the arid zone because of enormous variability in environmental conditions over time and in vegetation patterns over space, making replication logistically difficult. The most serious attempt to overcome these problems has been by Edwards, Dawson & Croft (1995) at Fowlers Gap. Six contiguous paddocks each of approximately 620 ha were stocked with either only red kangaroos, mainly sheep, or both red kangaroos and sheep, and diets examined on the basis of faecal pellet analysis over three years. The diets of the two herbivores overlapped by 52–73% at both high and low plant biomass availabilities, with grasses and forbs being the major items in the diets of both species. However, at times of lowest plant availabilities, the red kangaroos appeared to deprive sheep of preferred species, notably grasses, forcing them to include more chenopod shrubs in their diet. Whether this reduced the fitness of the sheep was not established.

Interactions between red kangaroos and cattle were reviewed by Dudzinski *et al.* (1982) in arid central Australia. Competition appeared to be less than between red kangaroos and sheep. Cattle and red kangaroos tend to eat different grass species and/or parts, and, importantly, there seems to be little spatial overlap between the two herbivores except during drought, when red kangaroos move from *Acacia* woodland with a shrub and perennial grass understorey out onto more open communities, and cattle continue to feed in these communities. There was no evidence that one species attracts or repels the other in spatial terms, and it was concluded that there was only limited competition for forage between the two species.

7.4 NUTRITION AND ECOLOGY

A number of the more complete studies will now be considered in order to explore in more detail the role of some of the digestive and metabolic features of the Macropodidae in their ability to exploit different nutritional environments.

7.4.1 The island wallabies

THE QUOKKA
Setonix brachyurus was formerly abundant throughout the south west of Australia. Following a catastrophic decline in the 1930s the mainland population was restricted to densely vegetated swamps where the quokka still occurs in low numbers. There are two extant island populations, one on Bald Island off the southern coast and the other on Rottnest Island near Perth. The quokka remains common on both islands.

Ecology
The quokka on Rottnest Island, freed from competition and predation after the island was isolated from the mainland some 7000 years ago, has undergone an expansion of its niche. The diet of the mainland quokka of sedges, perennial herbs and leguminous and myrtaceous shrubs (Storr 1964b) is not subject to fluctuations in digestible energy, nitrogen and water content to the same extent as that of the Rottnest Island quokka. Initially most of the habitats occupied by the Rottnest quokka were heavily vegetated with *Callitris*, *Melaleuca* and *Acacia* shrubs (Storr 1964a). The advent of human settlement on the island led to marked changes. Much of the shrub vegetation was cleared or thinned for sheep grazing, which enabled herbaceous plants, particularly exotic annuals, to flourish. Fre-

quent burning and severe browsing of regenerating shrubs led to the present situation in which much of the island is now dominated by the unpalatable shrub *Acanthocarpus* and the perennial grasses *Poa* and *Stipa* spp. In winter, germination of annuals and new growth of perennials provides a high-quality diet, but in summer only perennial shrubs and grasses provide food. The perennial shrubs become severely over-browsed and the quokkas are forced to eat more of the fibrous perennial grasses and the older, more fibrous leaves and twigs of perennial shrubs.

This suggests that there may be differences between the mainland and island quokka populations in aspects of their biology related to nutrition. Indeed, Shield (1965) showed that the mainland population breeds continuously throughout the year; in contrast, Rottnest quokkas have a season of birth restricted to six months. When female quokkas from Rottnest were maintained in captivity on good quality food *ad libitum* however, the annual anoestrus period decreased, and after a period of two years or more finally disappeared. Since the climate of the two areas, only 80 km apart, is similar, the breeding season difference between the two populations must be related to food supply.

Studies on adrenocortical function suggest that the Rottnest quokka undergoes an annual seasonal stress (Miller & Bradshaw 1979). The time of maximum environmental stress and mortality coincides with the end of the hot, dry summer–autumn period and particularly with the break in the season (the beginning of the winter rains). The animals lose up to 20% body mass, are dehydrated, there is a fall in body temperature and they suffer severe haemolytic anaemia (Miller & Bradshaw 1979). Shortages of digestible energy, nitrogen, water and vitamin E, hookworm (*Austrostrongylus thylogal*) infestation and *Salmonella* infection have variously been implicated, together with salt loading (Shield 1959; Storr 1964a; Barker 1974) and salt deficiency (Miller & Bradshaw 1979). It may well depend upon which part of the 1900 ha island is under consideration. For instance, one population of quokkas live on West End (Fig. 7.5) which entirely lacks fresh water. Others, living in the Lakes Area, have access to fresh water seepages around the salt lakes throughout the year. No quokka marked on West End has ever been recorded on the other side of Narrow Neck, a 200 m wide isthmus between the two areas. This leads to marked differences between the two populations in their access to free water and in the quality of their diet (Storr 1964a). West End animals show much greater fluctuations between winter and summer in haemoglobin concentrations and in body condition than animals from the Lakes Area (Tyndale-Biscoe 1973).

Nutrition

In an attempt to define the limiting factor(s) to the West End quokkas during summer and autumn, Wake (1980) conducted both short and

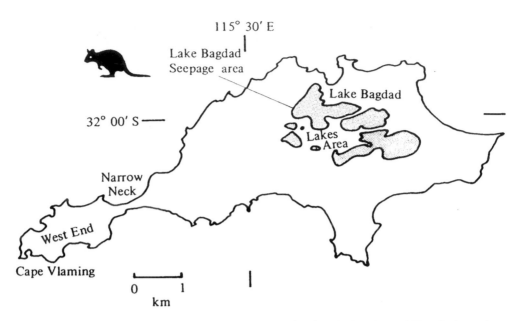

Figure 7.5 Rottnest Island, Western Australia, showing the locations of West End and the Lakes Area, the sites of several field studies of the quokka (*Setonix brachyurus*).

long-term supplementation experiments. Two short-term experiments were based on the technique used by Kinnear (1970) in his work with *Macropus eugenii* on the Abrolhos Islands. The treatments were control, urea-supplemented, starch-supplemented, and both urea- and starch-supplemented. The urea (200 mg of nitrogen in the early summer experiment, 93 mg in the later summer experiment) was dissolved in 2 mL distilled water and injected intraperitoneally. The starch (20g suspended in 20 mL water in both experiments) was administered by stomach tube. Animals were caught just after dark, weighed, and an initial blood sample taken; the early summer experiment involved 21 animals, and the late summer experiment 24. All animals were injected intraperitoneally with tritiated water to determine the size of the body water pool.

After receiving one of the four treatments the animals were kept for 12 hours in individual metabolism cages and urine and faeces collected separately. At the end of 12 hours a second blood sample was taken and the animals were released at point of capture. Changes in plasma urea concentrations and pool sizes over the 12 hour period were used to calculate a value for apparent utilisation of urea on each treatment. The results from the two experiments are summarised in Table 7.2.

The conclusions were that, in early summer, energy primarily and nitrogen secondarily limited microbial activity in the quokka forestomach; utilisation of injected urea was greatest with supplements of both energy and nitrogen. In late summer energy was the limiting factor, as shown by

Table 7.2. *Effect of short-term (12-hour) supplementation with urea, starch, or both on urea utilisation in quokkas* (Setonix brachyurus) *at West End, Rottnest Island*

	Treatment			
	Control	Urea	Starch	Urea + Starch
Early Summer				
Plasma urea (mg N dL^{-1}), T_0	16.0	20.8	24.0	19.6
Plasma urea (mg N dL^{-1}), T_{12}	27.3	32.8	28.9	27.0
Increment	11.3	12.0	4.9	7.4
Apparent utilisation of urea N (mg)	–	86	65	211
Late Summer				
Plasma urea (mg N dL^{-1}), T_0	21.7	20.4	20.4	22.5
Plasma urea (mg N dL^{-1}), T_{12}	28.3	30.5	18.7	22.1
Increment	6.6	10.1	− 1.7	− 0.4
Apparent utilisation of urea N (mg)	–	50	141	175

Source: After Wake 1980.

the negative increments in plasma urea nitrogen concentrations when starch was given alone. Presumably the difference between the two experiments was due to the fact that in early summer the quokkas were still in good condition and the food plants were not yet severely deficient in either readily fermentable carbohydrate or protein. As the summer progressed the decline in the level of readily fermentable carbohydrate in plants apparently assumed greater significance than the decline in protein level.

Wake (1980) then conducted long-term supplementation experiments in mid to late summer by distributing either starch, casein or water through part of the study area, each supplement in a different year. The supplements were marked with a fluorescein dye, which is quantitatively excreted in the urine, in order to identify those animals which took the supplement, and to obtain an estimate of the amount of supplement eaten. Wake (1980) also developed an index of body condition, the condition factor, which was the ratio of actual body weight to predicted body weight. Predicted body weight was derived from an allometric equation relating body weight to leg length. Thus a change in condition factor meant that the same proportion of the animal's weight had been gained or lost regardless of size. In Wake's (1980) study the extremes of condition factor were 0.64 and 1.45.

The energy supplement was taken readily. A total of 92 kg of starch was made available to the population, and almost all was consumed by quokkas – although other nocturnal animals such as rats and mice may also have consumed some. The supplements were always distributed just after dark and any excess was removed before sunrise to prevent losses to birds. A total of 156 quokkas was caught over six catching trips spanning a 14 week period between commencement of supplementation and soon after the beginning of the winter rains. On each occasion supplemented and unsupplemented animals were kept in individual metabolism cages for 12 hours of urine and faeces collection as before. A summary of results appears in Fig. 7.6.

The significantly higher condition factor of those quokkas which consistently ate the starch supplement and the increase in condition factor for some of those animals shows that readily fermentable energy was an important limiting factor in the diet of the quokka at West End during mid to late summer. This conclusion is supported by the differences shown in Fig. 7.6 for the other three measured parameters.

The long-term protein supplementation experiment was not successful. Only six of a total of 94 animals caught over four catching trips between late December and late March had high urinary fluorescein concentrations, confirming the observation that few animals ate significant amounts of the supplement despite the use of attractants such as vanilla essence and aniseed oil and of agar and water to increase its palatability. The water supplement was taken more readily. Quokkas were frequently observed drinking from water containers throughout the ten-week supplementation period, even during rain. The rain which fell just after water supplementation began would have alleviated any dehydration of both supplemented and unsupplemented populations and reduced the effects of the water supplement. However, results from 33 animals caught over two catching trips after supplementation commenced suggested that dehydration was a contributing factor, though not the primary cause, of weight loss in quokkas at West End in late summer. Condition factor increased with water supplementation, but not as much as after rain. Parallel increases in total body water (Fig. 7.7) indicate that the improvement in condition represented rehydration rather than an increase in body solids.

Storr (1964a) and Barker (1974) considered that the symptoms of starvation exhibited by quokkas in summer were a secondary effect of water deprivation. Water restriction reduced voluntary food intake in the tammar (Barker, Lintern & Murphy 1970). Storr (1964a) argued that the absence of free water at West End forced quokkas to eat such a high proportion of succulents to satisfy their water requirements that the high water content of the succulents limited their dry matter consumption. However, this need not necessarily lead to a reduced digestible energy intake since the carbohydrates of succulents are generally more digestible than those of perennial shrubs and grasses. Also, quokkas from the Lakes Area, with access to free water, still undergo seasonal anaemia and weight loss (Shield 1959; Barker 1974). Thus other factors such as a deficit of energy, and perhaps also protein, must be more important.

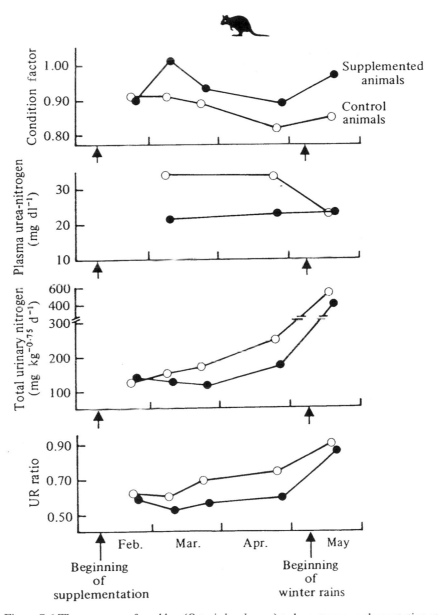

Figure 7.6 The response of quokkas (*Setonix brachyurus*) to long-term supplementation at West End, Rottnest Island. Condition factor is the ratio of actual body mass to predicted body mass. Predicted body mass is derived from an allometric eqation relating body mass to leg length. After Wake (1980).

A further factor contributing to the seasonal debility of Rottnest quokkas, not directly examined by Wake (1980), may be salt loading. Ramsay (1966) found that in field-caught quokkas held in metabolism cages, 12 hour urine volumes were always higher for 'coastal' (West End) animals than for 'inland' animals. Except in spring, the latter always showed lower urinary

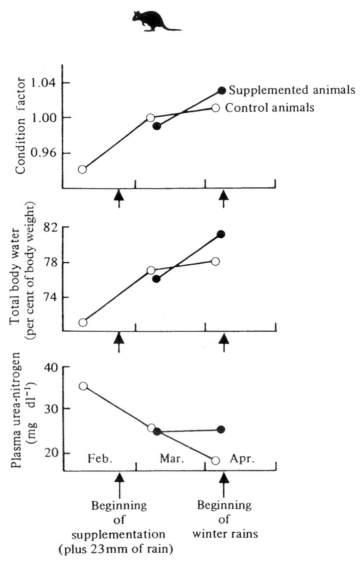

Figure 7.7 The response of quokkas (*Setonix brachyurus*) to long-term water supplementation at West End, Rottnest Island. After Wake (1980).

sodium levels despite the lower urine output, suggesting that the West End quokkas were salt loaded in late summer which would tend to exacerbate any dehydration resulting from a lack of free water. Ramsay (1966) interpreted her results as being due in part to the incidence of salt spray on plants in coastal situations, and in part to the high salt content of succulents eaten by quokkas at West End. Contrary to this, Miller & Bradshaw (1979) concluded that 'inland' quokkas subsist for a part of the year on a low-salt diet as indicated by raised levels of plasma aldosterone, but 'coastal' animals have some means of supplementing their salt intake.

Thus a number of possible causes have been proposed for the seasonal stress experienced by the quokka on Rottnest Island. As suggested by Ramsay (1966), the primary cause may well differ between the Lakes Area and West End, despite the similar nature of the seasonal anaemia which always develops (Barker 1974). Thus the conclusion from the short-term supplementation experiments of Wake (1980) that in early summer quokkas on West End were affected primarily by a shortage of digestible energy need not necessarily apply to the Lakes Area. Similarly, on West End the primary limiting factor may well differ between seasons and between years. It would seem that the most important factors involved are probably the levels of protein and readily fermentable carbohydrate in the plant material consumed, but other factors are clearly involved.

Other nutrients in short supply on Rottnest include several minerals and also possibly vitamin E. The limestone-derived sandy soils of Rottnest Island are poor in their status of most minerals. Thus early attempts to graze sheep on the island met with failure when the animals quickly developed a wasting disease now known to be a dual deficiency of copper and cobalt. This experience, together with the finding that many quokkas on Rottnest were emaciated and anaemic at the end of the annual summer drought, led Barker (1960, 1961a, b) to embark on a study of the role of several minerals in the nutrition of the Rottnest quokka.

Copper, molybdenum and sulphate   Grazing merino sheep require a minimum of 6 µg copper g^{-1} in dry pastures to avoid defective keratinisation of the wool (Underwood 1977). Copper deficiency can be caused by a low intake of copper, or it can be induced on normal copper diets containing moderate levels of molybdenum and inorganic sulphate. There are three essential steps in the induction of copper deficiency in ruminants: (1) reduction of sulphate to sulphide in the rumen; (2) the reaction at relatively neutral pH of this sulphide with molybdate in the rumen to produce thiomolybdate; and (3) reaction of the thiomolybdate with the copper to form the very insoluble copper thiomolybdate, $CuMoS_4$.

It appears from the work of Barker (1960, 1961a) that the same three-way copper-molybdenum-sulphate interaction occurs in quokkas. In an experiment with captive animals fed a diet containing 7.5 µg copper g^{-1}, drenching animals daily with 434 µg molybdenum and 1 g inorganic sulphate per 100 g dry matter intake significantly depressed blood and liver copper levels. However, even though the experiment continued for 20 weeks, copper deficiency anaemia was not induced.

Analysis of Rottnest plants showed that some species contained very little copper, moderate levels of molybdenum and high levels of inorganic sulphate (Barker 1961b). Therefore blood samples were taken from quokkas caught in the Lakes Area and at West End (Fig. 7.5) on nine collecting trips spaced over a period of 17 months (Barker 1961a). Despite higher copper levels in favoured food plants such as *Sporobolus* from the Lakes Area than in plants from West End, animals from the Lakes Area had lower blood copper levels on all but three collecting trips. This was correlated with higher plant molybdenum levels in the Lakes Area, as demonstrated by higher blood molybdenum levels in Lakes Area quokkas. This suggests that in this area blood copper levels were depressed by a higher molybdenum intake, as occurs in ruminants (Underwood 1977).

In the Lakes Area blood copper levels were related to seasonal changes in the blood haemoglobin concentration. Quokkas from West End, on the other hand, had higher and more uniform blood copper levels throughout the year, despite wider fluctuations in haemoglobin concentration. It thus appears, both from the field observations and the laboratory experiment, that induced copper depletion alone does not cause the seasonal anaemia on Rottnest Island, but may be one factor associated with it in the Lakes Area at least (Barker 1961a).

Two other indications came from this work. First, compared with the minimum copper requirements of merino sheep of 6 μg g^{-1} in dry feed, the quokka requires only 3 μg g^{-1} copper (Barker 1960), depending on molybdenum and inorganic sulphate levels in the feed. Second, female quokkas in both study areas in spring had lower blood copper and haemoglobin levels than did males. At other times of the year there was no sex difference. It appeared that the difference in spring blood copper levels could be attributed to the fact that in this season the females were suckling large pouch young. Copper levels in milk samples collected from lactating females in the spring of the following year ranged from 0.13 to 1.46 μg mL^{-1}. In all but two of the 25 lactating females sampled the copper level in milk was higher than that in the blood of the same animal. This indicates that the quokka, unlike the sheep and cow, but like the rat (Underwood 1977), can concentrate copper in the milk. (Both the quokka and the rat also concentrate iron in the milk, something that humans and domestic ruminants cannot do (Kaldor & Ezekiel 1962)). Barker (1962) considered that the maintenance of these milk copper levels could impose sufficient strain on the copper reserves of the females, after a prolonged period of lactation, to be responsible for the low levels of copper in the blood observed in the spring.

Cobalt and vitamin B$_{12}$ On the basis of ruminant studies, another possible factor in the emaciation and anaemia experienced by quokkas on Rottnest is a deficiency of cobalt. Plant cobalt levels on Rottnest are approximately 0.03 μg g^{-1} (Barker 1960), much lower than the cobalt requirements of sheep, viz. 0.08–0.11 μg g^{-1} (Underwood 1977). Barker (1960) found that captive quokkas were able to survive on diets containing as little as 0.01–0.03 μg g^{-1} of cobalt for at least 12 weeks. During this period liver cobalt levels fell from 0.3 – 0.7 μg g^{-1} to 0.08–0.19 μg g^{-1}. In sheep, liver cobalt levels fall from about 0.15 μg g^{-1} to 0.02 g g^{-1} in the deficient state (Underwood 1977). However, no symptoms typical of co-balt-deficient sheep appeared in the quokkas, and no anaemia developed. Although 12 weeks may not be long enough to deplete an animal's cobalt stores, the results nevertheless suggest that cobalt deficiency was not directly associated with the seasonal anaemia in the Rottnest quokka (Barker 1960).

In ruminants a cobalt deficiency is actually a vitamin B_{12} deficiency. The vitamin is synthesised by bacteria in the reticulo-rumen and in other parts of the gut. Thus on low-cobalt diets there are reductions in blood and liver not only in cobalt concentrations but in vitamin B_{12} levels as well. At the end of Barker's (1960) 12-week feeding period, vitamin B_{12} concentrations had fallen from 3–20 μg ml^{-1} to 1–2 μg ml^{-1} in plasma and from 0.8–1.5 μg g^{-1} dry weight of liver to 0.4–0.8 μg g^{-1}. By way of comparison, plasma B_{12} levels in quokkas on Rottnest ranged between 0.5 and 3.0μg ml^{-1}, being lower in juveniles than in adults, and lower in West End than in Lakes Area animals. This latter difference was not correlated with differences in plant cobalt levels (Barker 1960). Nor were seasonal differences in plasma B_{12} concentrations correlated with anaemia, since the lowest plasma B_{12} levels were recorded in spring, not autumn. Thus the field observations supported Barker's (1960) conclusion from his laboratory experiment that cobalt deficiency, like copper deficiency, was probably not directly linked with seasonal anaemia in the Rottnest quokka.

Selenium and vitamin E   For several years during studies on captive quokkas from Rottnest Island, workers observed development of a paralysis of the hind limbs in animals maintained on a diet of high-protein (17–21% crude protein) commercial sheep pellets consisting of bran, pollard, oat meal, linseed meal, whale meal, molasses, urea and mineral salts.

The disorder is characterised by weakness of the hind limbs which begins insidiously and progresses rapidly to complete paralysis. There is marked wasting of the muscles of the pelvic girdle and the disease invariably terminates in death (Kakulas 1961). The same disorder has been observed in other small wallabies such as *Macropus eugenii* and *Thylogale thetis* maintained on lucerne hay in captivity. Histologically, there are marked degenerative changes in the muscles of the hind limbs, lesions typical of a deficiency in vitamin E (α-tocopherol) in the diet.

Kakulas (1961) demonstrated convincingly that the lesions could be completely reversed by oral dosing of affected quokkas with 200–600 mg vitamin E daily for several days. In lambs, the condition (called 'white muscle disease' because of the pallor of the pelvic and femoral groups of muscles) can also be prevented by the addition of as little as 0.1μg selenium g^{-1} to vitamin E deficient diets. This is because both vitamin E and the selenium-containing enzyme glutathione peroxidase protect cells against oxidative damage caused by free radicals formed during cellular metabolism. Radicals are scavenged by vitamin E as a first line of defence. As a second line of defence, glutathione peroxidase destroys any peroxides formed from oxidation of unsaturated fatty acids by free radicals. These two defence mechanisms complement each other (McDonald *et al.* 1995). Selenium therefore has a sparing effect on vitamin E, often to the extent that it is an effective substitute, as has been found in ruminants and many other species (Underwood 1977). One species in which it is not is the rabbit. In further experiments with quokkas, Kakulas (1963a) demonstrated that supplementation of the commercial sheep pellets used before (Kakulas 1961) with selenium (0.5 g kg^{-1} body mass) was without any prophylactic effect. We have found the same with *M. eugenii* maintained on lucerne hay. These results indicate that the quokka (and the tammar) belongs to the small group of animals in which nutritional muscular dystrophy is not prevented by trace amounts of selenium.

During his experiments Kakulas noticed that the smaller the size of the enclosure, the higher was the incidence of nutritional muscular dystrophy in captive quokkas. In a formal experiment with four animals in enclosures of three different sizes (1.2, 9.0 and 30 m^2) replicated four times, Kakulas (1963b) confirmed that the size of the enclosure was an important factor in the development of muscular dystrophy in the quokka. Apparently the additional stress of crowding increases the vitamin E requirement significantly. In all cases, however, administration of the vitamin was completely effective in preventing paralysis in animals in similar and even smaller enclosures.

To ascertain whether the apparently high requirement for vitamin E by the quokka was a factor in the seasonal debility of the quokka on Rottnest Island, Kakulas (1966) measured the vitamin E status of plants and animals throughout the year. Results of these tests suggested that in spring the animals were receiving adequate vitamin E, but that during summer their vitamin E status was marginal. This correlated well with vitamin E levels in plants such as *Atriplex* sp. and *Sporobolus* sp. which were highest in winter and spring, but not with the results of muscle biopsy tests carried out concurrently on the same animals; the proportion of animals with muscle lesions was actually highest in winter. Also, any muscle lesions found were very mild. It would seem from these findings that vitamin E is not in itself an ecological problem for quokkas on Rottnest Island.

THE TAMMAR

Macropus eugenii has been the subject of numerous laboratory investigations, but only comparatively recently has its field ecology received any great attention. Like the quokka, the tammar wallaby was once widespread in south-west Western Australia, and, in addition, was common in South Australia. It is now largely confined to a number of offshore islands extending from Kangaroo Island (36° south latitude) in the east to the Abrolhos

Islands (28–29° south) in the west (Tyndale-Biscoe 1973). In fact, it is on these islands, at opposite ends of its range, that the tammar has been investigated most extensively. Three field studies will be considered here. The first arose from circumstantial evidence that tammars on the Abrolhos Islands supplement their water intake by drinking sea water, the second, on Garden Island, because of its proximity (10 km) to Rottnest Island, and the third from the need for an effective management plan for the tammar on Kangaroo Island where it is regarded as a pastoral pest.

The Abrolhos Islands

The first observations of an Australian marsupial were made by the Dutch navigator Francisco Pelsaert, after his ship the *Batavia* was wrecked on Houtman's Abrolhos in 1629; he described the tammar wallaby. The climate of the Abrolhos is maritime semi-arid (Kinnear, Purohit & Main 1968). The annual average rainfall is no more than 350 mm, whereas the annual evaporation is approximately 1400 mm. For only four winter months does the average rainfall exceed effective rainfall. The long dry summer is characterised by strong and persistent winds. The soils are sandy or rocky and fresh surface water is unavailable for long periods. In this harsh environment the tammar wallaby is subjected annually to a prolonged period of poor-quality food and a shortage of fresh drinking water.

The first aspect of the Abrolhos tammar's physiology to be investigated by Kinnear *et al.* (1968) and Purohit (1971) was its ability to survive when the only drinking water available was sea water. Kinnear *et al.* (1968) caught three tammars at the water's edge on a beach on one side of the Abrolhos Islands. These animals were taken to the University of Western Australia where, seven days after capture and without any acclimation period, they were placed in metabolism cages and offered a maintenance diet of chopped oaten hay supplemented with starch, sucrose, molasses, casein and minerals, and sea water for 30 days. The results were remarkable considering the absence of any acclimation period. A 10–15% weight loss occurred over the first three days, but then weight loss ceased and all animals slowly regained weight. Diarrhoea was never observed. Clearly tammars can survive for considerable periods by drinking sea water, although, as Purohit (1971) demonstrated, dry food and sea water alone will not support the tammar wallaby indefinitely either in the field or in the laboratory.

Additional experiments by Kinnear *et al.* (1968) sought to determine the physiological adaptations involved in the tammar's ability to drink sea water. When fed either a high- or a low-protein dry diet with 100% sea water the maximum urine condosity (i.e. the molarity of a sodium chloride solution having the same specific conductance) was 1.1, very similar to that in Merriam's kangaroo-rat (*Dipodomys merriami*), a desert rodent shown by Schmidt-Nielsen & Schmidt-Nielsen (1950) to be able to survive on a dry grain diet with 100% sea water. This indicated that, like *D. merriami*, the tammar had an excellent capacity for excreting electrolytes economically with respect to water. The absence of diarrhoea on the 100% sea water treatments indicated that tammars have a high capacity to absorb salt and water from the gut, thus avoiding purgation. Kinnear *et al.*(1968) also suggested that their results indicated that the tammar has a low minimal water turnover rate. This was confirmed by Denny & Dawson (1975a); under non-stress conditions these latter workers recorded a water turnover rate of 65 ml $kg^{-0.80}$ d^{-1} in the tammar compared with their mean for five macropodid marsupials of 90 ml $kg^{-0.80}$ d^{-1} (Table 1.5).

The urine-concentrating ability of the Abrolhos tammar is reflected in a high relative medullary thickness (RMT) of its kidneys, 6.8 ± 0.3. This is higher than either the quokka from Rottnest Island (5.1–5.8) (Brown 1964) or the tammar from Garden Island near Rottnest (5.9 ± 0.3, Kinnear *et al.* 1968) (Table 7.3). This raises the question of physiological differences between geographically isolated populations of the same species. As suggested by the lower RMT, Garden Island tammars appear unable to concentrate their urine to the same extent as Abrolhos tammars, since they lost weight on 100% sea water while Abrolhos tammars maintained weight (Kinnear *et al.* 1968).

The other aspect of the Abrolhos tammar examined by Kinnear (1970) was its ability to recycle urea from the blood to the digestive tract when faced with a low-protein diet. Body weights of Abrolhos tammars undergo a cyclic pattern, from a maximum in mid-winter to a minimum in mid-summer.

The weight loss in late spring is associated with the onset of the annual drought and a decline in plant protein levels. Concomitantly there was a fall in the urinary UR ratio. The concept of the UR ratio as an index of urea recycling was introduced in Chapter 5 in relation to the low urinary nitrogen loss and high urea recycling rate of koalas (Cork 1981). The UR ratio is the ratio of urea-nitrogen to total nitrogen concentrations in the urine. In Kinnear's (1970) tammars the UR ratio was 0.72 in winter, but only 0.30 in summer (Fig. 7.8). This suggests that, in summer, when the animals were protein deficient, urea was being retained by the kidneys, and instead of being excreted some was diffusing from the blood into the gut where a certain proportion was incorporated into microbial protein.

In order to test this hypothesis Kinnear & Main (1975) conducted a laboratory experiment in which one tammar wallaby and one red kangaroo (*Macropus rufus*) were infused duodenally with an almost nitrogen-free but otherwise complete nutrient mixture. By by-passing the stomach in this way recycling of urea from the blood was virtually prevented, since the stomach is the major site of microbial protein synthesis in the macropodid digestive tract (Chapter 6). In these two animals the UR ratio was 0.72 and 0.78, very similar to the mean winter value in the field population (Fig. 7.8). On this

Table 7.3. *Relative medullary thickness (RMT) of macropodid and potoroid marsupials. RMT is the medullary thickness × 10 divided by the cube root of the length, breadth and width of the kidney*

Species (number of animals)	Adult body mass (kg)	RMT	Habitat	Ref.
Dendrolagus matschiei (1)	9	3.9	Tropical rainforest	Yadav 1979
Thylogale thetis (4)	4–7	5.7 ± 0.2	Temperate rainforest	Hume & Dunning 1979
Setonix brachyurus	2–4	5.1 – 5.8	Mediterranean, summer drought	Brown 1964
Macropus irma (1)	8	5.8	Mediterranean, summer drought	Yadav 1979
Bettongia penicillata (2)	1–2	5.8, 5.9	Mediterranean, summer drought	Yadav 1979
Macropus eugenii (19)	6–8	5.9 ± 0.3	Mediterranean, summer drought	Kinnear *et al.* 1968
Macropus giganteus (1)	32–66	5.3	Mediterranean, summer drought	Yadav 1979
Macropus rufus	27–66	5.8	Semi-arid to arid	Dawson & Denny 1969
M. robustus erubescens	25–47	7.2	Semi-arid to arid	Dawson & Denny 1969
Peradorcus concinna (1)	1–2	7.3	Semi-arid, winter drought	Yadav 1979
Macropus eugenii (18)	6–8	6.8 ± 0.3	Maritime semi-arid	Kinnear *et al.* 1968
Lagorchestes hirsutus (2)	1–2	6.0, 7.2	Maritime arid	Yadav 1979
Lagostrophus fasciatus (1)	1–3	7.2	Maritime arid	Yadav 1979
Bettongia lesueur (2)	1–2	8.0, 8.8	Maritime arid	Yadav 1979
Lagorchestes conspicillatus (2)	2–5	8.4, 9.0	Maritime arid	Yadav 1979

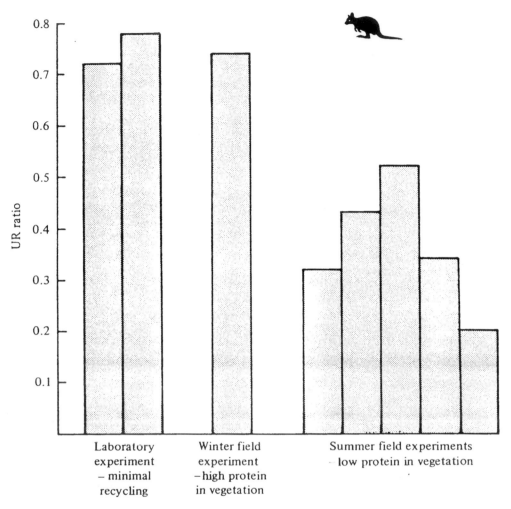

Figure 7.8 The proportion of total urinary nitrogen in the form of urea (the UR ratio) in captive tammar wallabies (*Macropus eugenii*) in which recycling of urea from the blood to the digestive tract was minimised, compared with free-living animals on the Abrolhos Islands, Western Australia, in winter when protein levels in the vegetation are maximal and in summer when the vegetation is low in protein content. After Kinnear & Main (1975).

basis Kinnear & Main (1975) argued that the Abrolhos tammars in winter were not recycling significant quantities of urea, presumably because their dietary protein intake was more than adequate to satisfy the nitrogen requirements for maximal microbial growth in the forestomach. However, in summer, from the low UR ratio it appeared that urea recycling made a substantial contribution to the animal's ability to utilise the low-protein vegetation available. Support for this contention comes from the stabilisation of body weight in the field between mid-summer and the beginning of the winter rains, suggesting that the animals must have been close to nitrogen equilibrium during the autumn period.

In another experiment Kinnear & Main (1975) captured tammars on the Abrolhos Islands at dusk and kept them in individual metabolism cages for 12 hours. Over this 12 hour period the fate of urea injected intraperitoneally was estimated from urinary urea output and changes in the size of the plasma urea pool. The use of this technique by Wake (1980) with quokkas on Rottnest Island has already been described (page 280). The disappearance of injected urea from the blood of tammars in early and late summer suggested to Kinnear and Main (1975) that at those times nitrogen was the primary limiting factor in the vegetation. In mid-summer, however, all the injected urea was either detectable in the blood or excreted in the urine. The lack of urea utilisation at this time implies that energy, rather than nitrogen, was primarily limiting. When Kinnear (1970) administered starch by stomach tube to these animals, urea utilisation improved dramatically, confirming that the supply of readily fermentable energy to the forestomach microbes was suboptimal in mid-summer.

Thus deficiencies of both energy and protein have been implicated in the seasonal weight loss of the Abrolhos tammar population, just as they were in Wake's (1980) study of the quokka on Rottnest Island. However, because of the tammar's well-adapted kidneys, dehydration and salt loading appear not to be important.

Garden Island

Garden Island has twice the annual rainfall and a much longer growing season than the Abrolhos Islands. Its natural vegetation is similar to that of Rottnest, only 10 km to the north, but Garden Island has been less altered by human activity. Bakker, Bradshaw & Main (1982) compared seasonal changes in body condition and water and electrolyte metabolism in two populations of tammars, one near the coast with no access to fresh water, the other in the vicinity of a continuously flowing fresh water bore. During a mild summer the availability of fresh water had no significant influence on body condition or water and electrolyte metabolism, and there were none of the signs of the seasonal debility seen so regularly in the Rottnest quokkas nearby. The coastal animals were able to maintain body condition by having very low water turnover rates and by balancing water influx and efflux. They did not drink sea water as Abrolhos tammars do, a conclusion based on direct observations and on their low plasma sodium concentrations at the end of summer.

Only after a particularly long summer drought was the availability of fresh water of substantial benefit. Then, compared with the bore population, coastal animals were in poor condition, and had high circulating levels of corticosteroids and antidiuretic hormone (ADH), indicative of stress related to water shortage. Body condition continued to decline, even after a mild summer, to a minimum in mid-winter. Bakker et al. (1982) thought that this was due to the high water content of the vegetation at this time, which prevented the tammars from ingesting enough energy to satisfy requirements, but firm evidence for this is lacking. Body condition peaked in early summer, when the animals are neither water deprived nor suffering from nutritional deficiency.

Kangaroo Island

Like Garden Island, and in contrast to the Abrolhos Islands, Kangaroo Island has an average rainfall of 800 mm and a growing season of about 6.5 months. However, the summers are still dry, and the tammars inhabiting Flinders Chase National Park on the western extremity of the island exhibit cyclical weight changes. Inns (1980) found that males underwent greater fluctuations in body mass than females (Fig. 7.9), and he suggested that mating activity by males in late summer contributed to this difference. However, there was considerable loss in body mass of both females and males between early summer and winter; in one year (1978) this amounted to 16% in females and 21% in males. This was reflected in seasonal changes in the amount of fat present within the peritoneal cavity (Fig. 7.10).

Figure 7.9 Seasonal changes in body mass of adult male and female tammar wallabies (*Macropus eugenii*) on Kangaroo Island, South Australia. After Inns (1980).

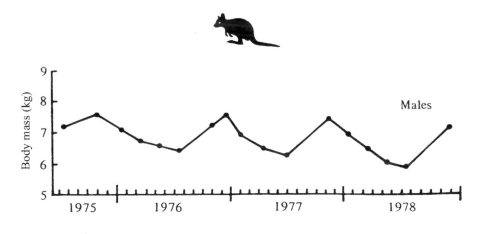

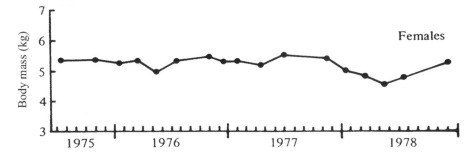

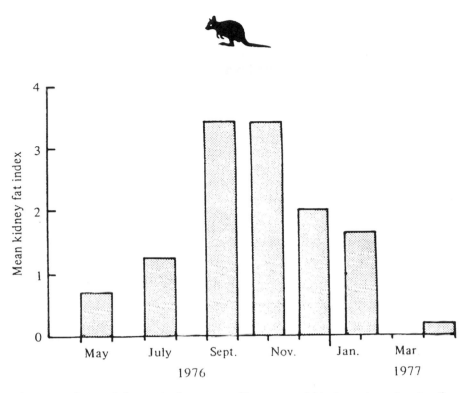

Figure 7.10 Seasonal changes in the amount of fat present within the peritoneal cavity of male tammar wallabies (*Macropus eugenii*) on Kangaroo Island, South Australia. The kidney fat index is based on a visual assessment of kidney and mesenteric fat. After Inns (1980).

Similarly, growth rates of young tammar wallabies after leaving the pouch for the last time in spring were rapid, but over the late summer and winter period there was little growth.

For the Kangaroo Island tammar there is no evidence that nitrogen is a limiting factor at any time of the year, as it is for the Abrolhos tammar. Barker (1971) caught tammars on Kangaroo Island at various times throughout the year and held them in individual metabolism cages without food or water for 24 hours. Urine and nitrogen excretion were both less in summer than in winter, but the UR ratio was never less than 0.70, indicating that nitrogen deficiency was probably not a problem. Inns (1980) presented radiotracking data which indicated that the tammars on Kangaroo Island expanded their home ranges in summer, presumably because they were searching for better quality food as the annual grasses dried off. In this regard Andrewartha & Barker (1969) observed that tammars foraged beneath *Acacia retinodes* trees during summer, suggesting that the animals may have been obtaining nitrogen from the seeds of this leguminous species.

Thus it is more likely that the Kangaroo Island tammar suffers from an inadequate energy intake over the late summer and early winter. The greater output of faecal dry matter at this time of year (Barker 1971) suggests that digestibility of the tammar's diet is low during the summer months. Greater energy expenditure, as suggested by an increase in the size of their home ranges in this season, would tend to exacerbate any energy deficiency. There was no evidence of severe dehydration in Inns' (1980) study.

Since nitrogen and water do not appear to be limiting factors on Kangaroo Island with its higher annual rainfall and longer growing season, tammar wallabies from this island may be expected to be less able to cope with shortages of nitrogen and water than Abrolhos tammars. No direct comparisons between Abrolhos and Kangaroo Island tammars have been reported, but they would be interesting in view of the possible physiological differences between the two populations already alluded to.

THE HARE-WALLABIES

Lagorchestes conspicillatus (spectacled hare-wallaby) appears to be declining throughout much of its mainland range, but is secure on Barrow Island off the north-west coast of Western Australia. There it shelters during the day under spinifex hummocks, coming out to feed at night. It does not drink, even when water is available. Bakker & Bradshaw (1989) compared the body condition and water and electrolyte metabolism of two populations in October, at the end of the dry season, and in March, after 171 mm of rain. The water content of the most frequently eaten food plants increased from 58% in October to 69% in March, and the protein content from 7.2% to 12.8% of dry matter over the same period. One population was in undisturbed *Triodia* hummock grassland in the north of the island, the other in a disturbed area in a large oilfield in the south. *Triodia pungens* was the main part of the diet in the north, whereas in the south several colonising species made up the larger part of the diet.

Body condition of males in the north improved between October and March, but that of both sexes in the south was consistent between seasons. All animals were in water balance, but water turnover rates in the northern animals in October were the lowest recorded for a macropod (Table 1.7). Water turnover rates more than trebled in the northern animals between the dry and wet seasons, but those in the southern animals barely changed. Plasma corticosteroid levels were higher, and plasma urea concentrations were lower in northern than in southern animals in October but not in March. Taken together these findings suggest that although the spectacled hare-wallaby is superbly adapted to surviving extremely desiccating conditions on Barrow Island, in undisturbed areas it suffers from shortages of both protein and water towards the end of the dry season. Laboratory findings of very low glomerular filtration rates, especially on a low-protein diet, and the ability to elaborate a highly concentrated urine (4000 mOsm kg^{-1}) (Bakker & Bradshaw 1983), are all consistent with the performance of free-living animals. A high relative medullary thickness (RMT) of the kidneys (8.7) (Table 7.3) reflects their concentrating power. Whether mainland *L. conspicillatus* show these adaptations to the same degree is unknown.

7.4.2 The arid-zone species

THE EURO

Barrow Island is the smallest island (233 km^2) off the coast of Australia to support an isolated population (about 1800) of one of the larger macropods, the euro *Macropus robustus isabellinus* (Short & Turner 1991). However, compared with *L. conspicillatus* on Barrow Island, we know little about its physiology. The most comprehensive field studies on the euro (*M. robustus erubescens*) are those of Ealey (1967a, b) and Ealey & Main (1967). More recent work on the nutritional requirements and digestion and metabolism in the euro has confirmed many of Ealey's predictions, and we now have a good understanding of the ways in which the euro survives in and exploits its often very harsh mainland environment.

Ealey's study area was the Pilbara district in the north west of Western Australia. It is an area of leached-out soils and poor vegetation. Low rainfall, low humidity and very high summer temperatures produce a harsh environment in which, in aboriginal times, the numbers of euros were low and their distribution patchy. The long hot summers and lack of surface water would have induced severe annual mortality, and restricted hardy survivors to rocky hills. Here they would have found caves in which to escape the desiccating heat, and patches of *Triodia pungens* (spinifex) on which to feed.

Ecology

The introduction of domestic sheep by Europeans into the Pilbara in 1866 (Ealey 1967a) disrupted the pattern of annual mortality in the euro population. The sinking of dams and wells every 5–8 km meant that water no longer acted as a population control (Ealey 1967b). Overgrazing by sheep meant that the more nutritious native grasses were soon eaten out and the durable spinifex encroached onto the plains, eventually to dominate them. Spinifex is a submaintenance diet for sheep, but not for the euro with its low maintenance protein requirement (Table 1.8). Consequently, euro numbers increased along with the sheep population. By the late 1920s sheep numbers had increased to nearly 800 000 but a drought in 1935–36 caused a drastic reduction in numbers, and numbers have since continued to fall. Meanwhile the euro continued to thrive, becoming more abundant than the sheep. Today, euros even inhabit the plains, sheltering under trees by day like red kangaroos. The additional water lost in cooling is readily recouped at the stock-waters that were so important in disrupting the old ecological regime in the first place (Newsome 1975).

Ealey & Main (1967) studied the seasonal cycle in the nutritional status of two euro populations in the Pilbara, one where there was only low-protein forage available (Mt. Edgar), the other where a mixed vegetation existed containing plants of high and low-protein content (Woodstock). Although euros are able to live and even to breed on low-protein vegetation, many thin and starving animals were seen by Ealey & Main (1967) during periods of protein shortage, even though there was an excess of edible vegetation. After a prolonged dry season spectacular mortality may occur. The animals which died were always emaciated. Thus it seems probable that nutrition now plays a major role in the regulation of euro density in areas of adequate water supply in the Pilbara.

Two types of mortality were observed by Ealey & Main (1967). The first was the regular seasonal die-off of a certain proportion of the population at Mt. Edgar where only the hardiest animals were able to survive semi-starvation almost every summer. Mt. Edgar euros consistently had lower blood haemoglobin levels in summer than did Woodstock euros, reflecting the low-protein status of the vegetation (Fig. 7.11).

The second type of mortality was the population crash seen in the Woodstock euros after a particularly dry summer. In most years, early summer storms and the subsequent monsoonal rains resulted in Wood-stock euros being in a good nutritional state to survive the latter part of the summer and autumn. A number of such years resulted in a gradual increase in the population, with little selection pressure against less fit individuals. When the euro density became high the more nutritious vegetation would be eaten very early in the season. The animals of this dense population could not attain a sufficiently good state of nutrition to last them over a dry summer. A prolonged dry season would result in a spectacular population crash, as occurred in 1954 (Ealey & Main 1967).

Physiology

Notwithstanding such population crashes, compared with the sheep the euro must be regarded as being extremely well adapted to an arid environment. The physiological bases of this adaptation have been reasonably well researched. The first of these is related directly to the generally low basal metabolic rate of marsupials (Chapter 1). Thus the adult euro has a maintenance energy requirement 27% below that of sheep (Table 1.2). Its maintenance nitrogen requirement of 160 mg truly digestible nitrogen. $kg^{-0.75} d^{-1}$ (Brown & Main 1967; Freudenberger & Hume 1992) is the lowest of the macropodid marsupials so far examined (Table 1.8), and less than half that of eutherians such as the sheep and horse. Similarly, the euro has a water turnover rate under non-stress conditions approximately 22%

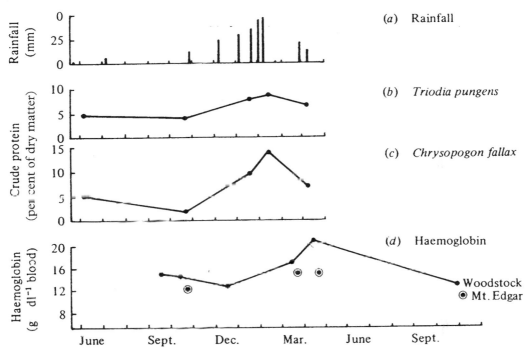

Figure 7.11 Effects of rain (a) on the protein content of (b) *Triodia pungens* (spinifex) and (c) *Chrysopogon fallax* in the Pilbara region of Western Australia. Note the difference in protein content of the two species in summer. Mt. Edgar vegetation consists mainly of *Triodia*; this is reflected in the lower haemoglobin concentrations in blood (d) of Mt. Edgar euros (*Macropus robustus erubescens*) compared with Woodstock euros. After Ealey & Main (1967).

below that of the eutherian mean (Table 1.6), although not different from many other marsupials studied.

These comparatively low nutrient requirements are all factors contributing to the ability of all arid-zone macropodid species to survive long periods of poor nutrition. The efficiency with which the euro utilises and retains both nitrogen and sulphur when fed poor-quality roughage was demonstrated by Hume (1974) in a comparison with red kangaroos and sheep (Table 7.4). All three species were in positive balance with respect to both elements on the good-quality chopped lucerne hay diet. On the medium-quality chopped oaten hay, however, the red kangaroos were unable to conserve enough nitrogen or sulphur and slipped into negative balance. On the poor-quality milled wheaten straw diet all three species were in negative nitrogen balance, but both the euro and the sheep re-

Table 7.4. *Intake and retention of the nitrogen and sulphur of three diets by euros* (Macropus robustus erubescens), *red kangaroos* (Macropus rufus) *and sheep*

	Chopped lucerne hay	Chopped oaten hay	Milled wheaten straw
Nitrogen			
Intake (g kg$^{-0.75}$ d^{-1})	1.51	0.50	0.11
Euro	1.46	0.43	0.06
Red kangaroo	2.56	0.64	0.14
Sheep			
Balance (g kg$^{-0.75}$ d^{-1})			
Euro	+0.37	+0.14	−0.19
Red kangaroo	+0.14	−0.12	−0.57
Sheep	+0.66	+0.11	−0.24
Sulphur			
Intake (mg kg$^{-0.75}$ d^{-1})			
Euro	102	50	34
Red kangaroo	104	43	20
Sheep	173	53	43
Balance (mg kg$^{-0.75}$ d^{-1})			
Euro	+11	+0	+1
Red kangaroo	+1	−3	−38
Sheep	+12	+14	+2

Source: After Hume 1974.

mained in slight positive sulphur balance. The wheat straw contained less than 2% crude protein, even lower than the spinifex analysed by Ealey & Main (1967) during the dry season in the Pilbara. Fibre digestibility was higher in the euro than in the red kangaroo on the lucerne hay diet, although there was no difference between the two macropods on the straw diet (Table 6.5).

Similarly, the euro displays a suite of characteristics that make it more able to withstand water shortages than can the conspecific eastern wallaroo (*M. robustus robustus*). These include lower voluntary water intakes, an increase in fibre digestibility and maintenance of nitrogen balance during water restriction, and lower faecal water efflux associated with a consistently lower faecal water content (54% versus 59% in the wallaroo) during water restriction (Freudenberger & Hume 1993). The euro's colon is 37% longer than that of the wallaroo, which explains at least part of its thrifty water economy.

These results serve to illustrate how well the euro is physiologically adapted to utilise low-protein, high-fibre forages such as spinifex in the arid zone. In addition to Hume's (1974) study a number of other investigations have compared the euro with the red kangaroo, the other large arid-zone macropod.

THE RED KANGAROO

In contrast to the sedentary euro, the red kangaroo (*Macropus rufus*) (Fig. 7.12) is a highly mobile species of the open plains. Frith (1964), Newsome (1965) and Bailey (1971) have all recorded the movements of red kangaroos to areas of better quality feed, and in response to storms 10–20 km distant. Thus it appears that whereas the euro is physiologically adapted to survival in arid regions where the quality of available forage is often low, the red kangaroo owes its survival in the arid zone more to behavioural responses.

Ecology

Newsome (1975) studied a red kangaroo population near Alice Springs in the Northern Territory over a period which covered the longest drought on record in central Australia. This drought began in 1958 and lasted until 1966. Red kangaroos were relatively uncommon in the Alice Springs area before Europeans settled there. The two limiting factors were probably water and protein, just as in the case of the euro of the Pilbara. Their drought refuges were the flood-outs at the ends of creeks. Here the red kangaroos found almost continuous shade under dense stands of trees, and

Figure 7.12 The red kangaroo (*Macropus rufus*) from inland Australia. (Ray Williams)

persistent green grass. After widespread good rains, when green grass would have been widely abundant, they could disperse and breed. During drought, however, few would have survived outside these drought refuges. Europeans released ruminant stock onto the grassy plains in the 1870s, and sank wells and dams every 5–15 km across them to provide their stock with water. This removed one factor limiting the size of the red kangaroo population. The cropping of the standing long dry grass by domestic stock created a subclimax grassland of generally shorter but greener, higher-protein food. This largely removed the second factor limiting the red kangaroo population. This meant that breeding could be prolonged with greater success, and more kangaroos could survive periodical droughts. The result was an enormous increase in red kangaroo numbers throughout the arid zone. However, more recently there has been a continuous decline in red kangaroo numbers in north-western Australia, including the Pilbara district (Ealey 1967a), presumably as a result of overgrazing of the more nutritious pasture species by red kangaroos, euros and domestic stock. The accompanying encroachment of spinifex onto the open plains in the Pilbara has excluded both sheep and the red kangaroo from much of that area. Only the euro, with its physiological capacity to utilise the coarse spinifex, has continued to thrive (Newsome 1975).

Reproduction

The significance of the abundance and quality of food to the red kangaroo is emphasised by the relationship in Fig. 7.13 between the percentage of adult females in breeding condition and the abundance of food (Newsome 1966).

In addition to a failure of post-partum ovulation during drought (Newsome 1964), other effects of reduced food supply on the red kangaroo have been documented by Frith & Sharman (1964). There was some intra-uterine mortality but the principal mortality occurred among pouch young at the end of pouch life, from 196 days of age onwards, and among the young-at-foot. The mortality was greatest in drought areas where 83% failed to reach maturity.

Physiology

Returning now to physiological comparisons between the red kangaroo and the euro, the data of Denny & Dawson (1975a) suggest no significant difference in water turnover rate in non-stress situations between the two species. Similarly, both species were 'camel-like' in their response to dehydration (when there was a 20% reduction in body weight) in that plasma volume was maintained, falling by only 8% in red kangaroos and 7% in euros (Denny & Dawson 1975b). Animals that are capable of maintaining a constant plasma volume have a greater chance of survival during dehydration because of the effects on blood viscosity (Schmidt-Nielsen 1964). Of the large mammals that have been studied only two can maintain plasma

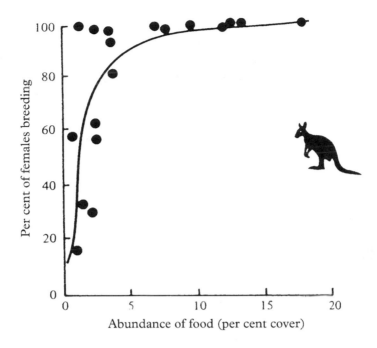

Figure 7.13 Relationship between the percentage of adult female red kangaroos (*Macropus rufus*) in breeding condition and the abundance of food. After Newsome (1966).

volume during dehydration as effectively as the euro and the red kangaroo. These are the camel, which loses less than 10% of its plasma volume during a 20% weight loss (Schmidt-Nielsen 1964), and the burro (*Equus asinus*), which loses 7% of its plasma volume during a 20% weight loss (Yousef, Dill & Mayes 1970). Other mammals which have been studied lose from 40% to 50% of their plasma volume for a similar level of dehydration.

The pattern of water loss from other body compartments differed between the two macropodid species, particularly gut water loss. This compartment contributed 56% of the total water loss of red kangaroos but only 22% of the loss from euros. The preferential maintenance of gut water in the euro at the expense of interstitial water could give that species a significant advantage in maintaining fermentation in the forestomach during drought. Freudenberger & Hume (1993) found that water restriction actually increased fermentation rate in the tubiform forestomach of the euro, but not in that of the eastern wallaroo, nor in other fermentation regions (sacciform forestomach, hindgut) of either species. Effects in the red kangaroo have not been measured.

Also of significance is the finding by Denny & Dawson (1977) that euro kidneys reabsorbed much more urea from the glomerular filtrate when dehydrated (89%) than did red kangaroo kidneys (69%). As a consequence plasma urea concentration in euros was almost twice that of red kangaroos when both species were dehydrated to 14% reduction in body weight; in hydrated animals plasma urea concentrations were similar between the species. Although not measured, recycling of urea to the gut in the dehydrated euro would be expected to be significantly greater than in the dehydrated red kangaroo.

The combination of the preferential maintenance of gut water and the expected greater urea recycling in the euro under conditions of water shortage must be of great ecological significance. Since euros are much less mobile than red kangaroos the continued efficient functioning of the digestive system during drought when only poor-quality feed is available must be important to the long-term survival of this sedentary species.

The ability of the euro to reabsorb more filtered urea than the red kangaroo was correlated with the relative medullary thickness (RMT) of their kidneys which was 7.2 in the euro and 5.8 in the red kangaroo (Dawson & Denny 1969). The significance of the RMT as an index of the concentrating ability of the kidney and as a measure of the animal's ability to withstand dehydration is discussed in Chapter 2. The RMTs of a range of macropodid marsupials are shown in Table 7.3. Although there is a marked difference in RMT between *Dendrolagus matschiei* from New Guinea at one end of the spectrum and *Lagorchestes conspicillatus* and *Bettongia lesueur* from Barrow Island off the north-west Australian coast at the other (Yadav 1979), within the range there is considerable overlap between habitat types. Thus minor differences in RMT must be interpreted with some caution, particularly because of the strong negative relationship between body size and urine concentrating ability (Beuchat 1990a, b). For instance, despite its lower RMT the red kangaroo has better urine concentrating abilities than the euro (Dawson & Denny 1969). This may be related to the greater heat loads and thus a greater need to conserve water in the open plains-dwelling red kangaroo than in the euro which traditionally shelters from the summer heat in caves and under rock ledges. Dawson & Brown (1970) also found that the red kangaroo has fur which gives greater protection from solar radiation in summer, and from heat loss in winter, than does that of the euro. This difference can again be related to the contrasting micro-habitats of the two species.

7.4.3 The forest kangaroos and wallabies

GREY KANGAROOS

In general grey kangaroos occur in areas of higher, more predictable rainfall than the arid-zone kangaroos. Their preferred habitat includes woodlands, shrublands, open forests and semi-arid mallee open scrubs (Russell 1974). There are two distinct species, *Macropus giganteus* (eastern grey kangaroo) and *M. fuliginosus* (western grey kangaroo). Their distributions overlap in western New South Wales (Fig. 7.14).

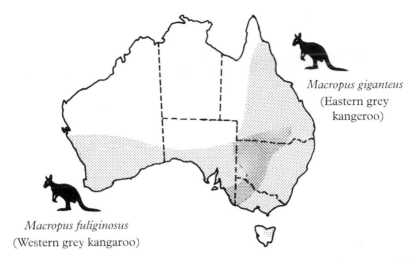

Figure 7.14 Distribution of the western grey kangaroo (*Macropus fuliginosus*) and the eastern grey kangaroo (*M. giganteus*). After Strahan (1995).

The eastern grey kangaroo

The eastern grey kangaroo is common throughout its range. Nevertheless relatively few studies have been made of its field biology, apart from movements and feeding patterns. The reports of Caughley (1964), Kirkpatrick (1965) and Bell (1973) all contain the remark that *M. giganteus* spends most of the day resting in shady open forest or woodland close to more open feeding areas. Partial clearing of woodland improves the habitat of *M. giganteus* by increasing the supply of grass (Calaby 1966).

The comprehensive studies carried out by Dellow (1979) on the digestive physiology of the eastern grey kangaroo are described in Chapter 6. Foley, Hume & Taylor (1980) established the maintenance nitrogen requirement of *M. giganteus* fed a series of diets based on chopped oaten hay with graded levels of casein. The value of 270 mg truly digestible nitrogen $kg^{-0.75}$ d^{-1} is close to the maintenance requirements of other macropods included in Table 1.8, with the exception of the euro which has the lowest requirement so far recorded for a macropod (Brown & Main 1967), and the red-necked pademelon and parma wallaby, which have the highest. The maintenance requirement of the sympatric wallaroo (*M. robustus robustus*), 240 mg truly digestible nitrogen $kg^{-0.75}$ d^{-1}, was similar to that of *M. giganteus* (and thus significantly higher than the other subspecies of *M. robustus*, the arid-zone euro). Analysis of the stomach contents of *M. giganteus* and *M. robustus robustus* showed that there were no significant differences between the two species in the nitrogen or fibre content of food consumed in summer or winter. From the results of their feeding experiments in the laboratory, Foley *et al.* (1980) concluded that nitrogen intakes were probably adequate to maintain both species in positive nitrogen balance throughout the year in the two study areas, one of which consisted largely of improved pasture species, but the other only of native pasture species.

Dellow *et al.* (1988) estimated the rate of production of short-chain fatty acids (SCFA) *in vitro* in *M. giganteus* shot in the field while grazing either native or improved pasture in late winter. Results were comparable with or slightly lower than estimates made by the same technique in the laboratory with animals maintained on chopped lucerne hay. Again in line with laboratory results SCFA production in the field animals was slower (mean of 21 mmol L^{-1} h^{-1}) in the tubiform region of the forestomach than in the sacciform region (29 mmol L^{-1} h^{-1}), and declined along the length of the tubiform region. SCFA production in the hindgut was slower still, 11 mmol L^{-1} h^{-1}.

A nutrient that is in short supply in some of the eastern grey kangaroo's range, namely the Snowy Mountains in south-eastern New South Wales, is sodium. Another sodium-deficient area is the Pilbara in north-western Western Australia.

Sodium is not a nutrient likely to be limiting to offshore island populations of quokkas, tammars or spectacled hare-wallabies. Urinary sodium/potassium ratios can be used as a guide to the sodium status of the diet. Thus when Main (1970) compared urinary Na/K ratios in free-living Rottnest quokkas and Abrolhos tammars with those of euros and red kangaroos in the Pilbara region of Western Australia he found much higher ratios in the island wallabies. Average Na/K ratios in the quokkas were 3.4–7.7 in coastal animals and 2.3–3.6 in inland animals, and in Abrolhos tammars average Na/K ratios ranged from 2.5 to 4.5. In contrast, in the Pilbara, euros averaged 0.5 and in red kangaroos 0.4. Analysis of the spinifex and other native grasses in the diet and which grow on deep sands in the Pilbara showed no sodium.

The grasses eaten by eastern grey kangaroos (and common wombats) in the Snowy Mountains (and to a lesser extent the grasslands surrounding Canberra nearby) are also very low compared with grasses eaten by these two herbivores in coastal Victoria (Table 4.3). Consequently, the urine of Snowy Mountain animals contained virtually no sodium. The adaptations to a constant sodium deficiency that Blair-West *et al.* (1968) found in the eastern grey kangaroo included enlarged adrenal glands, particularly the zona glomerulosa (the site of synthesis and secretion of aldosterone), elevated circulatory levels of aldosterone and structural changes for conserving sodium by the salivary glands. Animals from the Snowy Mountains had a much more extensive duct system of both the parotid and submandibular glands, and blood vessels were extraordinarily abundant around the striated (secretory) ducts. These structural differences are adaptations for conserving sodium. Although salivary concentrations of sodium and potassium were not measured by Blair-West *et al.* (1968), there would undoubtedly be a much lower Na/K ratio in the saliva of alpine *M. giganteus* than in that of coastal animals.

The generally higher urinary Na/K ratios in red kangaroos (1.2) and euros (0.9) at Fowlers Gap in western New South Wales (Dawson & Denny 1969) than in the Pilbara reflect the higher sodium status of the soil (3.6 mEq kg^{-1}, similar to soils in coastal Victoria (Table 4.3)) and consequently higher sodium levels in the grasses (96–22 mEq kg^{-1} dry matter, again in much the same range as grasses growing on the Victorian coast). The sodium content of *Atriplex* sp. (saltbush) averaged 2200 mEq kg^{-1} dry matter; halophytic chenopods, principally *Atriplex* and *Kochia* (bluebush), make a small contribution to the diet of both species of kangaroos at Fowlers Gap (Dawson & Ellis 1994, 1996).

Halophytic chenopods are much less available to euros and red kangaroos in the Pilbara. However, not all the soils are as sodium-deficient as the sands mentioned above. These variations in soil type are reflected in the sodium status of the herbivores. Main (1970) found that during a period of good rainfall and abundant plant growth about 30% of euros and red kangaroos had much lower body weights than the rest of the population. Further, these low body weights were correlated with very low urinary sodium excretion as indicated by very low sodium/potassium ratios in the bladder urine of shot animals. It was apparent that these animals were severely sodium-depleted.

Main (1970) tested this idea in the euros. By assuming that the daily excretion of creatinine by the euro was constant per unit of body weight (Fraser & Kinnear 1969), from the amount of creatinine in a known urine volume Main (1970) calculated that there were no gross differences in daily urine production between animals with average or with very low sodium concentrations in the urine. This confirmed that the low-sodium concentration of the urine was not due to dilution resulting from high urine flows.

The explanation for the very low urinary sodium excretion in some animals came from analysis of plant material in the mouth of euros shot while grazing. These animals appeared to be eating mainly spinifex (*Triodia*) and other native grasses which grew in deep sands and showed no sodium. On the other hand, animals with average sodium excretion were found to be eating mainly the grasses *Cenchrus* and *Eriachne*, which grew on loams and clays between the sand dunes and showed sodium concentrations between 20 and 450 mEq kg^{-1} dry matter. These levels are quite high compared with grasses from the sodium-deficient Snowy Mountains (Table 4.3). Thus, the differences in urinary sodium excretion, at least in the euros, were probably due to differences in the diet of this sedentary species. The low body weights of sodium-deficient euros and red kangaroos were probably due to the fact that these animals would be unable to rehydrate fully by drinking since to do this would mean increased sodium losses in the increased urine flow. Low water intakes would depress intakes of dry matter (and therefore energy), leading to poor body condition, even though food was abundant.

The western grey kangaroo

The Western Grey Kangaroo has received least attention of all the large kangaroos. Prince (1976) conducted a series of physiological experiments comparing *M. fuliginosus* with the euro and the red kangaroo in both the laboratory and field. Fermentation rates were measured in the forestomach of animals shot in the field. The sampling procedure covered most of the 24 hours of the day, enabling estimates to be made of total daily SCFA production in each species. There were two peaks of fermentation activity, corresponding to peaks of grazing activity in the early morning and late afternoon (Fig. 7.15).

Fermentation rates indicated that SCFA production rate was higher and total daily SCFA production greater in wild western grey kangaroos than in either euros or red kangaroos (Fig.7.16). This could be a reflection of either a basic physiological difference between *M. fuliginosus* and the two arid-zone kangaroos, or a higher nutritive value of pasture grazed by *M.*

Figure 7.15 The pattern of short-chain fatty acid (SCFA) production in the forestomach of free-living western grey kangaroos (*Macropus fuliginosus*), red kangaroos (*M. rufus*) and euros (*M. robustus erubescens*) determined *in vitro*. The two peaks in production correspond with peaks of foraging activity in the evening and early morning. After Prince (1976).

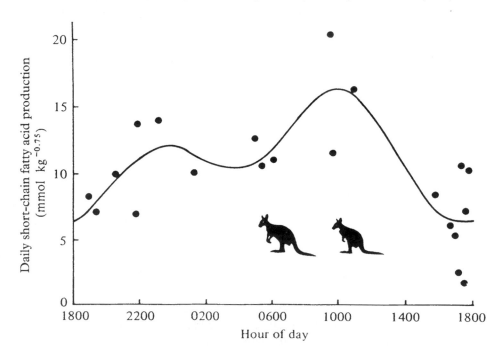

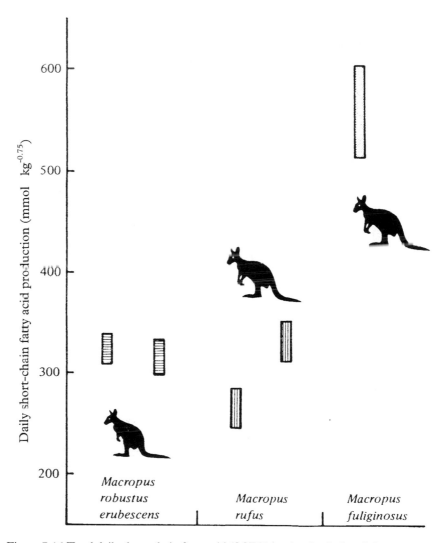

Figure 7.16 Total daily short-chain fatty acid (SCFA) production in free-living euros (*Macropus robustus erubescens*), red kangaroos (*M. rufus*) and western grey kangaroos (*M. fuliginosus*) measured by *in vitro* incubation. After Prince (1976).

fuliginosus in the south-west corner of Western Australia compared with that of arid-zone pastures. The higher nitrogen concentration in fore-stomach digesta of *M. fuliginosus* (Fig. 7.17) suggests that the western grey was consuming higher quality plant material than were the other two species. However, when all three species were maintained on common diets in captivity, dry matter digestibility of a low-nitrogen diet was highest in *M. fuliginosus*. This is good evidence that there is indeed a difference in digestive efficiency between the western grey kangaroo and the euro and red kangaroo.

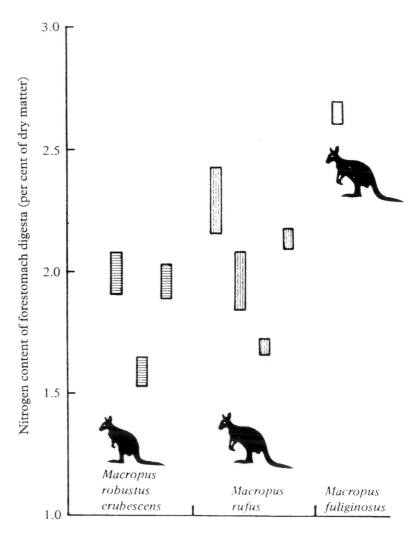

Figure 7.17 Nitrogen content of forestomach digesta in free-living euros (*Macropus robustus erubescens*), red kangaroos (*M. rufus*) and western grey kangaroos (*M. fuliginosus*). The high nitrogen content in *M. fuliginosus* is consistent with the high fermentation rate in Fig. 7.16. After Prince (1976).

Prince (1976) also found that total and inorganic phosphorus concentrations in forestomach digesta fluid of field animals were higher in *M. fuliginosus* than in the red kangaroo, although not the euro. He suggested that this difference could reflect differences between the euros and western greys on the one hand and red kangaroos on the other in the recycling of endogenous phosphorus to the stomach via salivary secretions, and hence differences in their capacity to maintain conditions favourable for microbial fermentation. This suggestion is in general agreement with differences in fibre digestion and nitrogen and sulphur retention found by Hume (1974) between euros and red kangaroos in the laboratory. It also finds support in the higher maximal flow rates, and higher rates of parotid urea clearance and higher rates of phosphate secretion at equivalent flow rates in eastern grey kangaroos than in euros and red kangaroos (Beal 1989, and Section 6.2.5). However, it would be useful to have the differences found by Prince (1976) in phosphorus concentrations in the field confirmed in captive animals maintained on common diets.

PADEMELONS

The range of the genus *Thylogale* extends from the Bismarck Archipelago (3° S latitude), through New Guinea (*T. brunii*) and the east coast of the Australian mainland (*T. stigmatica*, *T. thetis*), to Tasmania (*T. billardierii*) at 43° S latitude. The habitats occupied by *Thylogale* have not been described in detail but the primary requirement appears to be dense forest vegetation, either rainforest (closed forest) or wet sclerophyll forest (tall open forest).

Pademelons are most numerous at the forest edge, from which they move short distances at night to feed in open areas (Calaby 1966). Johnson (1980b) found that radio-tracked red-necked pademelons (*T. thetis*) rarely moved more than 70 m from their rainforest refuge when feeding in open fields at night. During the day they moved up to 500 m into the rainforest, apparently searching for food, since Johnson (1980b) often saw them feeding on ground vegetation. In fact he remarked that to refer to *T. thetis* as nocturnal is misleading. Although the population of *T. thetis* studied by Johnson (1980b) near Dorrigo in north-eastern New South Wales occupied habitat greatly modified by Europeans, Johnson (1980b) argued that the present mosaic of open pasture and dense rainforest probably resembles the situation obtaining before the advent of Europeans, when there were, even then, large, naturally occurring grasslands bordering rainforest, possibly as fire climaxes resulting from Aboriginal activities (Calaby 1966).

Although there is a winter trough in food availability, Johnson (1977) could find little seasonal variation in adult body mass in the *T. thetis* population he studied near Dorrigo. He also found that wild female pademelons bred continuously, but with a peak in autumn and spring. These peaks were spaced by the length of the average pouch life (181 days) and it is probable that females raise two young each year. The peaks in breeding at Dorrigo occurred at times which enabled females to carry small pouch young and to be under little nutritional stress from lactation during the winter trough in food availability.

During his observations of feeding behaviour Johnson (1977) found that in their diurnal home range within the rainforest pademelons browsed on shrubs such as *Helichrysum diosmifolia* throughout the year, and grazed on herbs such as *Viola* sp. and on grasses. During winter months *Solanum mauritianum* (wild tobacco) was also eaten, despite its appreciable alkaloid content, common within the genus. Presumably the trough in food availability in winter forced the animals to expand their diet spectrum. In their nocturnal home range of open pasture area, grasses comprised the bulk of food items taken. In both diurnal and nocturnal sections of their home range *T. thetis* seemed to avoid eating dry vegetation whenever possible (Johnson 1977).

These observations led Hume (1977a, b) to examine some nutritional and metabolic aspects of *T. thetis* in the laboratory. *T. thetis* is notable among the Macropodidae in having a maintenance nitrogen requirement at least twice that of most other species investigated so far (Table 1.8), and a rate of whole-body protein turnover that is 43% greater than in the tammar wallaby (Table 6.11). Its inability to tolerate saline drinking water (Hume & Dunning 1979) supports the contention that its kidneys are not adapted for conserving urea or water; *T. thetis* has a relative medullary thickness (RMT) of 5.7, compared with the tammar wallaby of 7.2. Thus, when Chilcott, Moore & Hume (1985) compared urea recycling under conditions of *ad libitum* and restricted water intakes, there was no change in urinary urea excretion or in urea recycling to the digestive tract in *T. thetis*. In contrast, in the tammar wallaby urea excretion decreased and urea recycling to the gut increased, at least on diets above 9% crude protein. Also consistent with these findings is the high water turnover rate reported by Dellow & Hume (1982b) in the red-necked pademelon compared with the tammar wallaby (Table 1.6). All these data suggest an animal ill-adapted to an environment where food quality or quantity may be seasonally limiting. The width of the nutritional niche occupied by *T. thetis* is constrained by its high nutrient requirements.

The only other nutritional information on the genus *Thylogale* is on *T. billardierii* (Tasmanian pademelon). From their measurements of FMR, Nagy, Sanson & Jacobsen (1990) calculated that free-living 6 kg *T. billardierii* in an enclosed natural area consumed about 200 g dry matter daily. This is similar to the intake by captive *T. thetis* (Dellow & Hume 1982a) fed chopped lucerne hay of similar digestibility to that of the fresh grass consumed by the free-living *T. billardierii*.

PARMA WALLABY

The preferred habitat of *Macropus parma* is similar to that of *T. thetis*. Most sightings of parma wallabies by Read & Fox (1991) were in wet sclerophyll forest with a moist or rainforest understorey. Hume (1986) found that the maintenance nitrogen requirements of *M. parma* and *T. thetis* were remarkably similar (Table 1.8), and at least double those of the four other macropodid species, all of which are from less mesic environments, that have been studied. Rates of whole-body protein turnover are similar in *M. parma* and *T. thetis* (Table 6.11). Voluntary intakes of dry matter and water by *M. parma* were also similar to those of *T. thetis*, but higher than in other macropods. Urea recycling actually decreased in response to water restriction, compared with the absence of change in *T. thetis* but an increase in *M. eugenii* (see above). Thus within the Macropodidae nutrient requirements seem to be linked closely with preferred habitat, regardless of phylogeny.

McArthur & Sanson (1993a) compared the effects of a tannin on digestive performance of *M. parma* (a grazer) and *T. billardierii* (a browser – despite the consumption of fresh grass in the study by Nagy *et al.* (1990) discussed above). The parma wallabies were affected more by adding a condensed tannin (quebracho) to a tannin-free basal diet than the Tasmanian pademelons, supporting the thesis that tannins should affect grazers more than browsers. In eutherian herbivores, browsers are protected against dietary tannins by the presence of tannin-binding salivary proteins (Austin *et al.* 1989). Likewise, McArthur, Sanson & Beal (1995) found that *T. thetis* produced more salivary protein, some of which was tannin binding, than *M. parma*. Faecal tannin recovery was higher in *M. parma* than in *T. billardierii*, but the fate of the absorbed tannins in the latter species is unknown; McArthur & Sanson (1993a) could not detect any costs of the tannins in the form of increased energy or nitrogen losses in the urine.

7.5 SUMMARY AND CONCLUSIONS

The relationships between dentition and diet among the family Macropodidae are now reasonably well understood, thanks to continuing studies of the dietary habits of various species, and to the functional and integrative approach to diet and dentition taken by Sanson (1989). Consideration of the natural diets of kangaroos and wallabies in this chapter, along with the constraints imposed on free-living animals by shortages of energy, protein and water, have provided the opportunity to extend the laboratory findings discussed in Chapter 6 to the animal in its natural habitat.

The effects of the regular summer drought on two small wallabies, the Rottnest Island quokka and the tammar wallaby on the Abrolhos, Garden and Kangaroo Islands, have been the subject of detailed investigation. The finding that quokkas in a more mesic habitat on the mainland close to Rottnest Island breed continuously throughout the year, while those on Rottnest breed only once, underscores the central role of food quality in the reproductive success of this species. The most definitive work on the trace mineral nutrition of marsupials has been carried out with the quokka. This is because of the generally poor nutritional status of the limestone-derived sandy soils of Rottnest, and of the failure of early attempts to graze domestic sheep on the island.

Other species from semi-arid and arid environments have also been well studied. The comparison between the euro and the red kangaroo illustrated two different ways, one largely physiological, the other largely behavioural, in which an animal can adapt to frequent shortages of food and water during extended periods of drought. Attempts to estimate field energy and water budgets are continuing and adding to our knowledge of the nutritional constraints operating, particularly in arid-zone species.

Our knowledge of the nutritional ecology of many forest-dwelling macropods is less advanced. The pademelons and parma wallaby have unusually high maintenance requirements for protein and seem profligate in their usage of water, but are probably rarely confronted with shortages of either nutrient in their mesic forest habitats. There is increasing support for the notion that within the Macropodidae nutrient requirements are linked closely with preferred habitat, regardless of phylogeny.

8 Foregut fermenters – the rat-kangaroos

The rat-kangaroos (family Potoroidae) are the other group of marsupials in which the main site of digesta retention and microbial fermentation is the forestomach. The one exception to this statement is the musky rat-kangaroo (*Hypsiprymnodon moschatus*), in which the stomach is simple and has no forestomach. In the kangaroos and wallabies (family Macropodidae) the forestomach is grossly tubiform in nature, but in the rat-kangaroos that have a forestomach it is sacciform. This difference in gastric morphology, together with their uniformly small size, limits rat-kangaroos to diets of low fibre content.

8.1 DENTITION

Sanson's (1989) classification of dental organisation in the family Macropodidae (Chapter 7) recognises three grades. The grazer grade is suited to shearing fibrous grasses and is found mainly in large kangaroos. The intermediate browser/grazer grade is less specialised for shearing, and has an enhanced grinding capacity as well; it is found mainly in rock-wallabies and other macropods of smaller body size. The browser grade is specialised for crushing and grinding low-fibre content plant material, and is found predominantly in small wallabies.

The fourth grade of dental organisation recognised by Sanson (1989) is a basal grade which is suited to diets of very low fibre content, and is occupied exclusively by members of the family Potoroidae. Being smaller than most macropodids, they are even more selective for low-fibre, concentrate foods. They take mainly roots and rhizomes, seeds and fungi, rather than above-ground structural parts of plants. Some invertebrates are consumed seasonally, and for this reason they are sometimes described as omnivores rather than herbivores. However, only the musky rat-kangaroo is truly omnivorous throughout the year. All other rat-kangaroos consume principally plant or fungal material, and have a complex forestomach and microbial fermentation. They are therefore classified here as concentrate selectors and foregut fermenters. Clearly they are only marginal herbivores.

The basal (potoroid) dental grade is well suited to foods such as seeds, fungal sporocarps and invertebrates that are large items compared with the size of a tooth. Such foods are relatively easily comminuted once the encasing material (testa, pericarp or exoskeleton) is ruptured. Prehension involves precise manipulation of individual items with the incisors. The emphasis in the cheek teeth is on the premolars, which in the Potoroidae are long and sectorial (well suited for cutting) (Fig. 8.1). The food item is opened by a long but relatively coarse shearing stroke involving the premolars, and the liberated contents are then simply crushed and ground by the low-crowned (bunodont) molars The whole cheek-tooth row is in occlusion simultaneously (Fig. 8.2) and the large premolar prevents any forward movement of the molars (molar progression). The phenomenon of molar progression, which is so pronounced in some of the Macropodidae, is covered in Chapter 7.

The musky rat-kangaroo differs from other rat-kangaroos in several respects, and is currently placed in its own subfamily, the Hypsiprymnodontinae. One way in which it stands out from the other rat-kangaroos (the subfamily Potoroinae) is in its dentition. *Hypsiprymnodon* has a first incisor (the largest of the incisors) which is relatively small, but a diastema between the incisors and first premolar that is greater (in both absolute and proportional terms) than that of any extant potoroine. The third premolar is tall and narrow (i.e. plagiaulacoid), is inserted obliquely to the molar row, and bears seven almost vertical ridges (Heighway 1939).

Figure 8.1 Skull of *Potorous tridactylus* (long-nosed potoroo).

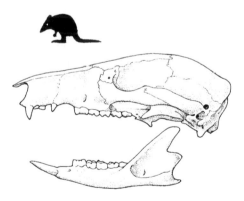

Figure 8.2 Lateral view of the lower jaw and upper incisors of the basal (potoroid) dental grade. The relative width of the arrows indicates relative occlusal emphases within the tooth row. The diet of hard individual items is manipulated by the large first incisors and penetrated by the premolars. The large premolar constrains the bunodont (low-crowned) molars to anterioposterior occlusion, and all teeth occlude. After Sanson (1989).

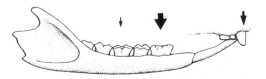

8.2 DIET STUDIES

8.2.1 Musky rat-kangaroo

The musky rat-kangaroo (*Hypsiprymnodon moschatus*) occurs only in closed tall forest (rainforest) in north Queensland (Johnson & Strahan 1982). It is one of only a few diurnally active marsupial species. It feeds on the forest floor, on fallen fruits and nuts of trees and palms, invertebrates and epigeal (above ground) fungi as well (Dennis & Johnson 1995). In contrast to the Potoroinae, *Hypsiprymnodon* has a manus (forepaw) that is not specialised for digging. The forepaws are used instead to turn over leaf litter in search of food. Food items are picked up with the mouth, transferred to the forepaws, and held there for consumption. When eating arthropods the head is turned to one side and the sectorial premolars are used to shear through the chitinous exoskeleton (Johnson & Strahan 1982). There is evidence that the main component of the diet is fruit, and it has been described as primarily frugivorous, but when fruits are less available it consumes more invertebrates, and epigeal fruiting bodies of fungi become an important dietary item in the late wet season (Dennis & Johnson 1995). Therefore it is best described as an omnivore.

8.2.2 Potoroos

There are two extant species of potoroos, *Potorous tridactylus* (long-nosed potoroo), which is still common along the seaboard of south-eastern Australia, including Tasmania, and *P. longipes* (long-footed potoroo), which is found only in a small area in east Gippsland in Victoria and in the southeast of New South Wales (Seebeck, Bennett & Scotts 1989). Both appear to be mainly mycophagous (fungus eating). Guiler (1971) found that *P. tridactylus* in Tasmania was dependent largely upon hypogeal (below ground) fungi as food, based on faecal analysis. Food in the stomach closely corresponded to that identified in the faeces in terms of both the species present and their proportions. The per cent by area of faecal fragments that were fungal ranged from 67% in summer to 92% in winter. The median claws of the manus of *P. tridactylus* are long, strong and well adapted for digging. The forepaws are also extensively used for holding and manipulating food. Other foods consumed were plant material (grasses and sedges, herbs, unidentified seeds and bryophytes) that made up 6% to 22% of total faecal fragment area, and insects (1% in winter to 21% in summer).

In south-western Victoria, Bennett & Baxter (1989) found *P. tridactylus* to be mainly mycophagous also, particularly in winter (Fig. 8.3). At other times of the year vascular plants, arthropods and fruit made up more of the diet. In east Gippsland and the south-east corner of New South Wales, Claridge, Tanton & Cunningham (1993) likewise concluded that *P. tridac-*

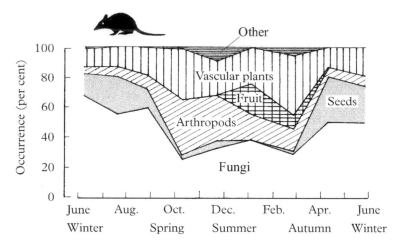

Figure 8.3 Diet of the the long-nosed potoroo (*Potorous tridactylus*) in south-western Victoria. Fungi are important throughout the year. After Bennett & Baxter (1989).

tylus was primarily mycophagous (33% to 82% of faecal fragments), with plant tissue (3% to 56%), seeds (1% to 29%) and invertebrates (1% to 16%) making up the rest.

Similarly, during most of the year, *P. longipes* was found to be primarily mycophagous (Seebeck *et al.* 1989). Hypogeal or semi-hypogeal fungi contributed at least 80% of faecal volume throughout the year. Other items identified in the faeces were vascular plant tissue, seeds and insects.

8.2.3 Bettongs

There are four species of *Bettongia* still extant. Most had extensive geographic ranges, but all are now confined to small remnant populations. Only the Tasmanian bettong (*B. gaimardi*) is considered common.

The burrowing bettong (*B. lesueur*) persists only on Bernier, Dorre and Barrow Islands off the coast of Western Australia (Short & Turner 1993). It is distinctive among the rat-kangaroos in that it digs burrows and lives in warrens. It also digs for part of its food (roots and tubers), as well as feeding on green leaves, fruits, flowers and seeds (Finlayson 1958).

The brush-tailed bettong (*B. penicillata*) persists in only three small areas of open eucalypt forest in the south west of Western Australia. In eucalypt forest near Manjimup, its diet consists largely of the fruiting bodies of hypogeal fungi (Lamont, Ralph & Christensen 1985), supplemented by bulbs, tubers, seeds, insects and gum exudates, probably from *Hakea* shrubs (Kinnear *et al.* 1979). The fungi appeared to be the main source of protein for the animals. Although the fungal sporocarps are protected by an indigestible outer cover (pericarp), this cover is removed and the inner 'kernel' along with some adhering spores is ingested. The kernel contained

8–10% crude protein, 42% lipid and less than 1% ash. As a protein source the kernel appeared to be deficient in lysine but very high in methionine, indicating a grossly imbalanced protein (Kinnear *et al.* 1979), but other analyses suggest a much better balance of amino acids (see Section 8.7). Modification of this protein by microbial fermentation in the sacciform forestomach may be important in enabling rat-kangaroos to utilise these hypogeal fungi (see below).

Hypogeal fungi also constitute an important food source after fire. *B. penicillata* return to feed on the fungi after bushfires. Being hypogeal, the fungi escape incineration. *B. penicillata* are able to maintain weight on this material, the only food source left, while tammar wallabies (*Macropus eugenii*) in the same area lose weight because grass, their principal food source, has disappeared (Christensen 1980).

In contrast, Sampson (1971) made no mention of fungi in the diet of *B. penicillata* in a reserve at Tutanning, a drier site than Manjimup. Scats contained the remains of both mono- and dicotyledonous leaves, seeds of several plant species, roots and tubers, bark and arthropods. Sampson (1971) felt that roots and tubers formed the bulk of the diet, but that quantitative evaluation of the diet of *B. penicillata* was not justified because of their feeding behaviour. In the case of seeds, they chewed the seed and spat out the epidermal material and hard parts, and swallowed only the soft, easily digested contents. When insects were eaten, they cracked the cuticle and sucked out the soft parts of the body, swallowing only after spitting out the hard parts. Thus scat analysis would lead to an underestimate of the contribution of seeds and insects to the diet.

Both Sampson (1971) and Christensen (1980) noticed that *B. penicillata* buried seeds, including the hard nuts of the quandong (*Santalum acuminatum*). Sampson (1971) suggested that they may dig up the seeds and eat the soft parts once the seeds germinate. This behaviour could be important for the dispersal of plants like the quandong that have hard seed coats, because it is unlikely that all buried seeds would be relocated and eaten.

The northern bettong (*B. tropica*) is restricted to open forest habitats on and adjacent to the Atherton Tableland in north Queensland (Winter and Johnson 1995). It is closely related to *B. penicillata*. Its diet consists mainly of hypogeal fruiting bodies of mycorrhizal fungi (McIlwee & Johnson 1998). It eats the soft spore-mass from within the fruiting body, leaving the tough pericarp behind at the digging site. It also eats roots and tubers, the leaves of grasses, lilies and forbs, and some invertebrates. When it eats the roots of cockatoo grass (*Alloteropsis semialata*), the fibrous parts are left behind at the digging site as a thoroughly chewed pellet (Winter & Johnson 1995).

The Tasmanian bettong (*B. gaimardi*) is the most common and the most studied bettong. It is also dependent on hypogeal fungi as its main food source (Rose 1986; Taylor 1992) (Fig. 8.4). Epigeal fungi are also eaten, as are fruits (mainly of *Astroloma humifusum*), leaves and stems of forbs and

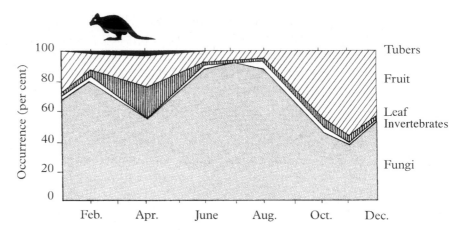

Figure 8.4 Diet of the Tasmanian bettong (*Bettongia gaimardi*). Note the heavy dependence on fungi in all seasons of the year. After Johnson (1994a).

grasses, monocot seeds (mainly of *Vulpia*, *Aira* and *Stipa*) and *Acacia* gum (Taylor 1992). The abundance of *B. gaimardi* was related to the extent of mycorrhizal root development, even though associated soils were of low fertility (Taylor 1993). Diggings were clustered around *Eucalyptus* and *Acacia* stems, and increased with the density of *A. dealbata* stems (Johnson 1994c). Body condition of *B. gaimardi* tended to increase with increased representation of fungi in the diet (Johnson 1994a).

The tolerance of three species of bettongs to sodium fluoroacetate (compound 1080), a potent toxin found in the leguminous genera *Gastrolobium* and *Oxylobium* in south-western Australia, but not in south-eastern Australia (see Chapter 2, section 2.9), was compared by Mead *et al.* (1985). Most mammals are fatally poisoned by less than 1 mg kg^{-1} body mass of 1080. The two plants contain 1080 at up to 2.65 g kg^{-1} fresh leaf, presumably for protection against herbivory. Mead *et al.* (1985) found that *Bettongia penicillata*, whose range includes these plants, are exceptionally resistant to 1080, being able to tolerate levels of 100 mg kg^{-1} body mass. In contrast, *B. lesueur* from Dorre Island, which has been separated from the mainland for at least 7000 years, is only moderately tolerant of the toxin, perhaps to 10 mg kg^{-1} body mass, while *B. gaimardi* from Tasmania is highly sensitive to 1080, succumbing to 1 mg per kg body mass. Mead *et al.* (1985) concluded that *Bettongia* radiated from the east into Western Australia, the western populations coevolving a degree of resistance dependent on the extent of exposure to fluoroacetate-bearing plants.

8.2.4 *Aepyprymnus*

Aepyprymnus rufescens (the rufous rat-kangaroo or rufous bettong) (Fig. 8.5) is the largest of the Potoroidae, at 2.5–3.5 kg. It is also the most

Figure 8.5 The rufous rat-kangaroo or rufous bettong (*Aepyprymnus rufescens*) is the largest of the rat-kangaroos. (Vertebrate Biocontrol Cooperative Research Centre, Canberra)

rhizophagous (Finlayson 1931). Schlager (1981) found that swollen tap roots of the introduced flat weeds *Hypochoeris*, *Centaurea* and *Taraxacum* were the main food items selected in northern New South Wales. Feeding is strictly a nocturnal activity, but by day the diggings are easily identified by the discarded tops of the weeds. When feeding on the rhizomes of blady grass (*Imperata cylindrica*), the forepaws are used to peel off the outer fibrous layers, and only the softer inner parenchyma is eaten. Other items seen being eaten by Schlager (1981) were hypogeal fungi associated with *Casuarina* roots, epigeal fungi, grass seeds and gum exudate from *Acacia* and *Melaleuca* trees. Some insects were eaten during spring and summer, and flower parts were sometimes eaten. *Aepyprymnus* was seen to shift its home range to take advantage of a seasonal flush of fungi. Calaby (1966) also reported them grazing on introduced clover and the grass *Paspalum dilatatum*. Thus *Aepyprymnus* has a wide dietary spectrum, without the same reliance on fungi observed for *Bettongia* spp.

8.2.5 Mycophagy and spore dispersal in Australian forests

This topic was introduced in the discussion of bandicoot diets in Chapter 3. It is treated here also because mycophagy appears to be most prevalent within the Potoroidae (Claridge & May 1994). However, it has been recorded for a variety of other marsupials (Table 8.1) as well as among

Table 8.1. *Australian marsupials known to consume sporocarpic fungi*

Species	% of samples containing spores	Mean proportion of fungus in stomach or faeces by volume	Ref.
Family Dasyuridae			
Phascogale tapoatafa	n.a.	n.a.	Troughton 1965
Family Peramelidae			
Isoodon obesulus	100	n.a.	Christensen 1980
Perameles nasuta	n.a.	++	Claridge 1993
Macrotis lagotis	100	++	Watts 1969
Family Vombatidae			
Vombatus ursinus	n.a.	n.a.	Troughton 1965
Family Burramyidae			
Burramys parvus	2	+	Mansergh *et al.* 1990
Family Petauridae			
Petaurus australis	15	n.a.	Craig 1985
Family Phalangeridae			
Trichosurus caninus	39	++	Seebeck *et al.* 1984
Trichosurus vulpecula	90	+	Statham 1984
Family Potoroidae			
Hypsiprymnodon moschatus	n.a.	n.a.	Dennis & Johnson 1995
Potorous tridactylus	100	+++	Bennett & Baxter 1989
Potorous longipes	100	+++	Seebeck *et al.* 1989
Bettongia lesueur	n.a.	n.a.	Burbidge 1995
Bettongia penicillata	100	+++	Christensen 1980
Bettongia tropica	100	+++	McIlwee & Johnson 1998
Bettongia gaimardi	100	n.a.	Taylor 1992
Aepyprymnus rufescens	100	n.a.	Schlager 1981
Family Macropodidae			
Setonix brachyurus	n.a.	n.a.	Christensen 1980
Thylogale thetis	n.a.	+	Jarman & Phillips 1989
Thylogale stigmatica	n.a.	n.a.	Jarman & Phillips 1989
Macropus eugenii	n.a.	n.a.	Christensen 1980
Wallabia bicolor	100	++	Hollis *et al.* 1986
Macropus fuliginosus	n.a.	n.a.	Christensen 1980

Note: + Less than 10%
  ++ 10–25%
  +++ Greater than 25%
n.a. no estimate available

rodents (especially *Pseudomys* and *Rattus*), feral pigs and foxes and invertebrates (especially beetles).

Several researchers have proposed a tripartite symbiotic interrelationship among hypogeal mycorrhizal fungi, mycophagous mammals and plants (e.g. Maser, Trappe & Nussbaum 1978). Most hypogeal fungi are either known or thought to establish mycorrhizal associations on (ectomycorrhizae) or in (endomycorrhizae) plant roots. Mycorrhizal fungi are known to perform a number of functions when in symbiosis, including the uptake and transfer of water and nutrients from the soil to the host plant and protection of the host's root system from soil pathogens. The vegetative part of fungi is the mycelium, a mass of microscopic, thread-like hyphae. Spores may be formed directly by the mycelium, or the mycelium may give rise to one or more sporocarps (fruiting bodies) in or on which spores are formed. Mycorrhizae of land plants are most abundant in the humus and soil immediately below the humus, and mycorrhizal activity is enhanced by favourable soil moisture conditions.

The fungi that are eaten by Australian mammals have one of two distinctive fruiting habits. Hypogeal species form sporocarps in the ground. In some species the sporocarps may be covered only by the litter layer, but in others (e.g. *Mesophellia* spp.) the sporocarps may be at depths of up to 50 cm. However, most hypogeal sporocarps are found between 1 cm and 15 cm deep. All truffles and false-truffles are hypogeal. Epigeal species form sporocarps above ground, as seen in agarics such as mushrooms and toadstools. Spores liberated from epigeal sporocarps are dispersed mainly by wind. In contrast, spores in hypogeal sporocarps are assumed to rely upon ingestion by animals for their dispersal. When a mammal feeds on hypogeal sporocarps, spores can be liberated and dispersed in three ways: by clouds of powdery spores being released into the air during feeding, by some spore-bearing tissue adhering to the animal's fur and later being brushed off by vegetation or by grooming or washed off by rain, and by ingestion by the animal and deposition in the faeces at some other location.

The fungal spores are essentially indigestible, yet there is evidence that passage of spores through the digestive tract is necessary for their viability. For instance, Lamont *et al.* (1985) found that roots of *Eucalyptus calophylla* (marri) seedlings inoculated with the faeces of *B. penicillata* developed seven different ectomycorrhizae, but similar seedlings inoculated with fresh spore tissue of two fungal species known to be associated with eucalypt roots (including *Mesophellia* sp.) failed to develop ectomycorrhizae. Similarly, Claridge *et al.* (1992) recorded no ectomycorrhizae on the roots of seedlings of two species of eucalypts inoculated with spores present in sporocarps of *Mesophellia pachythrix*, but ectomycorrhizae were established when spores present in the faeces of *P. tridactylus* were used as the inoculum. However, Malajczuk, Trappe & Molina (1987) were successful in culturing ectomycorrhizae on the roots of *E. diversicolor* (karri) seedlings from spores in the sporocarps of two species of *Mesophellia*.

Whether passage through the digestive tract is an essential pretreatment for spore germination or not, deposition in faeces is almost undoubtedly essential for the dispersal of otherwise immobile fungi. As these mycorrhizal fungi play important roles in the nutrition of their host trees, the health of Australian forest ecosystems may be particularly reliant on the presence of adequate numbers of spore dispersers such as rat-kangaroos and bandicoots. Forest regeneration following logging or fire may be compromised by the loss of these small marsupials.

8.3 THE POTOROID DIGESTIVE TRACT

A major difference between the two potoroid subfamilies is in their gastrointestinal tract morphologies. That of the musky rat-kangaroo (Hypsiprymnodontinae) is simple, while those of the Potoroinae are complex.

8.3.1 Musky rat-kangaroo

Both Carlsson (1915) and Heighway (1939) remarked on the relatively enormous parotid salivary glands of *Hypsiprymnodon*. The glands consist of the main mass of secretory tissue that extends from the ear to the clavicle, a thin extension that passes medially over the mandibular gland to meet a similar extension from the other parotid, and a separate island of glandular tissue just before the parotid duct enters the buccal cavity; Carlsson (1915) referred to this as an accessory parotid gland.

The simple stomach of *Hypsiprymnodon* (Fig. 8.6) has been described by Carlsson (1915) and by Heighway (1939). Carlsson (1915) thought that it was more similar to the stomach of *Trichosurus* (the brushtail possums) than to any of the other potoroids. The simple stomach, together with other features such as its pronounced plagiaulacoid premolars, is responsible for its separate subfamilial status within the Potoroidae.

Although Heighway (1939) agreed with Carlsson (1915) in describing the stomach as sac-like, she drew attention to a fairly distinct groove on either side of the oesophageal opening. The grooves are most prominent on the lesser curvature, and disappear before they reach the greater curvature. Internally the grooves correspond to ridges which she used to divide the stomach into cardiac, oesophageal and pyloric 'compartments'.

The gastric mucosa is entirely glandular with a closely pitted appearance. In addition, the mucosa of the oesophageal 'compartment' is organised into a thick, heavily plicated structure, with numerous glandular crypts opening on and between the folds. From its structure and position, Heighway (1939) considered this to represent the cardiogastric gland described in Chapter 2 for *Caenolestes*, in Chapter 4 for the wombats and in Chapter 5 for the koala.

(*a*)   External features   (*b*)   Stomach in section

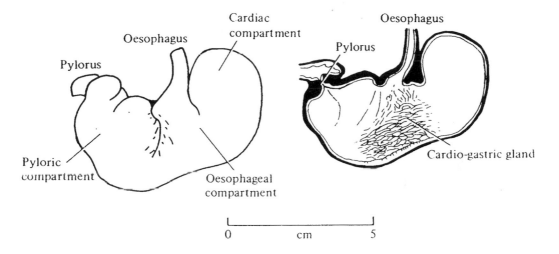

Figure 8.6 The simple stomach of the musky rat-kangaroo (*Hypsiprymnodon moschatus*): (a) external features, (b) in longitudinal section. After Heighway (1939).

Heighway (1939) thought that the duodenum was devoid of lymphoid tissue (Brunner's glands) and of longitudinal folds, but that the mucosa exhibited a 'closely villous surface'. The mucosa of the rest of the small intestine was villous and folded, with a large number of small lymphoid nodules. In an adult specimen of 25 cm body length, Carlsson (1915) reported lengths for the small intestine of 90 cm, the caecum of 4.8 cm and the colon of 64 cm. The caecum was described by Heighway (1939) as a 'simple blind sac', but capacious. Both the caecum and proximal colon were weakly haustrated, and the taeniae were barely discernible. The musky rat-kangaroo has only three species of nematodes, one in the small intestine and two in the caecum and colon (Beveridge & Spratt 1996).

8.3.2 The Potoroinae

The stomach of the Potoroinae is much more complex than that of *Hypsiprymnodon*. The complete gastrointestinal tract of *Aepyprymnus* is shown in Fig. 8.7, and its stomach in Fig. 8.8.

The potoroine stomach is a large organ that, in captive *Aepyprymnus*, contains 50% of the total contents of the gastrointestinal tract (Hume & Carlisle 1985). As in the Macropodidae, functionally it consists of two regions, an enlarged forestomach where food is stored and fermented, and

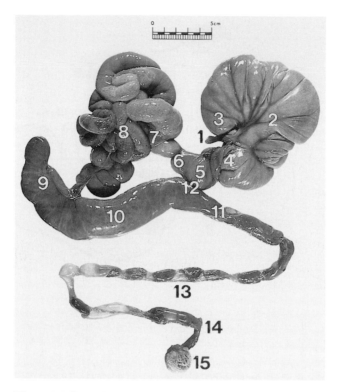

Figure 8.7 Gastrointestinal tract of the rufous rat-kangaroo or rufous bettong (*Aepyprymnus rufescens*). The sacciform region of the forestomach (2) has been rotated through 180° to expose the tubiform forestomach (4) and hindstomach (5) regions. Except for the small intestine (8) and the gastrocolic ligament (12), mesenteric attachments have been cleared to allow good visualisation of the tract. 1. oesophagus; 3. blind sac of sacciform forestomach; 6. pyloric sphincter; 7. duodenum; 8. jejunum and ileum; 9. caecum; 10. proximal or ascending colon; 11. proximal colic flexure; 13. distal colon; 14. rectum; 15. cloaca. From Hume & Carlisle (1985).

a hindstomach where hydrochloric acid and pepsinogen are secreted. The forestomach can be divided into a sacciform region with its blind sac, and a tubiform region into which the oesophagus enters (Langer 1980). The two regions are delimited *in situ* by permanent dorsal and ventral folds close to the cardia (the opening of the oesophagus). In all potoroine species the sacciform forestomach is by far the largest region of the stomach, and is extensively haustrated. It occupies the left side of the abdominal cavity, with the blind sac extending across to the right side. The short tubiform forestomach lies in a medial position. It is only weakly haustrated. There is no gastric sulcus in the tubiform forestomach. Another permanent fold separates the tubiform forestomach from the hindstomach, which lies on the right side, dorsal to the blind sac of the sacciform forestomach.

Langer (1980) reported that the relative volumes of the three stomach regions (sacciform forestomach, tubiform forestomach and hindstomach)

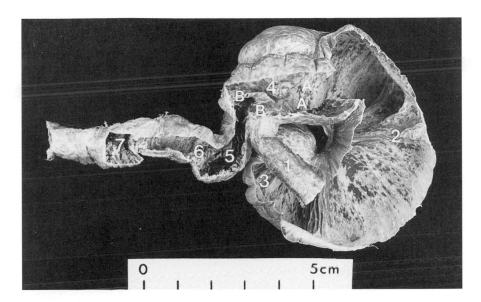

Figure 8.8 Stomach of *Aepyprymnus rufescens*. Preserved and dried preparation in which parts of the wall have been resected to display internal features. 1. oesophagus; 2. sacciform region of the forestomach showing haustrations; 3. blind sac of the sacciform forestomach; 4. tubiform forestomach; 5. hindstomach; 6. pyloric sphincter; 7. duodenum. AA. permanent fold between sacciform and tubiform forestomach regions; BB. permanent fold between tubiform forestomach and hindstomach. From Hume & Carlisle (1985).

were 75%, 14% and 11% respectively in *Potorous tridactylus* (long-nosed potoroo) and 73%, 22% and 5% in *Bettongia lesueur* (burrowing bettong).

The duodenum and the remainder of the small intestine lie primarily caudal and ventral to the hindstomach (Hume & Carlisle 1985). The caecum lies caudal and left of most of the small intestine, but its blind end is quite mobile and can be found on either the left or right side of the abdominal cavity. It does not have haustra or taeniae (Richardson 1989). The proximal or ascending colon is of similar diameter to the caecum, and lies on the right side of the abdominal cavity. The loosely coiled transverse colon lies close to the forestomach, and is attached to the sacciform fore-stomach by the gastrocolic ligament. Faecal pellets form in the transverse colon, distinguishing this colonic region from the proximal colon. The descending colon lies on the left side of the abdominal cavity, and passes more-or-less directly caudal to the cloaca. In *Aepyprymnus* the small intestine contained 15% of total gut contents and the large intestine or hindgut contained 35% (8% in the caecum, 18% in the proximal colon and 9% in the distal colon) (Hume & Carlisle 1985). Thus the capacity of the caecum and proximal colon in *Aepyprymnus* is about half that of the forestomach, suggesting that it may be an important secondary area of microbial fermentation. Richardson (1989) similarly concluded that in *Bettongia penicillata* (brush-tailed bettong) the relatively long caecum and colon were also probably important fermentation sites.

The forestomach of the Potoroinae is lined entirely with mucus-secreting cardiac glandular mucosa (Langer 1980). Brunner's glands (which also secrete mucus) form a glandular collar immediately distal to the pyloric sphincter, and in *Potorous* and *Caloprymnus* (the extinct desert rat-kangaroo) empty by an intricate duct system onto both gastric and intestinal epithelium (Krause 1972).

8.3.3 The helminth fauna

As in the Hypsiprymnodontinae, the helminth fauna of the Potoroinae is sparse (Beveridge & Spratt 1996). *Potorous* has two species of cestodes in the small intestine, one strongyloid and one trichostrongyloid nematode in the forestomach and one oxyuroid nematode in the caecum and proximal colon. A second strongyloid nematode occurs in the forestomach of *Aepyprymnus*.

There is no information available on the microbiology of the potoroid gastrointestinal tract.

8.3.4 Digesta passage

Wallis (1994) has made a detailed study of the rate of passage of inert digesta markers through the gastrointestinal tracts of three species of potoroine marsupials, *Potorous tridactylus*, *Bettongia penicillata* and *Aepyprymnus rufescens*. He used ^{51}Cr-EDTA as a solute marker, and ^{103}Ru-Phenanthroline as a particle-associated marker. Animals were offered grain-based diets containing either 15% or 28% neutral-detergent fibre (NDF) *ad libitum*, and the markers were given as a pulse dose either immediately before they commenced feeding (at dusk), when gut fill should have been minimal, or after they had consumed about 40% of their typical total daily intake. The curves describing the patterns of appearance of the markers in the faeces were remarkably similar for all three species, and were not significantly altered by either the fibre content of the diet or time of dosing in relation to initiation of feeding. Marker concentration in the faeces showed a delay period corresponding to the transit time of the marker, followed by a sharp rise to a peak concentration and then an exponential decline. There were no differences in passage rate between the two markers. Transit time (measured as 5% excretion time) averaged 6–17 hours and 8–17 hours for the solute and particle markers respectively. Mean retention times averaged 20–31 hours and 20–37 hours. Wallis (1994) thought that because the bulk of values for mean retention time fell between 24 and 30 hours digesta flow in the three rat-kangaroos was entrained to their circadian rhythm. In contrast to *Hypsiprymnodon*, potoroine marsupials are strictly nocturnal in their activity patterns. They take refuge in grass nests during daylight hours, emerging after dusk to forage. Foraging takes place in short bursts throughout the night, and they disappear again before sunrise.

Of course the mean retention times quoted above are for the whole gut (i.e. mouth to cloaca) and represent the net result of several events within the gastrointestinal tract. Radiography has been used to trace the movement of suspensions of barium sulphate and of radio-opaque particles through the stomach and intestines of all three of the species used by Wallis (1994) in his passage rate studies. In the studies by Hume & Carlisle (1985), *Aepyprymnus* and *Potorous tridactylus* were fed *ad libitum*, and dosed between 0800 and 0900 hours with either barium sulphate coated onto slices of sweet potato or radio-opaque particles incorporated into pellets consisting of 45% cracked wheat, 45% cracked corn and 10% crushed dry dog-kibble. The animals were first examined fluoroscopically while restrained in a standing position in a Perspex box for 20 minutes, then kept in their holding cages except for short periods when radiographs were taken with the animal recumbent either on its back or on its left side. Fig. 8.9 shows a series of radiographs of *Aepyprymnus* taken 5 min, 2 hours, 24 hours and 48 hours after dosing.

In both species barium sulphate given orally entered the sacciform forestomach within 20 minutes of dosing, but in the same time the hindstomach and duodenum were clearly outlined. Contrast medium reached the hindgut within 1–2 hours, but was retained there for at least 24 hours. Some of the contrast medium that entered the sacciform forestomach was retained there for at least 24 hours. A proportion of radio-opaque particles given orally also bypassed the sacciform forestomach, but those that entered this gastric region were retained there for up to 93 hours. These results suggested that both the forestomach and the caecum and proximal colon were important retention sites for digesta in rat-kangaroos. They also show that the short tubiform forestomach and the consequent close juxtaposition of cardia, sacciform forestomach and hindstomach mean that ingesta from the oesophagus can flow either into the fermentation region (the sacciform forestomach) or through the hindstomach into the small intestine. What controls the direction of flow is not fully understood, but the degree of fill of the small intestine and hindstomach appear to be involved, at least in *Aepyprymnus* (Hume *et al*. 1988). There does not seem to be any sorting of ingesta in the tubiform forestomach on the basis of particle size.

In similar studies with *Bettongia penicillata*, Richardson (1989) also found that some marked ingesta entered the sacciform forestomach while other contrast medium entered the hindstomach and small intestine. Radio-opaque pellets passed through the gastrointestinal tract much more slowly than barium sulphate. It also took longer for large (1 mm × 2–3 mm) pellets than small (1 mm × 1 mm) pellets to initially pass through the pylorus, to completely clear the stomach and to clear the entire tract. These results

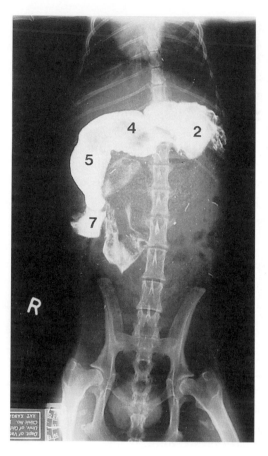 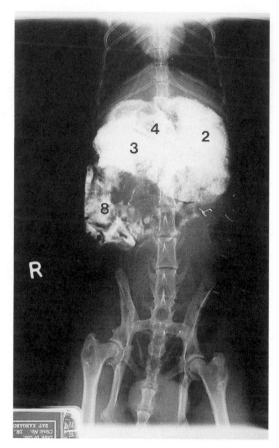

(a)   (b)

Figure 8.9 Series of ventro-dorsal radiographs of *Aepyprymnus rufescens* taken 5 min (a), 40 min (b), 6 h (c) and 48 h (d) after a single oral dose of barium sulphate. (a) At 5 min the contrast medium has outlined the cranial area of the sacciform forestomach (2) and tubiform forestomach (4), and some has bypassed the sacciform forestomach and entered the hindstomach (5) and duodenum (7). (b) At 40 min after dosing the sacciform forestomach (2) is completely outlined, and its blind sac (3) overlies the hindstomach. Contrast medium now outlines the jejunum and ileum (8). At 6 h the density of contrast medium in the sacciform forestomach (2, 3) has decreased, and the small intestine (8), caecum (9) and proximal colon (10) are clearly outlined. (d) By 48 h the stomach and small intestine are clear of contrast medium, but the caecum (9), proximal colon (10), proximal colic flexure (11) and faecal pellets in the distal colon (13) and rectum (14) are all clearly defined. From Hume & Carlisle (1985).

confirm a lack of particle sorting in the tubiform forestomach. More likely the pyloric sphincter is the site for sorting of different sized particles, as it is in most vertebrate stomachs (Stevens & Hume 1995).

A study of *Potorous tridactylus* by Frappell & Rose (1986) also confirmed that the hindgut was an important site for digesta retention, but compared with Hume & Carlisle (1985) and Richardson (1989) it appeared to under-

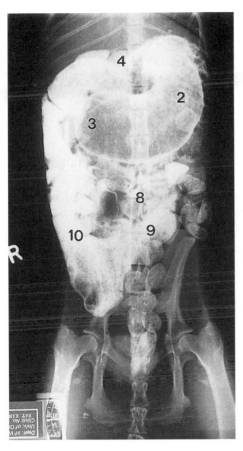

(c)

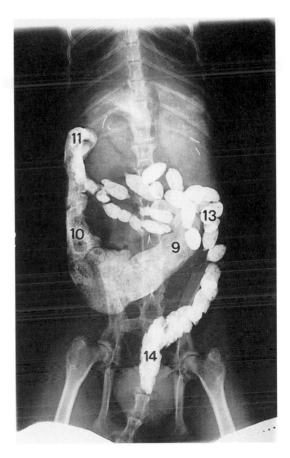

(d)

estimate retention times in the forestomach. The only major difference between the studies appeared to be that the *Potorous* used by Frappell & Rose (1986) were fasted for 12 hours before dosing. However, according to Frappell (1984), they were also sedated with ketamine hydrochloride before dosing. This led Hume *et al.* (1988) to examine the effects of both fasting and sedation on digesta movements in rat-kangaroos. In *Potorous*, fasting for 12–15 hours resulted in no change in the direction of initial dispersion of barium sulphate; the sacciform forestomach was still clearly outlined within 5 minutes of dosing. However, the time to reach the hindgut was significantly delayed. These effects of fasting were completely reversed when fasted animals were sedated; most contrast medium bypassed the sacciform forestomach, and the time to reach the hindgut was reduced from 660 minutes to 115 minutes. This effect was clearly due to the greatly increased gut motility observed fluoroscopically in sedated animals. Similar increases in gut motility upon sedation were seen in *Aepyprymnus*.

An interesting effect of fasting in *Aepyprymnus* was a change in the direction of initial dispersion of the contrast medium, from the sacciform forestomach to the hindstomach. This led Hume *et al.* (1988) to postulate the following daily sequence of events in the potoroine digestive tract: when animals begin to feed at night forestomach fill is at a minimum and the hindstomach and small intestine are virtually empty. Because of the close juxtaposition of the cardia and hindstomach, most ingesta initially bypass the sacciform forestomach and pass through the hindstomach into the small intestine. With further feeding, the small intestine and then the hindstomach fill, which diverts most ingesta into the sacciform fore-stomach. At dawn forestomach fill is maximal. During daylight hours while the animal is resting, forestomach fill declines, partly because of microbial fermentation, partly because of digesta flow out of the sacciform region, through the short tubiform region, into the hindstomach and small intestine. The effect is to maintain a more-or-less continuous flow of material through the small intestine despite a markedly discontinuous feeding pattern. Performance of plug-flow reactor (PFR) systems such as the small intestine is most efficient when flow is continuous.

In this proposal the primary role of the sacciform forestomach is storage of ingesta for later digestion in the small intestine, with microbial fermentation of only secondary importance. This is consistent with the generally highly digestible nature of the concentrate foods on which rat-kangaroos depend (these are more efficiently digested directly in the small intestine than fermented in the forestomach) and with the need for storage of food rapidly ingested in short and intermittent feeding bouts, particularly if bouts are restricted in number by the presence of predators.

Two other factors which impinge on this proposal are the extent of microbial fermentation in the forestomach and the relative importance of the forestomach and hindgut as areas of microbial activity. There is only limited information on both.

8.3.5 Fermentation

That microbial fermentation does occur in the potoroine forestomach was suggested by Kinnear *at al.* (1979). In six animals the pH ranged from 5.5 to 5.7 in the sacciform region, in contrast to a mean of 2.8 in the hind-stomach. In these animals short-chain fatty acids (SCFA) were present at a mean concentration of 106 mmol L^{-1}. However, in two other animals, SCFA concentrations were only 34 and 37 mmol L^{-1}, which correlated with higher pH values, 6.8 and 6.5 respectively. This suggests that despite relatively large parotid salivary glands (3.12 g kg^{-1} body mass), buffering of the forestomach contents is limited. This is supported by Beal's (1992) analysis of parotid gland function in *Potorous* and *Aepyprymnus*. Although there were some differences between the two genera, the composition of the parotid saliva of both was high in protein content (mainly amylase) and low in secretory capacity for bicarbonate and phosphate anions (about half that of red kangaroos). On the basis of the relatively low buffering capacity of potoroine parotid saliva, Beal (1992) questioned the importance of fermentation in the forestomach to digestion in these animals.

Wallis (1990) measured the rate of production of SCFA *in vitro* in captive *Potorous* and *Aepyprymnus* fed a finely ground pelleted diet based on corn, wheat and cornflour, with ground oat hulls added to provide 27% NDF. Animals were killed at intervals between 4 and 9 hours after dark, the time at which they commenced feeding. In four *Potorous*, SCFA concentrations and production rates were consistent over this time period, but in *Aepy prymnus* there was negligible SCFA present in the forestomach at 7 and 9 hours after dark. The pH of forestomach contents was low (4.1–4.5) in all four animals. Although animal numbers were minimal, a low buffering capacity may have been responsible for the lack of discernible microbial activity in two of the four *Aepyprymnus*.

In the other two *Aepyprymnus*, forestomach SCFA production rate was lower than in the hindgut. This was also the case for *Potorous* and the wild *Aepyprymnus* (Table 8.2). This is in marked contrast to the relatively low rates of production in the macropod hindgut (Chapter 6). However, in terms of total daily production, the forestomach was 3 to 5 times greater than the hindgut because of its greater fluid volume. Nevertheless, this is much less than the 10 to 30 times differential in the macropods (see Table 6.4).

The low molar proportion of acetate and the high proportion of propionate in the forestomach of both captive and wild animals suggests a diet high in non-structural carbohydrates and a microbial population dominated by non-cellulolytic species. Conversely, the higher acetate/propionate ratio in the hindgut is consistent with a substrate higher in structural polysaccharides after digestion of soluble components in the stomach and small intestine, and a higher proportion of cellulolytic bacteria. Additionally, disposal of reducing equivalents by reductive acetogenesis could be expected to be significant in the potoroid hindgut, as it is in the hindgut of humans, some rodents and termites (Drake 1994; Wolin & Miller 1994). Based on its relatively large size and rapid fermentation rate, the potoroine hindgut does appear to be an important site of microbial digestion. This makes sense if there is significant bypass of the forestomach fermentation region by resistant starch, pectin and non-starch storage carbohydrates, as well as structural carbohydrates. These would not be digested in the small intestine, but some would provide a readily fermentable substrate for the hindgut microbiota, particularly on the typical potoroid diet of seeds, roots and fungi.

Table 8.2. *Concentration, molar proportions and rate of production* in vitro *of short-chain fatty acids (SCFA) in the forestomach and hindgut (caecum plus proximal colon) of captive rufous rat-kangaroos (rufous bettongs) (*Aepyprymnus*) and long-nosed potoroos (*Potorous*) fed a concentrate diet, and wild rufous rat-kangaroos*

	Captive *Aepyprymnus*	Captive *Potorous*	Wild *Aepyprymnus*
Body mass (kg)	2.94	0.86	2.66
Forestomach SCFA			
Total concentration (mmol L^{-1})	83	77	128
Molar proportion (%)			
Acetate	44	46	47
Propionate	37	44	36
Butyrate[a]	14	6	9
Valerate[a]	5	4	8
Production			
mmol L^{-1} hr^{-1}	30	54	27
mmol d^{-1}	84	57	97
% of MER[b]	13.5	16.4	–
Hindgut SCFA			
Total concentration (mmol L^{-1})	88	97	111
Molar proportion (%)			
Acetate	62	70	81
Propionate	27	22	14
Butyrate[a]	8	6	5
Valerate[a]	3	2	0
Production			
mmol L^{-1} hr^{-1}	44	79	48
mmol d^{-1}	16	17	38
% of MER[b]	2.3	4.1	–

Note: [a] Negligible amounts of iso-acids present in all samples.
[b] MER = maintenance energy requirement.
Source: From Wallis 1990.

8.4 FOOD INTAKE AND DIGESTION

In his studies of the digestive physiology of potoroine marsupials, Wallis (1990) noticed that, on the type of diet used in the fermentation study above, digestibility of fibre was low and highly variable. For instance, on the 29% NDF diet, both *Potorous* and *Aepyprymnus* digested 23% of the NDF, including 10% of the ADF, with coefficients of variation of 8% and 9% for NDF and 16% and 23% for ADF. On the same diet, coefficients of variation associated with apparent dry matter digestibilities of 70% and 68% were less than 2% in both genera. Although forestomach pH was not measured in this experiment, it suggested that microbial activity may be highly variable in the forestomach depending on such factors as rate of ingestion, the salivary secretory capacity of individuals, and the balance between cellulolytic and non-cellulolytic bacteria. On a 51% NDF diet based on coarsely ground alfalfa (lucerne) and corn, apparent digestibility of dry matter was lower (58% and 59%), but fibre digestibility was much higher and more consistent among animals (56% and 58% for NDF, 36% and 38% for ADF). These results suggest a more stable microbial population in the sacciform forestomach.

Notwithstanding the highly variable performance of rat-kangaroos on the high-concentrate diet, this diet may better represent the type of diet that these concentrate-selectors consume in the wild. More detailed studies on the digestive physiology of free-living rat-kangaroos are needed.

A comparison of digestive performance between potoroine and macropodid marsupials fed the lucerne-based diet described above is summarised in Table 8.3. It includes the two genera described above, plus *Bettongia penicillata* and *Macropus robustus robustus* (eastern wallaroo). The results show that although apparent digestibility of dry matter was a little lower in the three rat-kangaroos than the wallaroo, digestibility of NDF and ADF was actually greater in all three rat-kangaroos, despite their small body size. This difference is explained by the gross anatomies of the fermentation

Table 8.3. *Intake and digestibility of a pelleted diet of 75% chopped lucerne (alfalfa) hay and 25% crushed corn, and containing 51% neutral-detergent fibre (NDF) by eastern wallaroos* (Macropus robustus robustus) *and three potoroine marsupials* (Aepyprymnus rufescens, Bettongia penicillata *and* Potorous tridactylus)

n	*Macropus robustus* (8)	*Aepyprymnus rufescens* (3)	*Bettongia penicillata* (2)	*Potorous tridactylus* (5)
Body mass (kg)	17.6	2.7	1.1	0.9
Dry matter				
Intake (g kg$^{-0.75}$ d^{-1})	56.8	36.9	43.4	35.4
Apparent digestibility (%)	63.5	58.8	56.8	57.8
Digestible intake (g kg$^{-0.75}$ d^{-1})	36.1	21.7	24.5	20.6
Fibre				
NDF digestibility (%)	39.3	59.6	55.5	57.9
ADF digestibility (%)	29.3	36.0	36.0	37.8

Note: NDF = neutral-detergent fibre (hemicelluloses, cellulose and lignin).
ADF = acid-detergent fibre (cellulose and lignin).
Source: From Freudenberger, Wallis & Hume 1989.

chambers involved. The sacciform forestomach of the rat-kangaroos has many of the performance characteristics of a continuous-flow, stirred-tank reactor (CSTR) (see Chapter 2). These include relatively long retention times and high extent of conversion of reactants to products. In contrast, the mainly tubiform forestomach of the wallaroo performs largely as a modified PFR (see Chapter 4), which features shorter retention times and, consequently, lower extent of conversion of reactants to products. The larger body size and therefore reactor volume of the wallaroo do not completely compensate for its tubiform gastric anatomy.

On the other hand, dry matter intake of the experimental diet was much greater by the wallaroos than by the three rat-kangaroos, a difference that can also be explained by the different performance characteristics of the two reactor types involved. The longer retention times of a CSTR-type forestomach limit the amount of material a rat-kangaroo can process, even when food is available *ad libitum*. Thus, despite high fibre digestibilities on an experimental diet of 51% NDF, potoroine marsupials are limited to natural diets of low fibre content; this is why they are concentrate selectors.

8.5 NITROGEN METABOLISM

The maintenance nitrogen requirements of the three potoroid marsupials that have been determined are remarkably uniform, at least on grain-based diets. Truly digestible requirements were 199 mg N per $kg^{0.75}$ d^{-1} for *Potorous tridactylus*, *Bettongia penicillata* and *Aepyprymnus* (Table 1.8). Level of fibre in the diet (from 10% to 30% NDF) had no effect on the estimates. The total nitrogen requirements of female *Aepyprymnus* at peak lactation were about four times those for maintenance of adult males.

The uniformity in maintenance requirements among the three genera is somewhat surprising in view of the large difference in requirements of wallabies from contrasting environments. Hume (1986) found that for maintenance of nitrogen balance, *Macropus parma*, a wallaby from moist forest environments, required 477 mg truly digestible N $kg^{-0.75}$ d^{-1}, whereas *M. eugenii*, a wallaby of similar size from maritime arid habitats, required only 230 (Table 1.8). Of the three potoroids compared by Wallis & Hume (1992), the range of *Potorous* is restricted to moist forest and coastal heath, while that of *Aepyprymnus* includes both moist and dry forest habitats, and *B. penicillata* is an animal primarily of open woodlands. Nevertheless, their maintenance nitrogen requirements are all below those of the parma wallaby.

8.6 ENERGY METABOLISM

Consistent with maintenance nitrogen requirements, Wallis & Farrell (1992) found no significant differences among *Potorous tridactylus*, *Betton-*

gia penicillata and *Aepyprymnus* in their resting metabolic rates (Table 1.1). All are slightly above the Dawson & Hulbert (1970) 'marsupial mean' basal metabolic rate of 204 kJ $kg^{-0.75}$ d^{-1}. Activity at night doubled the daytime fasting heat production of *Potorous* and *Bettongia*, but the increase in *Aepyprymnus* was only 25%. Consequently, the maintenance energy requirement of captive *Aepyprymnus* (386 kJ DE $kg^{-0.75}$ d^{-1}) was lower than that of *Potorous* and *Bettongia* (529 and 540) (Table 1.2). The combined heat production of fed female *Aepyprymnus* with pouch young remained quite constant for the first two-thirds of pouch life, but then rose sharply by 20% in response to rapid growth of the pouch young. The female only entered negative energy balance during the last week of pouch life (Wallis & Farrell 1992). Milk consumption rates by young *B. penicillata* increase from about 1 mL d^{-1} at 35 days of age to about 23 mL d^{-1} at 90 days, just before permanent emergence from the pouch (Merchant, Libke & Smith 1994). Milk lipid content, and thus its energy content also, increases continuously over this period, and beyond, as it does in *Potorous tridactylus* (Crowley, Woodward & Rose 1988) and in macropodid marsupials too. Thus the period around permanent pouch exit is a period of maximum drain on the mother's resources, as depicted in Fig. 1.4 for nitrogen.

When Rübsamen, Hume & Rübsamen (1983) measured metabolic rates of *Aepyprymnus* in respirometry chambers, they found a minimum metabolic rate of 240 kJ $kg^{-0.75}$ d^{-1}, a value very similar to the fasting metabolic rate reported by Wallis & Farrell (1992), between ambient temperatures of 25° and 35°C. Maintenance of a stable body temperature above 30°C was achieved mainly by panting and saliva spreading. Below the lower critical temperature the animals conserved body heat by reducing their effective surface area by hiding head, limbs and tail under the body, and below 18°C they visibly shivered. However, shivering was never observed in free-living *Aepyprymnus* in winter, even at temperatures as low as −4°C. Thus temperature regulation in the wild probably involves the use of a well-insulated nest during the day and heat generated by activity at night. Between ambient temperatures of 6° and 30°C, the temperature in the wall of occupied nests always exceeded ambient except at 30°C. The maximum difference was at 6°C ambient, when the lowest nest temperature was 19°C. This would promote significant energy savings on cold days. At night the most frequently observed activity was hopping rather than a slow quadrupedal gait, which would have generated enough heat to maintain body temperature even in cold, windy conditions. Similarly, measurements of oxygen consumption by long-nosed potoroos on treadmills at ambient temperatures between 5° and 25°C convinced Baudinette, Halpern & Hinds (1993) that heat generated during locomotion can largely substitute for heat that would otherwise be required for thermoregulation.

Wallis & Green (1992) examined seasonal energy budgets of free-living

Aepyprymnus at Drake in northern New South Wales. Field metabolic rates (FMRs) were similar in summer and winter, even though mean minimum temperatures in winter were 20°C below those in summer. They were also similar in males and females even though several of the females carried large pouch young or were suckling young-at-foot. FMRs were a higher multiple of BMR (3.3–3.4) than in all macropodid species that have been measured (1.8–2.5) (Table 1.4). An explanation for the relatively high FMR of *Aepyprymnus* is not at hand, although Wallis & Green (1992) thought that the energetic costs of nest building and of digging for hypogeal foods may be major contributors.

FMRs of long-nosed potoroos in southern Victoria were also similar in males and females, even though all females carried pouch young or had young-at-foot (Wallis, Green & Newgrain 1997). They were however about 30% higher for both sexes in early spring than in mid summer, even though mean minimum temperatures differed by only 4°C between the two seasons. No measurements were made in winter, but the authors felt that the seasonal difference they found in FMRs was probably due to a combination of the small body size of *Potorous tridactylus* and the cold, wet winters but mild summers experienced in southern Victoria.

8.7 THE NUTRITIVE VALUE OF FUNGI

The importance of fungi in the diets of *Potorous* and *Bettongia* was emphasised in Section 8.2. Johnson (1994a) showed that when production of hypogeal fungal sporocarps was high, *B. gaimardi* was almost exclusively mycophagous, while at lower levels of production the animals became more herbivorous by adding dicotyledonous leaves and fruit to their diet. Body condition of both sexes increased with increasing proportions of fungus in the diet, suggesting that fungi are a primary food source of high value to this rat-kangaroo.

There is some confusion in the literature over the nutritive value of fungi to mycophagous marsupials. Kinnear *et al.* (1979) published an amino acid profile for the edible kernal of hypogeal fungi consumed by *B. penicillata* that is grossly imbalanced, especially with respect to methionine, which was 36 times higher than that of edible truffles (Manozzi-Torini 1976) and 44 times higher than that of hypogeal fungi consumed by *Aepyprymnus* at Drake, northern New South Wales. Beckmann (1986) found that several unidentified hypogeal fungi eaten by *Aepyprymnus* were well-balanced sources of amino acids compared with truffles. Similarly, Wallis *et al.* (1997) concluded that the inner gleba of hypogeal fungi consumed by *Potorous tridactylus* was also a well-balanced source of essential amino acids. Claridge & Cork (1994) have provided a description of the other chemical components of the sporocarps of two hypogeal fungi, *Mesophellia glauca* and *Rhizopogon luteolus*, consumed by long-nosed

Table 8.4. *Chemical composition of the dry matter of sporocarps of two hypogeal fungi eaten by the long-nosed potoroo* Potorous tridactylus

Constituent	Mesophellia glauca	Rhizopogon luteolus
Water content (% of fresh matter)	7.1	7.6
Ash	2.9	2.0
Neutral-detergent fibre (total cell walls)	46.6	18.9
Gross energy (kJ g^{-1})	25.0	19.9
Total nitrogen	1.63	1.60
Protein nitrogen	0.67	0.51
Non-protein nitrogen	0.78	0.78
Cell-wall nitrogen	0.19	0.32

Source: From Claridge & Cork 1994.

potoroos (*P. tridactylus*) (Table 8.4). Both fungi were marginal in total nitrogen (1.6%), and 59–68% of this was in non-protein form or associated with cell walls. Another notable feature is the high NDF content of *Mesophellia* (47%), probably chitin, suggesting a low availability of energy as well as nitrogen.

Despite the questionable nutritional value of the two fungi, *P. tridactylus* maintained positive nitrogen balance and high intakes of digestible energy. Digestibilities were high for dry matter (80–86%), energy (76–93%) and nitrogen (72%). Claridge & Cork (1994) concluded that the good performance of potoroos on these fungi was made possible by their fore-stomach fermentation system. Microbial fermentation allows more effective utilisation of fungal structural carbohydrates, and of crude protein that is either of poor quality or is protected by incorporation into cell walls, than would be possible in simple-stomached mammals. This is probably why rat-kangaroos can use hypogeal fungi as their sole source of nutrients, whereas bandicoots, with a simple stomach and limited hindgut fermentation, use fungi to a lesser extent.

8.8 SUMMARY

The rat-kangaroos are interesting because they occupy one end of the spectrum in body size, diet, dentition and digestive strategy within the superfamily Macropodoidea. *Hypsiprymnodon* is the extreme because of its simple stomach and omnivorous diet. The other rat-kangaroos are foregut fermenters, but despite high fibre digestibilities their gastric morphology and small body size force them to be concentrate selectors, feeding on either hypogeal fungi or plant material of low fibre content. Their ability to utilise fungi of doubtful nutritional value has been demonstrated, and this explains the widespread occurrence of mycophagy among the Potoroinae. A three-way symbiotic association between hypogeal mycorrhizal fungi, rat-kangaroos and plants may be an important requirement for the long-term health of Australian eucalypt forests.

9 Evolution of marsupials and of digestive systems

This chapter provides an opportunity to review and compare the foraging and digestive strategies of the various groups of marsupials in the context of their possible evolution and radiation on the South American and Australian continents. Only a few marsupial families are known only by their modern representatives, and the available fossil record provides at least a basic outline of the evolutionary history of marsupials from the Miocene to the Recent (Clemens, Richardson & Baverstock 1989). Events prior to the Miocene are much less certain and consequently are the subject of continuing debate. The following account is based primarily on reviews by Clemens *et al.* (1989) and Woodburne & Case (1996), and should be considered in the context of the dynamic nature of the views of palaeontologists as additional fossil, molecular and other evidence comes to light.

Also, the bulk of the fossil evidence is based on fragments of teeth (often a single tooth), as these are the toughest, most enduring parts of the vertebrate skeleton. Although much can be learnt from teeth, it is wise to remember that there is abundant evidence that the gut and behaviour of an animal can compensate for the dietary adaptations of teeth (Sanson 1991). This point is discussed in Chapter 2.

9.1 OUTLINE OF THE ORIGINS OF SOUTH AMERICAN AND AUSTRALASIAN MARSUPIALS

A possible chronology of events is shown in Fig. 9.1. Although the date of 135 million years ago (mya) (early Cretaceous) has been assumed for the separation of marsupials and eutherians, teeth of the earliest undoubted marsupials found to date, in western North America, are from the late Cretaceous, 83 mya (Fox 1987). Although earlier (approximately 100 mya) fossils, barely 1 mm long, from Texas, have been attributed to the Marsupialia, their affinities are unclear. Marsupials were relatively abundant in western North America in the late Cretaceous, where they underwent a 'modest evolutionary radiation' (Clemens *et al.* 1989). By 70 mya three families were distinct. The central group was the Didelphidae, a morphologically conservative family that persists to the present and includes 70 of the 78 extant marsupials in the Americas (Rich 1991).

The other two families in the North American late Cretaceous, the Pediomyidae and the Stagodontidae, are thought to have descended from

Figure 9.1 Selected phyletic, dispersal, geologic and climatic events in the development of the South American and Australasian marsupial faunas.

Age	mya[1]	Event
Recent	0.01	Extinction of Australia's megafauna
Pleistocene	2.5	(Dispersal of Didelphidae from South America to Central and North America)
Pliocene	4.5	(Arrival to eutherian murid rodents in Australia)
Miocene	10	Increasing aridity and radiation of the Macropodinae. (Extinction of marsupials in North America and Europe)
	25	Open forest habitats, and basal evidence for Diprotodontia families
Oligocene	34	Final separation of South America from Antarctica, glaciers on Antarctica, and seasonal climates and open forests in Australia
Eocene	40	Dasyuromorphians still mainly plesiomorphic (primitive), but likely time of appearance of modern bandicoots
	42	Climate cooling in southern oceans and continents. Australian climate still non-seasonal, with closed forests dominated by *Nothofagus* (Antarctic beech)
	45	Maximum of non-seasonal warm climates, and derived South American taxons in Antarctic Peninsula; the fauna was too derived to be ancestral to Australian marsupials
	50	Radiation of Diprotodontia underway in Australia, with possum families distinct
	52	Final separation of Australia from Antarctica by a marine barrier, and probably vicariant separation of South America from Antarctica. Peak in marsupial family diversity in Australia with appearance of modern Peramelina, Dasyuromorphia and Diprotodontia
	55	Beginning of derived South American taxons in Antarctic Peninsula that were too derived to be ancestral to Australian marsupials
Paleocene	60	Appearance of Notoryctemorphia in Australia
	62	Presence of microbiotheriids (Australidelphia) and derived Ameridelphia (myaulestids, didelphids, caenolestids) in Bolivia
	63	Dispersal to Australia fundamentally over because of the sinking of the South Tasman Rise
	64	Appearance of syndactyly in Australia in the Diprotodontia at least
	65	Dispersal from Antarctica of derived microbiotheres to South America, and to Australia of basal Peramelina, Dasyuoromorphia and Diprotodontia
Cretaceous		(Marsupials dispersed from Europe to Africa and Asia (Kazakhstan), but no evidence of a significant radiation)
	66	Dispersal of marsupials to Antarctica from South America
	67	Likely entry to marsupials to South America from North America, and the split between Australidelphians and Ameridelphians
	(69–65)	(Dispersal of marsupials from North America to Europe)
	70	Basal split between Didelphidae and other marsupials in North America
	83	Earliest evidence of marsupials, in North America

Note: [1] mya = millions of years ago

primitive didelphids, as have all subsequent and more derived marsupials, including present-day Australian forms.

Marsupials appear to have dispersed from North America to both South America and Europe at the end of the Cretaceous or the beginning of the Paleocene (65 mya). They radiated in Europe through the Paleocene, Eocene and Oligocene epochs but became extinct in the middle Miocene. During this time they dispersed into Africa and Asia (as far as Kazakhstan). Although it was thought that there was no significant radiation in either Africa or Asia, recent reports of fossils of late Cretaceous mammals, thought to be aberrant marsupials, in Mongolia (Szalay & Trofimov 1996), and of a single upper molar from a tiny marsupial in middle Miocene sediments in Thailand (Ducrocq *et al.* 1992) may result in a revision of early marsupial history.

Marsupials appear to have disappeared from North America in the Miocene, to be followed several million years later by the reappearance of marsupials dispersing from South America in the Pleistocene. Five species in four genera are found today as far north as southern Mexico, but only one, *Didelphis virginiana*, ranges into the United States and southern Canada (Clemens *et al.* 1989).

According to Woodburne & Case (1996), the split into the two main groups of marsupials, the cohort Australidelphia and the cohort Ameridelphia (Fig. 9.2), took place in South America soon after their entry from North America. This was quickly followed by dispersal of both cohorts into Antarctica when South America, Antarctica and Australia were all connected as part of Gondwana. Central to the thesis of Woodburne and Case (1996) is the idea that Antarctica was a site of marsupial radiation, with dispersal from Antarctica as early as 65 mya of derived microbiotheriids to South America and of basal Dasyuromorphia, Peramelina and Diprotodontia to Australia.

Dispersal to Australia was virtually complete soon after (63 mya) because of the sinking of the South Tasman Rise (Fig. 9.3), with final separation of Australia from Antarctica about 52 mya.

There is evidence for the presence of derived microbiotheriids (Australidelphia) and derived didelphids, caenolestids and myaulestids (Ameridelphia) in Bolivia 62 mya. Radiation continued in the Antarctic Peninsula, with the first appearance there 58 mya of other South American taxons that were too derived to be ancestral to Australian marsupials. These derived South American taxons reached a maximum in the Antarctic Peninsula about 45 mya, when warm climates prevailed. By 42 mya, cooling of the southern oceans and continents was evident, and by the time of final separation of South America and Antarctica 34 mya the latter land mass had continental glaciers.

In Australia, the Notoryctemorphia (Fig. 9.2) are thought to have appeared by 60 mya (although the earliest fossils are from about 25 mya),

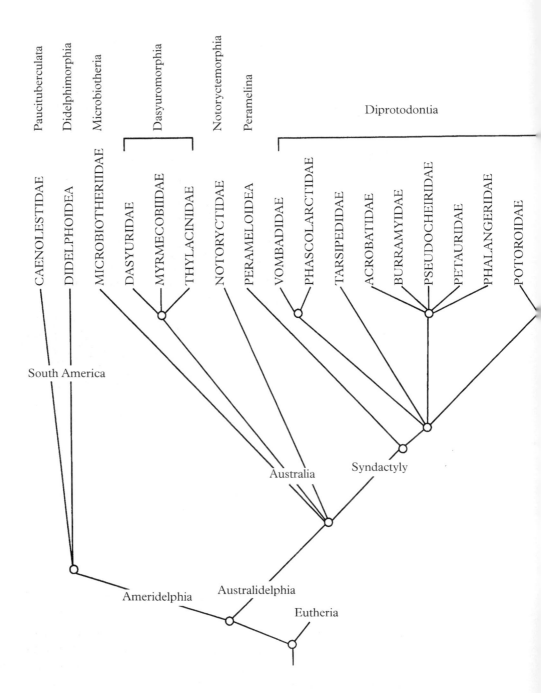

Figure 9.2 Pylogenetic tree of marsupial orders and families to show the basic dichotomy into the two cohorts Ameridelphia and Australidelphia. After Woodburne & Case (1996).

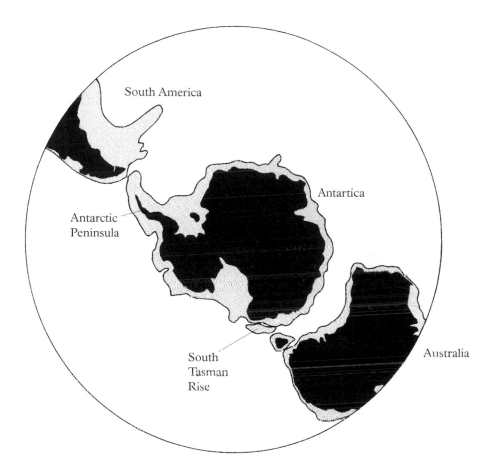

Figure 9.3 Gondwana, showing the connections between South America, Antarctica and Australia at 50 million years ago. After Woodburne & Case (1996).

and radiation of the Diprotodontia was well under way by 50 mya, with possum families distinct by 30 mya. Climates were warm and non-seasonal, and even at 42 mya Australia still had non-seasonal, though cooler, climates with closed forests dominated by Antarctic beech (*Nothofagus*). Modern bandicoots are thought to have appeared at about this time (40 mya), but most Dasyuromorphs were probably still pleisiomorphic (primitive).

By the time South America finally separated from Antarctica (34 mya), Australian climates were drying and becoming more seasonal, and open forests appeared. Increasing aridity in the Miocene (20–10 mya) coincided with the appearance of grasslands and the radiation of the Macropodinae. Eutherian murid rodents arrived in Australia from Asia 4.5 mya (Pliocene).

9.2 AUSTRALIAN MARSUPIALS IN THE MIOCENE

Although the available fossil record of Australian marsupials extends back into the Oligocene (38 mya) the earliest presently known extensive marsupial fossil assemblages are from the Miocene. By that time the basic part of the latest radiation had occurred, with marsupial families specialised for each major ecological niche present (Clemens *et al.* 1989). The Miocene marsupial fauna consisted largely of arboreal and terrestrial browsers and their predators.

Towards the end of the Miocene (8–10 mya) a number of events occurred that produced the drying trend mentioned above. Australia drifted north towards the Asian block and the New Guinean highlands started to rise. (The Great Dividing Range in eastern Australia was already in place, having risen about 90 mya at the time when New Zealand split off from Australia.) These events created a rainshadow for rain from the northeast, which was exacerbated by less rain from the west as the continent drifted into drier latitudes. Also, sea temperatures fell sharply towards the end of the Miocene, which further reduced precipitation.

This drying trend led to the replacement of closed forest (rainforest) by open forest, woodland and ultimately grassland in the interior. At the same time, fossil assemblages in central Australia changed from a domination by arboreal and terrestrial browsers to the newly evolving terrestrial grazers. The trend toward increasing aridity continued through the Pleistocene to a peak about 17 000 years ago (Clemens *et al.* 1989). At about this time the Australian megafauna disappeared.

During the Pleistocene the New Guinea highlands and the Great Dividing Range continued to rise, and the continued northerly drift of Australia brought its northern parts into a wetter zone. The cooler, wetter conditions so created provided refuge for groups from central Australia, including phalangerid possums and browsing macropods, that persists to the present day.

Further south, the uplifted Australian Alps created an alpine niche that was filled by the mountain pygmy-possum (*Burramys parvus*) and the broad-toothed rat (*Mastacomys fuscus*). Thus a wide range of habitats became available to the Australian fauna in the closing stages of the Tertiary Period. This is reflected in the distribution of the member species of seven extant marsupial families (Table 9.1).

To date, no marsupial has established itself on the Asian mainland, but cuscuses (Phalangeridae) are found as far west as Sulawesi in Indonesia.

9.3 THE AUSTRALIAN MEGAFAUNA

During the Pliocene and Pleistocene the body sizes of many vertebrates throughout the world tended to increase, and by the late Pleistocene North

Table 9.1. *Distribution of seven Australian marsupial families among major habitats*

| Family | Number of species | | | | | | |
	Closed forest	Open forest	Woodland	Open scrub	Low shrubland	Hummock grassland	Tussock grassland
Dasyuridae	10	14	14	10	10	18	14
Peramelidae	4	7	4	1	0	2	4
Vombatidae	0	1	1	0	1	0	0
Petauridae	6	7	5	0	0	0	0
Phalangeridae	4	1	1	0	0	0	0
Potoroidae	1	6	0	1	1	1	2
Macropodidae	9	13	10	5	0	3	7
Number of species per 1000 km² of habitat	0.49	0.18	0.03	0.03	0.04	0.08	0.04

Source: From Clemens, Richardson & Baverstock 1989.

America, South America, Eurasia, Africa, Australia and some islands each supported a megafaunal community. In the absolute sense, megafauna refers to Australian mammals of 40 kg or more. In the relative sense, members of the megafauna reached a significantly larger size than closely related forms (Murray 1991). Prolific speciation in the Pleistocene culminated in a diverse Australian megafauna of 19 genera and 35 species.

A typical Pleistocene marsupial community from temperate Australia about 50 000 years ago included the marsupial lion (*Thylacoleo carnifex*) at 45 kg, the giant wombat *Phascolonus* (150 kg), the marsupial tapir *Palorchestes* (300 kg), the diprotodontians *Diprotodon* (1150 kg), *Zygomaturus* (450 kg) and *Nototherium*, the potoroid *Propleopus* (45 kg), and the macropodids *Protemnodon* (a giant wallaby of 50 kg), *Simosthenurus* (50 kg), *Sthenurus* (50 kg) and *Procoptodon* (150-200 kg) (the short-faced kangaroos). In addition, there were some very large members of the living genera *Wallabia* and *Macropus* (Macropodidae) and *Sarcophilus* (Dasyuridae). At 1150 kg, *Diprotodon* was clearly the most magnificent member of the Australian megafauna.

More than 50 species of the Australian megafauna became extinct at the end of the Pleistocene (50 000 years ago) and prior to the Holocene, although reliable terminal dates are few in number. Australia was one of the most severely affected of the continental faunas in that no mammal greater than about 60 kg in mass survived among the living fauna encountered by the Europeans (Murray 1991). The proposed cause or causes of this mass extinction event are many, and include climate change, epidemic diseases, and hunting and ecological disruption by Aborigines. Many of the morphological specialisations of the Australian megafauna appear to have been related to the marginal conditions of a semi-arid climate (over 8 million years of cyclic drought) and the increasing senility of Australian soils. Thus they are unlikely to have entirely succumbed to the climate of the late Pleistocene. On the other hand, they appear to have evolved few specialised defensive and offensive features and were probably less fleet than most of the eutherian megafauna. For these reasons, it has been concluded by Murray (1991) and Flannery (1994b) that the cause of extinction of the Australian megafauna is most likely gradual populational attrition by human predation at a time of increased aridity and evidence of fire.

9.4 LIKELY FORAGING AND DIGESTIVE STRATEGIES OF EXTINCT MARSUPIALS

Most living South American marsupials are included in the family Didelphidae. Their early success in the Cretaceous period has been attributed in part to the acquisition of tribosphenic molars, which allow for both shearing and grinding actions, as opposed to only one or the other actions in their progenitors. This double action of the molars allowed these early marsupials to utilise a much wider range of foods than their basically insectivorous ancestors. For instance the Stagodontidae, which arose from the Didelphidae in the Cretaceous period, reached a body mass of about 40 kg. They had short, deep and heavy jaws, and premolars as well as lower molars with well-developed shearing surfaces. They seem to have evolved to become specialised carnivores.

The diets of living didelphids range from insectivory/carnivory to frugivory, but most are omnivorous (see Chapters 2, 3). It is probable that, in this morphologically conservative group, the simple gastrointestinal tract found in *Didelphis virginiana* (Fig. 3.1) is little changed from the earliest didelphids from western North America. A caecum most likely was present in these earliest forms, derived via a stem therian from a reptilian ancestor (Hume & Warner 1980).

There appears to be little variation in gastrointestinal tract morphology among extant South American marsupials, but more detailed information is needed in this area. Subtle variations in dentition and /or digestive tract morphology may be expected to accompany specific dietary niches. The stomach of *Caenolestes*, with its cardiogastric gland (Figs. 2.6, 2.7), is an exception to the simple didelphid model. Caenolestids also differ from didelphids and other marsupials by a pair of lower incisors that are enlarged into dagger-like teeth, molars that decrease in size posteriorly, and in some species a blade-like last premolar or first molar. All these features suggest an active predatory lifestyle, as seen in the dietary habits of modern species (see Chapter 2, Section 2.3.1).

The microbiotherid *Dromiciops* is of interest because it has a number of features (e.g. greatly simplified molars, enlarged and fully ossified auditory bullae and articular facets of the tarsus) that resemble those otherwise found only in Australian marsupials (Szalay 1982). These derived morphological features, along with the current distribution of *Dromiciops* in *Nothofagus* forests in southern Chile, support the idea that the Microbiotheres were an early offshoot of the radiation of Australian marsupials in Antarctica (Clemens *et al.* 1989). Further support for this idea is the fact that only two groups of marsupials have no hindgut caecum, *Dromiciops* and the Australian Dasyuromorphia.

The South American radiation of marsupials includes no herbivorous branches. The earliest faunal assemblages reconstructed from the Paleocene and Eocene of southern South America indicate an advanced radiation of the Marsupialia into insectivorous/carnivorous and omnivorous trophic niches (Eisenberg 1989). Another major mammalian group present at that time, the Xenarthra, was differentiating into insectivorous, omnivorous and herbivorous forms. At the same time, the Notoungulata and the Litopterna were filling the terrestrial large herbivore niche. Their place was taken by the Artiodactyla (deer, peccaries, camelids) in the Pliocene, and later by the Perissodactyla (tapirs, horses) and Proboscidea. Although horses, proboscideans and many of the larger camelids did not persist beyond the Pleistocene, the continual presence of large eutherian herbivores appears to have prevented the radiation of marsupials into this trophic niche.

The Borhyaenidae, the 'dog-like' marsupials, appear in the South American fossil record from the late Paleocene to the Pliocene (a span of 50 million years), and included both large and small carnivores (Clemens *et al.* 1989). Their decline coincides with the appearance of modern eutherian carnivores in the Pliocene, at the time of completion of the Panamanian land bridge (Stehli & Webb 1985). An alternative explanation is that the decline of the borhyaenids in the late Tertiary was brought about by competition from large carnivorous birds, the phororhacoids (Marshall 1978). Whatever the cause of the borhyaenids' demise, the modern marsupial fauna of South America lacks large herbivores and large carnivores, but fills the small carnivore/insectivore and omnivore niches.

The earliest Australian marsupials were also no doubt small insectivores. Except for the absence of a caecum the dasyurid digestive tract is probably virtually unchanged from its didelphid ancestor, and is remarkably similar to that of *Didelphis* (see Figs. 2.10 and 2.11 versus 3.1). The absence of a caecum in modern dasyurids is considered an example of secondary loss of a primitive organ (Hume & Warner 1980).

In the Pleistocene the appearance of marsupial carnivores of larger body size allowed for selection of larger prey. Thus *Thylacoleo carnifex*, at 45 kg, appears to have selected animals of large body size (though probably not as large as *Diprotodon*), while *Thylacinus cynocephalus* (Tasmanian tiger), at up to 35 kg, selected prey of small to medium body size. This led Case (1985) to suggest that these two Pleistocene marsupial carnivores could have coexisted within a single community because of non-overlapping dietary niches. The unique dentition of *Thylacoleo* was discussed in Chapter 2, Section 2.3.2. Its gross dental morphology indicates flesh eating and no bone crushing; the relatively large striations on the teeth of *Thylacoleo* are attributed to fragments of bone or grit in the fur of the prey (Wells *et al.* 1982). Tooth microwear patterns of the thylacine are also consistent with a diet of meat (Jones & Stoddart 1998).

Modern bandicoots appeared early in the fossil record, about 40 mya. The omnivorous dietary niche of extant bandicoots probably arose early, and may even have its origins in the first didelphids in North America. At some stage in the evolution of these early marsupial insectivores, there presumably arose some advantage in ingesting plant material such as seeds and nuts, fruits and succulent shoots, perhaps during periods of low insect prey availability. The consequent passage through the gut of more indigestible bulk would have had at least two significant effects (Hume & Warner 1980): First, there would have been increased secretion of fluid throughout the length of the gut, beginning with saliva (especially watery parotid saliva), and of mucus from the oesophageal wall, the cardiac and pyloric glands of the stomach, and from the small and large intestine. Second, there would have been increased epithelial sloughing because of the increasingly abrasive nature of the diet. From these events there would follow selective advantage in having some means of slowing digesta passage, resulting in more time for resorption of water, electrolytes and metabolites from these secretions to proceed in order to avoid excessive loss. The need for resorption would be greatest at the distal end of the gut. Selection pressure would thus be expected to be greatest for enlargement of the hindgut, principally the caecum, where digesta retention would have been longest.

In this scenario, microbial fermentation in the caecum would have followed soon after, but the initial stimulus to hindgut enlargement was probably related to recovery of secretions from more proximal parts of the gastrointestinal tract. The possession of only a small microbial population in the hindgut of these early omnivores would seem to have had little adaptive advantage (Hume & Warner 1980), and only when microbial activity made a significant contribution to the host's energy or nutrient economy would it be of any real advantage to the host, and subject to selection in response to progressive changes in the nutritional environment of the animal. As discussed in Chapter 3, the advantage of microbial fermentation in the hindgut, and the evolution of a mechanism to selectively retain microbes in the caecum of at least two extant bandicoot species (Moyle *et al.*1995; McClelland 1997), is an increased ability to exploit unpredictable nutritional environments.

Omnivory is clearly a highly successful dietary strategy, as evidenced by the large number of didelphids in South America that feed on both plant and animal material, and the evolution of a suite of arboreal omnivorous marsupials in Australia (Chapter 3) in addition to the terrestrial bandicoots and bilbies.

9.5 EVOLUTION OF HERBIVOROUS MARSUPIALS

Increasing plant intake and increasing body size among early Australian marsupials would have increased the supply of substrates to the hindgut for microbial fermentation. Among hindgut fermenters, two alternative strategies have evolved (Hume 1989): the first is colon fermentation, which involves separation of the colon into two functional areas, the distal section which has remained the major site for resorption of water and electrolytes (Stevens & Hume 1995), and the proximal section where digesta are retained by antiperistalsis and by structural modifications, principally haustration of the wall but occasionally by permanent folds as well. The proximal colon, together with the caecum, if present, is where microbial fermentation can make a significant contribution to the energy economy of the animal, up to 33% in the wombats (see Chapter 4). The digestive strategy of the giant wombats of the Pleistocene (*Phascolonus*) was no doubt similar to that of extant vombatid species. The colon digestive strategy is well suited to large body size, as *Phascolonus*, at 150 kg, was, because of its tolerance to highly fibrous plant materials. This type of hindgut fermentation strategy is covered in Chapter 4. The hypselodont (open-rooted) teeth of the Vombatidae are thought to have evolved during the late Tertiary in response to a general drying of Australian climates and expansion of savanna, grassland and desert habitats (Clemens *et al*. 1989). The giant wombats of the Pliocene and Pleistocene must have been wide-ranging herbivores, feeding on low-quality forage. The low energy and nutrient requirements of modern wombats probably evolved in these earlier, much larger forms. It is also likely that giant wombats did not dig burrows. Thus burrowing must have evolved recently among wombats in the course of body-size reduction of large ancestral forms (Johnson 1998), perhaps as a response to water stress in open habitats. The combination of low maintenance requirements, open-rooted teeth, colon digestive strategy and burrowing habits helps to explain the success of modern wombats relative to other large herbivores in low-productivity environments.

The second hindgut digestive strategy, which is found only in relatively small mammals, is seen in caecum fermenters (Chapter 5). In this strategy the caecum does not function as a simple extension of the proximal colon, but instead as a fermentation chamber in which microbes along with other small particles as well as fluid and solutes can be selectively retained, which facilitates the elimination of larger particles. This is a means of utilising fibrous diets without the encumbrance of an overly large gut, and

High-quality forage . . . . Low-quality forage
(low fibre, high protein) . . . . (high fibre, low protein)

Plant roots, tubers, bulbs, seeds, fungi (invertebrates)	Temperate grasses, herbs in growing season, shoots of woody shrubs, fruits	Tropical grasses in wet season, temperate grasses, herbs in dry season	Tropical grasses in dry season, arid-zone grasses and shrubs

Small body size . . . . Large body size
Crushing/grinding . . . . Cutting
dentition   dentition

Simple stomach Large sacciform forestomach . . . . Large tubiform forestomach

Hypsiprymnodon			*M. giganteus*
Potorous	*Thylogale*		*M. rufus*
Bettongia	*Setonix*	*Wallabia*	*M. fuliginosus*
Aepyprymnus	*Lagorchestes*	*M. rufogriseus*	*M. r. robustus*
	Dorcopsis	*Onychogalea*	*M. r. erubescens*
	Dendrolagus	*Petrogale*	
	Lagostrophus		
	Dorcopsulus	*Peradorcus*	

Figure 9.4 Nutritional niches occupied by kangaroos, wallabies and rat-kangaroos. There is a continuum of forage qualities from high on the left to low on the right. Examples of plant materials at four points on this continuum are shown, as are the general trends in species body size, dentition and relative sizes of the sacciform and tubiform regions of the forestomach. The superfamily Macropodoidea utilises the complete range of forage qualities.

is important for small animals because of their relatively high mass-specific nutrient requirements. The caecum digestive strategy is thought to be more derived than the colon digestive strategy (Hume & Warner 1980), and has allowed the relatively recent radiation of arboreal marsupials that feed on tree foliage. Species that feed on the leaves of *Eucalyptus* species are discussed in Chapter 5.

The other recent radiation has been the kangaroos. The earliest members of the superfamily Macropodoidea were the rat-kangaroos. The most primitive extant rat-kangaroo, *Hypsiprymnodon* (musky rat-kangaroo), is an omnivore with simple dentition and gut morphology (see Chapter 8) that lives only in closed forests that are reminiscent of the ancestral habitat of rat-kangaroos in the Oligocene, about 30 mya. Although measurements have not been made, *Hypsiprymnodon* probably has some fermentation in its small caecum, as occurs in other terrestrial omnivores such as the bandicoots.

Hypsiprymnodon's closest relative is possibly *Propleopus*, the gigantic Pleistocene rat-kangaroo. *Propleopus* had similar dental features to *Hypsiprymnodon*, with a large plagiaulacoid, sectorial premolar and tubercular, bunodont molars, only the teeth were much larger. The large premolars were possibly used to cut flesh, while the bunodont molars suggest a browsing herbivorous diet (Sanson 1991). It was therefore probably omnivorous, as is *Hypsiprymnodon* today (Chapter 8).

All other extant members of the Macropodoidea are foregut fermenters. As new nutritional niches appeared during the late Miocene and early Pliocene a diversity of morphological and physiological adaptations appeared to exploit these niches. The result is the range of body sizes, natural diets and digestive strategies that we see in the kangaroos, wallabies and rat-kangaroos today. Fig. 9.4 illustrates the range of nutritional niches exploited by extant Macropodoidea.

The consequences of small body size are seen in the Potoroinae. Their relatively high metabolic requirements limit them to plant and fungal material that is low in fibrous cell walls. Although they have a large forestomach in which microbial fermentation occurs (Wallis 1990), it probably evolved primarily as a storage organ and only secondarily for fermentation (Hume *et al.* 1988). Nevertheless, forestomach fermentation may provide 16–21% of the daily maintenance requirement for energy, and in addition detoxify some plant secondary metabolites and provide microbial amino acids and B vitamins that may be deficient in the diet (Freudenberger, Wallis & Hume 1989).

9.5.1 Forest herbivory

In contrast to the rat-kangaroos, wallabies and kangaroos feed primarily upon the aerial parts of plants. This dietary shift is associated with increases in body size, dental specialisations and shifts in the proportions of sacciform and tubiform regions of the forestomach (see Chapters 6 and 7). The early wallabies were inhabitants of wet forests. Some of their digestive features may be retained by extant genera such as *Thylogale* (pademelons). Pademelons have a lophodont, browsing grade dentition which maximises occlusal surface area and results in a crushing action that is suited to soft, non-abrasive plant materials. The pademelon forestomach is mostly sacciform, which was probably true of ancestral macropodids. However, about 40% is tubiform, which Freudenberger *et al.* (1989) proposed to be a derived feature to facilitate the passage of less digestible forage. Also, tropical browse often contains toxic plant secondary metabolites, and detoxifying microbial associations in the forestomach of early forest-dwelling wallabies may have been critical to their success.

9.5.2 Forest ecotone herbivory

As Australian climates began to dry in the Oligocene and Miocene, forests receded, which created ecotones between forests and developing grasslands. These ecotones are nutritionally diverse and currently support several large wallabies, including *Wallabia bicolor* (swamp wallaby), *Macropus* spp. (red-necked wallaby, black-striped wallaby, agile wallaby) and *Petrogale* spp. (rock-wallabies). Like pademelons, swamp wallabies and some rock-wallabies have the ancestral browsing grade dentition (Table 7.1), but wallabies in the genus *Macropus* have the more derived dentition of the intermediate browser/grazer grade, which has a reduced area of contact, resulting in more of a cutting and less of a crushing action. Common to most if not all forest-edge wallabies is an enlarged tubiform region of the forestomach which allows for increased passage of less digestible but abundant grasses.

The soft tissues of the gastrointestinal tracts of the giant Pleistocene genera *Sthenurus*, *Simosthenurus* and *Procoptodon* are not preserved in the fossil record. On the basis of dental and skull morphology, *Sthenurus* is considered to have been a grazer, while *Simosthenurus* and *Procoptodon* are thought to have been browsers (Tedford 1966). However, Sanson (1991) has noted that the molar patterns of *Sthenurus* are very similar to those of *Wallabia bicolor* (swamp wallaby), which is a browser, while those of *Procoptodon* are fundamentally the same as those of *Macropus giganteus* (eastern grey kangaroo), a grazer. These differences in opinion highlight the difficulties in interpreting the diet of extinct species from only their teeth and skulls.

A number of lines of evidence (molecular, immunological, anatomical and behavioural) have recently come together to support the idea that tree-kangaroos are closely related to and probably evolved from rock-wallabies, possibly more than 5 mya. Flannery *et al.* (1996) surmised that a proto tree-kangaroo/rock-wallaby, which had developed the special ability of making long and precise downward leaps found it advantageous to ascend sloping tree trunks, perhaps to avoid predators. Natural selection for strong, curved claws, ability to reach and grasp above their head with their fore-limbs, and ability to twist their feet so that the soles, facing each other, were able to grasp a tree trunk, would have led to the ability to climb trees. This would have further improved their chances of escaping both terrestrial predators and helminth parasites (Beveridge & Spratt 1996). It would also have allowed a dietary shift from the leaves of herbs and grasses to tree foliage, for which competition may have been less intense. A low basal metabolic rate (Table 1.1) and slow reproductive rate (Flannery *et al.* 1996) would have allowed them to reduce their intake of leaves and their protective agents, the plant secondary metabolites. However, it appears that *Eucalyptus* foliage is too heavily defended by bacteriocides to be utilised as food by foregut fermenters such as tree-kangaroos. Only hindgut fermenters such as the koala, greater glider, and, to a lesser extent, ringtail and brushtail possums have successfully exploited eucalypt leaves as food. In these marsupials the plant secondary metabolites are largely absorbed from the stomach and small intestine, detoxified in the liver and excreted in the urine, thereby protecting the bacteria resident in the caecum and proximal colon.

Tree-kangaroos reached New Guinea at least 2 mya, along with the forest wallabies *Dorcopsis* and *Dorcopsulus*, extinct giant wallabies of the genus *Protemnodon*, pademelons (*Thylogale* spp.) and the agile wallaby (*Macropus agilis*). In the extensive New Guinean tropical montane forests the tree-kangaroos radiated to fill a variety of niches. In the face of less competition, some even became largely terrestrial again (Flannery *et al.* 1996).

9.5.3 Grassland herbivory

The dramatic increase in body size of macropodids during the Pliocene and Pleistocene corresponded with the expansion of grasslands of low nutritive value. At this time there may have been evolutionary pressure to maximise body size and thus minimise the ratio of metabolic requirements to gut capacity, so that this new food resource could be exploited. The grazing kangaroos and wallabies tend to have the most reduced sacciform region and the most expanded tubiform region of the forestomach. This, in combination with an often greater body size, allowed them to exploit seasonally poor quality but abundant grasses.

As well as being high in fibre, mature grasses are often low in protein (see Table 4.1), and this could have limited their exploitation by kangaroos. However, this limitation would have been ameliorated by urea recycling to the gastrointestinal tract, as has been demonstrated in present-day wallabies (Chapter 6). For instance, Kennedy & Hume (1978) found that incorporation of endogenous urea into microbial protein in the gut of tammar wallabies (*Macropus eugenii*) fed a low-nitrogen diet equalled their dietary nitrogen intake. Forest and forest-edge wallabies also benefit from urea recycling, but this is of less nutritional importance in the presence of abundant green forage which is not usually limiting in nitrogen content.

The hindgut of macropodid marsupials also appears to have increased in length from the ancestral browsers to large grazing kangaroos (Osawa & Woodall 1992). This may have been primarily in response to increasing water stress (Woodall & Skinner 1993) as drought became more common, but would have also provided additional capacity in the proximal colon for microbial degradation of plant cell walls (Freudenberger *et al.* 1989).

9.5.4 **Arid-zone herbivory**

Arid-zone herbivory would also have been assisted by reductions in metabolic rate and thus water turnover rate (see Chapter 1). For instance, Freudenberger & Hume (1993) showed that even in captivity the arid-zone euro (*Macropus robustus erubescens*) required only two-thirds as much water as the closely related eastern wallaroo (*M. r. robustus*) from more mesic forested habitats. When given limited water, euros produced drier faeces (43% dry matter) than eastern wallaroos (30% dry matter); this was probably achieved by greater water resorption in a 65% longer colon. Euros may be able also to preferentially maintain gut water at the expense of interstitial water during periods of water shortage (Denny & Dawson 1975b), and thus better maintain microbial fermentation of dry grasses. A 33% lower maintenance requirement for truly digestible nitrogen than in eastern wallaroos is another important aspect of the euro's ability to exploit arid-zone food supplies.

9.6 **CONCLUSION**

Although mostly speculative, it has been possible in this chapter to describe the digestive systems of various extant groups of marsupials in terms of their likely origins. The description of the gastrointestinal tracts of extinct marsupials is problematic because of the absence of soft tissues in the fossil record. Although there is some skull material in the more recent fossil record, the earlier record is based almost entirely on fragments of teeth. Speculation on digestive strategies based solely on teeth are subject to considerable potential error because of the opportunities for the digestive tract and/or foraging strategy of an animal to compensate for lack of differences in dental specialisations. Nevertheless, by considering a combination of information from the paleontological, climatological and physiological literature, it is still possible to make a guess at how the wide array of digestive strategies seen in present-day marsupials may have come to be.

(10) Future directions

This book has concentrated on what we know about marsupial nutrition at the end of the twentieth century rather than on what we don't. However, it must be pointed out that the number of species included in analyses of digestive function are minimal, and especially for physiological studies a wider range of species from divergent habitats is needed. In any future research the comparative approach is to be encouraged in order to establish the common principles involved and at the same time the extent to which parameters vary within dietary groups, and why.

Because one of the main aims of the book is to compare metabolic rates and nutrient requirements among marsupials, a large part of Chapter 1 is devoted to compilations of comparative data. In this regard, West, Brown & Enquist (1997) provide a plausible basis for expecting that most metabolic functions of animals should be related to body mass by some multiple of the one-quarter power, and that for whole-body parameters the power function should be close to three-quarters because most animals are three-dimensional. Thus BMR in marsupials scales to the 0.75 power of body mass (Hayssen & Lacey 1985). Nevertheless, a range of empirically derived power functions for other metabolic parameters have appeared in the literature, perhaps because some animals are less three-dimensional than others. For instance, Nagy (1987, 1994) has found that the field metabolic rate (FMR) of marsupials as a group scales to the 0.58 power of body mass. Although Green (1997) suggested a different power function for macropodoid (0.69) and non-macropodoid marsupials (0.52), there is no clear biological basis for lumping all non-macropodoids together, and for this reason the exponent 0.58 recommended by Nagy (1987, 1994) is used in this book.

Likewise, different exponents for water turnover rates measured in captive (0.80) and free-living marsupials (0.71) appear in Tables 1.6 and 1.7 respectively. However, in all cases the data presented in tables are sufficient for the reader to recalculate the values on the basis of any other power function.

There is a continuing need to increase the data set on FMRs and water turnover rates of marsupials at both ends of the body size spectrum. For instance, in the carnivores there is a lack of data on both the smallest forms such as *Ningaui* and *Planigale* and some of the larger forms such as *Dasycercus (Dasyuroides) byrnei* (kowari) and *Dasyurus maculatus* (spotted-tailed quoll). There is also almost no information on FMRs in didelphid marsupials from South America.

Among omnivorous marsupials there is a shortage of data on energy and water flux in bandicoots from both cool and tropical climates, from the smallest potoroid, *Hypsiprymnodon* (musky rat-kangaroo), and from the extremes of body size of arboreal species such as *Acrobates* at the lower end and *Petaurus australis* (yellow-bellied glider) at the other. Among herbivorous marsupials there is a shortage of information on energy and water flux in the large kangaroos as well as in some of the smaller wallabies and hare-wallabies, and in rock-wallabies and tree-kangaroos.

As Green (1997) pointed out, a larger data set would increase our confidence in allometric descriptors of FMR and water turnover rate in marsupials, and would allow for better comparisons of the relative impacts of season and reproductive status on energy and water fluxes. The measurement of milk intake in free-living marsupials (for instance, by simultaneous measurement of sodium turnover in pouch young and the sodium content of milk) should also be extended, as it would help to explain differences in the energetic costs of reproduction between various marsupial species.

An expanded data base would also be expected to lead to a reduction in the coefficients of variation associated with estimates of FMR and field water turnover rate (WTR). These often exceed 25% (Green 1997). In addition to the effects of season and reproductive status, differences in activity levels of individuals could account for much of this variation. The use of activity loggers during measurements of FMR should prove useful in resolving the influence of differences in activity levels between individual animals on energy and water flux in the field.

The role of protein supply in the nutrition of marsupials has been under-researched. Protein shortages could well limit the exploitation of energy-rich foods such as plant exudates by some of the arboreal omnivores if pollen and /or invertebrates are not available. Because differences in seasonal conditions and activity levels have little effect on protein turnover, estimates of the maintenance nitrogen requirements of captive animals are directly applicable to non-reproducing adult animals in the field. However, there is a need for more information on the nitrogen requirements of young, growing marsupials (both pouch young and independent juveniles) and of lactating females. In cases such as the rat-kangaroos (Chapter 8) there is currently confusion in the literature over the biological value of the protein contained in hypogeal fungi. The majority of more recent papers dispute the unbalanced amino acid profile of hypogeal fungal protein reported by Kinnear *et al.* (1979), and there is a need to resolve this issue by further, careful analyses.

In Chapter 2 it was noted that increased emphasis on analysis of relationships between body size, jaw geometry, dental morphology and diet is leading to improved understanding of some of the subtle differences that exist among carnivorous marsupials in prey selection and thus community structure. Although our knowledge of marsupial diets has expanded enormously over the 17 years since this book's predecessor appeared, there remains a great and urgent need to quantify dietary preferences for many species. In this regard it is important to distinguish between anecdotal observations on feeding and those studies which carefully measure and compare food intake with food availability (Lee & Cockburn 1985). This is especially important for many South American species if conservation measures are to be maximally effective.

The problem of deriving the diet of extinct forms on the basis of structural features such as skull morphology (if the fossil is available) or dental structure, in the absence of gastrointestinal tract information, was highlighted by Sanson (1991). Several cases in which dietary differences are not reflected in dental patterns, because numerous modifications of the digestive tract and/or feeding behaviour compensate, are evident in extant species. There is no reason to believe that similar compensations did not operate in extinct species also.

Although the gastrointestinal tract of carnivorous marsupials shows little variation in structure between species, more information is needed on functional aspects, especially given the subtle dietary specialisations that have recently been described. *Caenolestes*, from South America, with its prominent cardiogastric gland, would be a good candidate for such investigations. Among Australian species, further information on food processing times and efficiencies in the larger dasyurids in relation to their FMRs would be welcome. There also remains the question of how much energy and protein is derived from the exoskeletons of invertebrate prey by the smaller dasyurids and carnivorous didelphids and caenolestids. The site and activities of chitinase and chitobiase, the enzymes involved in chitin breakdown, in these species would be of particular interest.

It is for the omnivorous marsupials that the greatest expansion in dietary information has come, especially for Australian species (Chapter 3). More needs to be done on the diet and ecology of South American didelphids across a range of environments. Studies of gastrointestinal tract function in marsupial omnivores lag behind those on diet. Although *Didelphis virginiana* is held in many laboratories, little research has been conducted on its digestive performance. The finding that in at least two bandicoot species (*Perameles nasuta* and *Isoodon macrourus*) there is a colonic separation mechanism that results in selective retention of fluid, bacteria and other fine particles, helps to explain how these marsupials exploit nutritionally unpredictable environments. The structural basis for the separation mechanism is so far obscure, and deserves detailed physiological study.

As in the small carnivores/insectivores, knowledge of the nutritional value of invertebrate exoskeletons to omnivorous marsupials such as bandicoots and bilbies would be particularly useful. Although it appears that the protein inside pollen grains is digested mainly in the small intestine, the exine layer passes through the short digestive tract of small marsupial omnivores such as *Tarsipes* (honey possum) virtually intact. This means that no volume reduction occurs, and the implications of this for the overall digestive process deserve further consideration. One of the principal energy substrates of these marsupials is gums, which are probably digested mainly by microbial fermentation. The large caecum of gumivores such as *Petaurus breviceps* (sugar glider) and *Gymnobelideus leadbeateri* (Leadbeater's possum) has been noted by several workers, but no functional studies on gum digestion, or on the effect of different levels of gum in the diet on the overall digestive process in these species have been conducted. The findings by Smith & Green (1987) and van Tets (1996) of low maintenance nitrogen requirements in the sugar glider and eastern pygmy-possum need to be extended to other omnivorous marsupials. Several small omnivorous marsupials are probably major pollinators of plants in the Myrtaceae and Proteaceae, but quantitative data on the benefits to the plants from the proposed co-evolutionary relationships are still needed.

The two main factors that define an animal's nutritional niche (its requirements for energy and nutrients, and the way in which it extracts those required nutrients from its environment) are well illustrated in Chapter 4, on the wombats. The studies of Barboza (1989) on the two southern species show how powerful the comparative approach can be in the fields of digestive physiology and nutritional ecology. Nevertheless, the several indications of low metabolic rates in the wombats described in that chapter are supported by only one direct measurement of energy flux, that of Wells (1978a) on the BMR of *Lasiorhinus latifrons* (southern hairy-nosed wombat). Similar measurements are needed in *Vombatus ursinus* (common wombat). There are no measurements of FMR in any wombat species. Data on water flux in free-living wombats are restricted to *L. latifrons* (Wells 1973). A more complete description of the energy and water metabolism of free-living wombats would be of great value. For instance, it would allow measurements of the energy costs of burrowing to be compared with estimates of likely energy and water savings from using burrows as refuges, within the framework of the total energy and water budgets of the animals.

In Chapter 5 the value of *Eucalyptus* foliage as food is described for a number of arboreal folivores, some of which feed mainly or exclusively on eucalypt leaves. Earlier work in the 1980s on digestive and metabolic features of the arboreal folivores is now being refined in terms of the factors likely to be responsible for differences in the degree to which different species specialise on *Eucalyptus*, and of the factors likely to determine the carrying capacity of eucalypt forests for these species. Further, detailed work is needed in both areas in order to better predict the ecological consequences of various forestry practices and of changes in the composition of the earth's atmosphere consequent upon global warming.

The three chapters devoted to the kangaroos and wallabies (Chapters 6 and 7) and rat-kangaroos (Chapter 8) describe the current state of knowledge for what are perhaps the most researched groups of marsupials. However, because of the large number of species and the wide range in body sizes and dietary niches occupied by macropodoid marsupials it is essential that the digestive and metabolic studies described are extended to other species from different habitats. Also, as mentioned above, several groups are poorly researched; these include the hare-wallabies, rock-wallabies, tree kangaroos and New Guinean forest wallabies. Further quantitative studies on fermentation in the forestomach and hindgut of rat-kangaroos are needed in order to better understand the relative importance of the two fermentation areas of the potoroine digestive tract.

The unfolding story of marsupial origins and their evolution in Gondwana and its constituent parts (Chapter 9) provided an opportunity to speculate on how the digestive systems of various groups of marsupials may have evolved to exploit the wide range of nutritional habitats that we find in South America, Australia and New Guinea today. Future collaborative efforts between paleontologists, molecular geneticists, anatomists, physiologists and ecologists can be expected to shed much more light on the ways in which the marsupial faunas of these three land masses came to be as they are today, and what the future may hold for them in terms of their conservation.

Appendix

A partial classification of extant and recently extinct marsupial species mentioned in the text. It is based on the classification used by Woodburne & Case (1996) to family level and by Eisenberg (1989), Redford & Eisenberg (1992), Strahan (1995) and Flannery (1995) to species level. Common names of American species are taken from Eisenberg (1989) and Redford & Eisenberg (1992), those of Australian species from Strahan (1995) and those of New Guinean species from Flannery (1995).

Class	Mammalia
Subclass	Metatheria (Marsupialia)
Cohort	Ameridelphia
Order	Didelphimorphia
Family	Didelphidae
	Caluromys derbianus
	Caluromys philander, bare-tailed woolly opossum
	Caluromysiops irrupta, black-shouldered opossum
	Chironectes minimus, water opossum (yapock)
	Didelphis albiventris, white-bellied opossum (white-eared opossum)
	Didelphis aurita
	Didelphis marsupialis, black-eared opossum (common opossum)
	Didelphis virginiana, Virginia opossum
	Glironia venusta, bushy-tailed opossum
	Lestodelphis halli, Patagonian opossum
	Lutreolina crassicaudata, little water opossum (thick-tailed opossum)
	Marmosa cinerea
	Marmosa fuscata, dusky mouse opossum
	Marmosa microtarsus
	Marmosa robinsoni, murine opossum
	Metachirus nudicaudatus, brown four-eyed opossum
	Monodelphis brevicaudata
	Monodelphis dimidiata, eastern short-tailed opossum
	Monodelphis domestica, gray short-tailed opossum
	Philander (Metachirops) opossum, gray four-eyed opossum
	Thylamys elegans, Chilean mouse-opossum
	Thylamys velutinus
Order	Paucituberculata
Family	Caenolestidae
	Caenolestes fuliginosus
	Caenolestes obscurus, rat opossum
	Lestoros inca
	Rhyncolestes raphanurus, Chilean shrew opossum

Cohort	Australidelphia
Order	Microbiotheria
Family	Microbiotheriidae

Dromiciops australis, Monito del Monte

Order	Dasyurida
Family	Dasyuridae

Antechinomys laniger, kultarr
Antechinus agilis, agile antechinus
Antechinus bellus, fawn antechinus
Antechinus flavipes, yellow-footed antechinus
Antechinus melanurus
Antechinus minimus, swamp antechinus
Antechinus stuartii, brown antechinus
Antechinus swainsonii, dusky antechinus
Dasycercus (Dasyuroides) byrnei, kowari
Dasycercus cristicauda, mulgara
Dasykaluta rosamondae, little red kaluta
Dasyurus geoffroii, western quoll
Dasyurus hallucatus, northern quoll
Dasyurus maculatus, spotted-tailed quoll (tiger quoll)
Dasyurus viverrinus, eastern quoll
Ningaui ridei, wongai ningaui
Ningaui timealeyi, Pilbara ningaui
Ningaui yvonneae, southern ningaui
Phascogale calura, red-tailed phascogale
Phascogale tapoatafa, brush-tailed phascogale
Planigale gilesi, Giles' planigale
Planigale ingrami, long-tailed planigale
Planigale maculata, common planigale
Planigale tenuirostris, narrow-nosed planigale
Pseudantechinus macdonnellensis, fat-tailed pseudantechinus
Pseudantechinus woolleyae, Woolley's pseudantechinus
Sarcophilus harrisii, Tasmanian devil
Sminthopsis crassicaudata, fat-tailed dunnart
Sminthopsis dolichura, little long-tailed dunnart
Sminthopsis granulipes, white-tailed dunnart
Sminthopsis griseoventer, grey-bellied dunnart
Sminthopsis hirtipes, hairy-footed dunnart
Sminthopsis leucopus, white-footed dunnart
Sminthopsis longicaudata, long-tailed dunnart
Sminthopsis macroura, stripe-faced dunnart
Sminthopsis murina, common dunnart
Sminthopsis ooldea, Ooldea dunnart
Sminthopsis virginiae, red-cheeked dunnart
Sminthopsis youngsoni, lesser hairy-footed dunnart

Family	Myrmecobiidae

Myrmecobius fasciatus, numbat

Family	Thylacinidae
	Thylacinus cynocephalus, thylacine (Tasmanian tiger)
Order	Notoryctemorphia
Family	Notoryctidae
	Notoryctes caurinus, northern marsupial mole
	Notoryctes typhlops, southern marsupial mole
Order	Peramelina
Family	Peramelidae
	Chaeropus ecaudatus, pig-footed bandicoot
	Isoodon auratus, golden bandicoot
	Isoodon macrourus, northern brown bandicoot
	Isoodon obesulus, southern brown bandicoot
	Macrotis lagotis, bilby (greater bilby)
	Macrotis leucura, lesser bilby
	Perameles bougainville, western barred bandicoot
	Perameles eremiana, desert bandicoot
	Perameles gunnii, eastern barred bandicoot
	Perameles nasuta, long-nosed bandicoot
Family	Peroryctidae
	Echymipera clara, dimorphic echymipera
	Echymipera kaluba, spiny echymipera
	Echymipera rufescens, rufous spiny bandicoot
Order	Diprotodontia
Family	Phascolarctidae
	Phascolarctos cinereus, koala
Family	Vombatidae
	Lasiorhinus krefftii, northern hairy-nosed wombat
	Lasiorhinus latifrons, southern hairy-nosed wombat
	Vombatus ursinus, common wombat
Family	Burramyidae
	Burramys parvus, mountain pygmy-possum
	Cercartetus caudatus, long-tailed pygmy-possum
	Cercartetus concinnus, western pygmy-possum
	Cercartetus lepidus, little pygmy-possum
	Cercartetus nanus, eastern pygmy-possum
Family	Petauridae
	Dactylopsila trivirgata, striped possum
	Gymnobelideus leadbeateri, Leadbeater's possum
	Petaurus australis, yellow-bellied glider
	Petaurus breviceps, sugar glider
	Petaurus gracilis, mahogany glider
	Petaurus norfolcensis, squirrel glider
Family	Pseudocheiridae
	Hemibelideus lemuroides, lemuroid ringtail possum
	Petauroides volans, greater glider
	Petropseudes dahli, rock ringtail possum
	Pseudocheirus canescens, lowland ringtail possum

Pseudocheirus mayeri, pygmy ringtail possum
Pseudocheirus occidentalis, western ringtail possum
Pseudocheirus peregrinus, common ringtail possum
Pseudochirops archeri, green ringtail possum
Pseudochirulus herbertensis, Herbert River ringtail possum

Family   Tarsipedidae
Tarsipes rostratus, honey possum

Family   Acrobatidae
Acrobates pygmaeus, feathertail glider
Distoechurus pennatus, feathertailed possum

Family   Phalangeridae
Phalanger gymnotis, ground cuscus
Phalanger intercastellanus (orientalis), southern common cuscus
Phalanger vestitus, Stein's cuscus
Spilocuscus maculatus, common spotted cuscus
Trichosurus caninus, mountain brushtail possum (bobuck)
Trichosurus vulpecula, common brushtail possum
Trichosurus vulpecula johnstoni, coppery brushtail possum
Wyulda squamicaudata, scaly-tailed possum

Family   Potoroidae
Aepyprymnus rufescens, rufous bettong (rufous rat-kangaroo)
Bettongia gaimardi, Tasmanian bettong
Bettongia lesueur, burrowing bettong (boodie)
Bettongia penicillata, brush-tailed bettong (woylie)
Bettongia tropica, northern bettong
Caloprymnus campestris, desert rat-kangaroo
Hypsiprymnodon moschatus, musky rat-kangaroo
Potorous longipes, long-footed potoroo
Potorous tridactylus, long-nosed potoroo

Family   Macropodidae
Dendrolagus bennettianus, Bennett's tree-kangaroo
Dendrolagus goodfellowi, Goodfellow's tree-kangaroo
Dendrolagus lumholtzi, Lumholtz's tree-kangaroo
Dendrolagus matschiei, Huon tree-kangaroo
Dorcopsis luctuosa, grey dorcopsis
Dorcopsulus vanheurni, small dorcopsis
Lagorchestes conspicillatus, spectacled hare-wallaby
Lagorchestes hirsutus, rufous hare-wallaby (western hare-wallaby)
Lagostrophus fasciatus, banded hare-wallaby
Macropus agilis, agile wallaby (sandy wallaby)
Macropus antilopinus, antilopine wallaroo
Macropus dorsalis, black-striped wallaby
Macropus eugenii, tammar wallaby (Kangaroo Island wallaby)
Macropus fuliginosus, western grey kangaroo
Macropus giganteus, eastern grey kangaroo
Macropus irma, western brush wallaby
Macropus parma, parma wallaby

Macropus parryi, whiptail wallaby (pretty-face wallaby)
Macropus robustus erubescens, euro
Macropus robustus isabellinus
Macropus robustus robustus, eastern wallaroo
Macropus robustus woodwardi
Macropus rufogriseus (= *M. bennetti*), red-necked wallaby (Bennett's wallaby)
Macropus rufus, red kangaroo
Onychogalea fraenata, bridled nailtail wallaby
Onychogalea unguifera, northern nailtail wallaby
Peradorcus concinna, little rock-wallaby (Nabarlek)
Petrogale assimilis, allied rock-wallaby
Petrogale brachyotis, short-eared rock-wallaby
Petrogale inornata, unadorned rock-wallaby
Petrogale lateralis, black-footed rock-wallaby
Petrogale penicillata, brush-tailed rock-wallaby
Petrogale rothschildi, Rothschild's rock-wallaby
Petrogale xanthopus, yellow-footed rock-wallaby
Setonix brachyurus, quokka
Thylogale billardierii, Tasmanian pademelon (red-bellied pademelon)
Thylogale brunii, dusky pademelon
Thylogale stigmatica, red-legged pademelon
Thylogale thetis, red-necked pademelon
Wallabia bicolor, swamp wallaby

References

Adrian, J. (1976). Gums and hydrocolloids in nutrition. *World Review of Nutrition and Dietetics*, **25**, 189–216.

Agar, N. S. & Baker, M. L. (1996). Role of catalase in H_2O_2-induced oxidant stress in marsupial erythrocytes. *Comparative Haematology International*, **6**, 32–4.

Agar, N. S., Gay, C. A., Gallagher, C., Steele, V. R. & Spencer, P. (1996). Glucose-6-phosphate dehydrogenase and oxidant sensitivity of erythrocytes of three species of wombats. *Comparative Haematology International*, **6**, 225–31.

Alexander, R. McN. (1991). Optimisation of gut structure and diet for higher vertebrate herbivores. *Philosophical Transactions of the Royal Society, London*, **B333**, 249–55.

Algar, D. (1986). An ecological study of macropodid marsupial species in a Reserve. Ph.D. Thesis, University of Western Australia, Perth.

Algar, D., Arnold, G. W. & Grassia, A. (1988). Effects of nitrogen and season on western grey kangaroo hematology. *Journal of Wildlife Management*, **52**, 616–19.

Allen, C. B. (1994). Extreme ecology: flood, drought and competition between native and introduced herbivores. *Ecological Society of Australia*. Unpubl. abstract.

Anderson, T. J. C., Berry, A. J., Amos, J. N. & Cook, J. M. (1988). Spool and line tracking of the New Guinea spiny bandicoot, *Echymipera kaluba* (Marsupialia: Peramelidae). *Journal of Mammalogy*, **69**, 114–20.

Andrewartha, H. G. & Barker, S. (1969). Introduction to a study of the ecology of the Kangaroo Island wallaby, *Protemnodon eugenii* (Desmarest), within Flinders Chase, Kangaroo Island, South Australia. *Transactions of the Royal Society of South Australia*, **93**, 127–32.

Archer, M. (1974). Regurgitation or merycism in the western native cat, *Dasyurus geoffroii*, and the red-tailed wambenger, *Phascogale calura* (Marsupialia, Dasyuridae). *Journal of Mammalogy*, **55**, 488–552.

Archer, M. (1976). The dasyurid dentition and its relationship to that of didelphids, thylacinids, borhyaenids (Marsupicarnivora) and peramelids (Peramelina:Marsupialia). *Australian Journal of Zoology (Supplementary Series)*, **39**, 1–34.

Archer, M. & Dawson, L. (1982). Revision of marsupial lions of the genus *Thylacoleo gervais* (Thylacoleonidae, Marsupialia) and thylacoleonid evolution in the late Cainozoic. In *Carnivorous Marsupials*, ed. M. Archer, pp. 477–94. Sydney: Surrey Beatty and the Royal Zoological Society of New South Wales.

Arnold, J. & Shield, J. W. (1970). Oxygen consumption and body temperature of the chuditch (*Dasyurus geoffroii*). *Journal of Zoology (London)*, **160**, 391–404.

Arnould, J. P. (1986). Aspects of the diet of the eastern pygmy possum (*Cercatetus nanus*). B.Sc. honours thesis, Monash University, Clayton, Victoria.

Attiwill, P. M. & Leeper, G. W. (1987). *Forest Soils and Nutrient Cycles*. Melbourne:

Melbourne University Press.

Atzert, S. P. (1971). A review of monofluoroacetate (Compound 1080): its properties, toxicology and use in predator and rodent control. *Special Scientific Report – Wildlife* No. 146, US Department of Interior, Fish and Wildlife Services, Bureau of Sport Fisheries and Wildlife, Washington, DC.

Austin, P. J., Suchar, L. A., Robbins, C. T. & Hagerman, A. E. (1989). Tannin-binding proteins in saliva of deer and their absence in saliva of sheep and cattle. *Journal of Chemical Ecology*, **15**, 1335–47.

Bailey, P. T. (1971). The red kangaroo, *Megaleia rufa* (Desmarest), in north-western New South Wales. I. Movements. *CSIRO Wildlife Research*, **16**, 11–28.

Bailey, P. T., Martenesz, P. N. & Barker, R. (1971). The red kangaroo, *Megaleia rufa* (Desmarest), in north-western New South Wales. II. Food. *CSIRO Wildlife Research*, **16**, 29–39.

Baker, M. L., Canfield, P. J., Gemmell, R. T., Spencer, P. B. S. & Agar, N. S. (1995). Erythrocyte metabolism in the koala, the common brushtail possum and the whiptail wallaby. *Comparative Haematology International*, **5**, 163–9.

Baker, S. K., Dehority, B. A., Chamberlain, N. L. & Purser, D. B. (1995). Inability of protozoa from the kangaroo forestomach to establish in the rumen of sheep. *Annals Zootechnie*, **44** (suppl.), 143.

Bakker, H. R. & Bradshaw, S. D. (1983). Renal function in the spectacled hare-wallaby, *Lagorchestes conspicillatus*: effects of dehydration and protein deficiency. *Australian Journal of Zoology*, **31**, 101–8.

Bakker, H. R. & Bradshaw, S. D. (1989). Rate of water turnover and electrolyte balance of an arid-zone marsupial, the spectacled hare-wallaby (*Lagorchestes conspicillatus*) on Barrow Island. *Comparative Biochemistry and Physiology*, **92A**, 521–9.

Bakker, H. R., Bradshaw, S. D. & Main, A. R. (1982). Water and electrolyte metabolism of the tammar wallaby *Macropus eugenii*. *Physiological Zoology*, **55**, 209–19.

Ballard, F. J., Hanson, R. W. & Kronfeld, D. S. (1969). Gluconeogenesis and lipogenesis in tissue from ruminant and nonruminant animals. *Federation Proceedings*, **28**, 218–31.

Barbosa, A. J. A., Noguira, J. C., Redins, C. A., Nogueira, A. M. M. F., Van Noorden, S. & Polak, J. M. (1990). Histochemical and ultrastructural studies on the enterochromaffin-like cell in the gastric mucosa of the opossum *Didelphis albiventris* (Marsupialia). *Cell and Tissue Research*, **262**, 425–30.

Barbour, R. A. (1977). Anatomy of marsupials. In *The Biology of Marsupials*, ed. B. Stonehouse and D. Gilmore, pp. 237–62. Baltimore: Baltimore University Press.

Barboza, P. S. (1989). The nutritional physiology of the Vombatidae. Ph.D. Thesis, University of New England, Armidale.

Barboza, P. S. (1993a). Digestive strategies of the wombats: feed intake, fiber digestion and digesta passage in two grazing marsupials with hindgut fermentation. *Physiological Zoology*, **66**, 983–99.

Barboza, P. S. (1993b). Effects of restricted water intake on digestion, urea recycling and renal function in wombats (Marsupialia: Vombatidae) from contrasting habitats. *Australian Journal of Zoology*, **41**, 527–36.

Barboza, P. S. & Hume I. D. (1992a). Digestive tract morphology and digestion in the wombats (Marsupialia: Vombatidae). *Journal of Comparative Physiology*, **B162**, 552–60.

Barboza, P. S. & Hume, I. D. (1992b). Hindgut fermentation in the wombats: two marsupial grazers. *Journal of Comparative Physiology*, **B162**, 561–6.

Barboza, P. S., Hume, I. D. & Nolan, J. V. (1993). Nitrogen metabolism and requirements of nitrogen and energy in wombats (Marsupialia: Vombatidae) *Physiological Zoology*, **66**, 807–28.

Barboza, P. S. & Vanselow, B. A. (1990). Copper toxicity in captive wombats (Marsupialia: Vombatidae). *1990 Proceedings of the American Association of Zoo Veterinarians*, pp. 204–6.

Barker, J. M. (1961). The metabolism of carbohydrate and volatile fatty acids in the marsupial, *Setonix brachyurus*. *Quarterly Journal of Experimental Physiology*, **46**, 54–68.

Barker, R. D. (1987). The diet of herbivores in the sheep rangelands. In *Kangaroos: Their Ecology and Management in Sheep Rangelands*, ed. G. Caughley, N. Shepherd and J. Short, pp. 69–83. Cambridge: Cambridge University Press.

Barker, S. (1960). The role of trace elements in the biology of the quokka (*setonix brachyurus*, Quoy & Gaimard). Ph.D. Thesis, University of Western Australia, Perth.

Barker, S. (1961a). Studies on marsupial nutrition. III. The copper-molybdenum-inorganic sulphate interaction in the Rottnest quokka, *Setonix brachyurus* (Quoy & Gaimard). *Australian Journal of Biological Sciences*, **14**, 646–58.

Barker, S. (1961b). Copper, molybdenum and inorganic sulphate levels in Rottnest plants. *Journal of the Royal Society of Western Australia*, **44**, 49–52.

Barker, S. (1962). Copper levels in the milk of a marsupial. *Nature*, **193**, 292.

Barker, S. (1968). Nitrogen balance and water intake in the Kangaroo Island wallaby, *Protemnodon eugenii* (Desmarest). *Australian Journal of Experimental Biology and Medical Science*, **46**, 17–32.

Barker, S. (1971). Nitrogen and water excretion of wallabies: differences between field and laboratory findings. *Comparative Biochemistry and Physiology*, **38A**, 359–67.

Barker, S. (1974). Studies on seasonal anaemia in the Rottnest Island quokka, *Setonix brachyurus* (Quoy & Gaimard) (Marsupialia: Macropodidae). *Transactions of the Royal Society of South Australia*, **98**, 43–8.

Barker, S., Brown, G. D. & Calaby, J. H. (1963). Food regurgitation in the Macropodidae. *Australian Journal of Science*, **25**, 430–2.

Barker, S., Lintern, S. M. & Murphy, C. R. (1970). The effect of water restriction on urea retention and nitrogen excretion in the Kangaroo Island wallaby, *Protemnodon eugenii* (Desmarest). *Comparative Biochemistry and Physiology*, **34**, 883–93.

Barkley, L. J. & Wittaker, J. O. Jr. (1984). Confirmation of *Caenolestes* in Peru with information on diet. *Journal of Mammalogy*, **65**, 328–30.

Barnard, E. A. (1969). Biological function of pancreatic ribonuclease. *Nature*, **221**, 340–4.

Barnes, R. D. (1977). The special anatomy of *Marmosa robinsoni*. In *The Biology of Marsupials*, ed. D. Hunsaker, II, pp. 387–413. New York: Academic Press.

Barnett, J. L., How, R. A. & Humphreys, W. F. (1979a). Blood parameters in natural populations of *Trichosurus* species (Marsupialia: Phalangeridae). I. Age, sex and season variation in *T. caninus* and *T. vulpecula*. *Australian Journal of Zoology*, **27**, 913–26.

Barnett, J. L., How, R. A. & Humphreys, W. F. (1979b). Blood parameters in natural populations of *Trichosurus* species (Marsupialia: Phalangeridae). II. Influence of habitat and population strategies of *T. caninus* and *T. vulpecula*. *Australian Journal of Zoology*, **27**, 927–38.

Bartholomew, G. A. & Hudson, J. W. (1962). Hibernation, estivation, temperature regulation, evaporative water loss, and heart rate of the pigmy possum, *Cercatetus nanus*. *Physiological Zoology*, **35**, 94–107.

Basden, R. (1966). The composition, occurrence and origin of lerp, the sugar secretion of *Eurymela distincta* (Signoret). *Proceedings of the Linnean Society of New South Wales*, **9**, 44–6.

Bauchop, T. (1979). The rumen anaerobic fungi: colonizers of plant fibre. *Annales de Recherches Vétérinaires*, **10**, 246–8.

Baudinette, R. V., Halpern, E. A. & Hinds, D. S. (1993). Energetic cost of locomotion as a function of ambient temperature and during growth in the marsupial *Potorous tridactylus*. *Journal of Experimental Biology*, **174**, 81–95.

Beal, A. M. (1984). Electrolyte composition of parotid saliva from sodium-replete red kangaroos (*Macropus rufus*). *Journal of Experimental Biology*, **111**, 225–37.

Beal, A. M. (1986). Effects of flow rate, duration of stimulation and mineralocorticoids on the electrolyte concentrations of mandibular saliva from the red kangaroo (*Macropus rufus*). *Journal of Experimental Biology*, **126**, 315–39.

Beal, A. M. (1989). Differences in salivary flow and composition among kangaroo species: implications for digestive efficiency. In *Kangaroos, Wallabies and Rat-kangaroos*, ed. G. Grigg, P. Jarman and I. Hume, pp. 189–95. Sydney: Surrey Beatty.

Beal, A. M. (1990a). Secretion rates and composition of parotid saliva in the koala (*Phascolarctos cinereus*). *Journal of Zoology, London*, **221**, 261–74.

Beal, A. M. (1990b). Composition of sublingual saliva in the koala (*Phascolarctos cinereus*). *Comparative Biochemistry and Physiology*, **97A**, 185–8.

Beal, A. M. (1991a). Influence of flow rate and aldosterone administration on mandibular salivary composition in the koala (*Phascolarctos cinereus*). *Journal of Zoology, London*, **223**, 265–79.

Beal, A. M. (1991b). Characteristics of parotid saliva from the common wombat (*Vombatus ursinus*). *Journal of Zoology, London*, **224**, 403–17.

Beal, A. M. (1992). Relationships between plasma composition and parotid salivary composition and secretion rates in the potoroine marsupials, *Aepyprymnus rufescens* and *Potorous tridactylus*. *Journal of Comparative Physiology*, **B162**, 637–45.

Beal, A. M. (1995a). Flow-composition relations for sublingual saliva of the common wombat (*Vombatus ursinus*). *Comparative Biochemistry and Physiology*, **111A**, 641–6.

Beal, A. M. (1995b). Secretion of electrolytes, protein and urea by the mandibular gland of the common wombat (*Vombatus ursinus*). *Journal of Comparative Physiology*, **B164**, 629–35.

Beckmann, E. A. (1986). Hypogeous fungi as a food source for the rufous rat-kangaroo *Aepyprymnus rufescens*. B. Nat. Res. Thesis, University of New England, Armidale.

Belcher, C. A. (1995). Diet of the tiger quoll (*Dasyurus maculatus*) in East Gippsland, Victoria. *Wildlife Research*, **22**, 341–57.

Bell, H. M. (1973). The ecology of three macropod marsupial species in an area of open forest and savannah woodland in north Queensland, Australia. *Mammalia*, **37**, 527–44.

Bennett, A. F. & Baxter, B. J. (1989). Diet of the long-nosed potoroo, *Potorous tridactylus* (Marsupialia: Potoroidae), in south-western Victoria. *Australian Wildlife Research*, **16**, 263–71.

Bensley, R. R. (1902). The cardiac glands of mammals. *American Journal of Anatomy*, **2**, 105–65.

Bergman, E. N., Reid, R. S., Murray, M. G., Brockway, J. M. & Whitelaw, F. G. (1965). Interconversions and production of volatile fatty acids in the sheep rumen. *Biochemical Journal*, **97**, 53–8.

Betts, T. J. (1978). Koala acceptance of *Eucalyptus globulus* Labill. as food in relation to the proportion of sesquiterpenoids in the leaves. In *The Koala. Proceedings of the Taronga Symposium*, ed. T. J. Bergin, pp. 75–85. Sydney: Zoological Parks Board of New South Wales.

Beuchat, C. A. (1990a). Metabolism and the scaling of urine concentrating ability in mammals: resolution of a paradox? *Journal of Theoretical Biology*, **143**, 113–22.

Beuchat, C. A. (1990b). Body size, medullary thickness and urine concentrating ability in mammals. *American Journal of Physiology*, **258**, R298–R308.

Beveridge, I. & Spratt, D. M. (1996). The helminth fauna of Australasian marsupials: origins and evolutionary biology. *Advances in Parasitology*, **37**, 135–254.

Birney, E. C., Jenness, R. & Hume, I. D. (1979). Ascorbic acid biosynthesis in the mammalian kidney. *Experientia*, **35**, 1425.

Birney, E. C., Jenness, R. & Hume, I. D. (1980). Evolution of an enzyme system: ascorbic acid biosynthesis in monotremes and marsupials. *Evolution*, **34**, 230–9.

Blackbourn, L. J. (1991). The diet and ranging of the black-footed rock-wallaby at Simpson's Gap, Alice Springs. B. Nat. Res. Thesis, University of New England, Armidale, New South Wales.

Blackhall, S. (1980). Diet of the eastern native-cat, *Dasyurus viverrinus* (Shaw), in southern Tasmania. *Australian Wildlife Research*, 7, 191–7.

Blair-West, J. R., Coghlan, J. P., Denton, D. A., Nelson, J. F., Orchard, E., Scoggins, B. A., Wright, R. D., Myers, K & Junqueira, C. L. (1968). Physiological, morphological and behavioural adaptation to a sodium deficient environment by wild native Australian and introduced species of animals. *Nature*, **217**, 922–8.

Bolton, B. L. & Latz, P. K. (1978). The western hare-wallaby, *Lagorchestes hirsutus*

(Gould) (Macropodidae), in the Tanami Desert. *Australian Wildlife Research*, **5**, 285–93.

Bradley, A. J., McDonald, I. R. & Lee, A. K. (1980). Stress and mortality in a small marsupial (*Antechinus stuartii*) (Macleay). *General and Comparative Endocrinology*, **40**, 188–200.

Bradshaw, S. D., Morris, K. D., Dickman, C. R., Withers, P. C. & Murphy, D. (1994). Field metabolism and turnover in the golden bandicoot (*Isoodon auratus*) and other small mammals from Barrow Island, Western Australia. *Australian Journal of Zoology*, **42**, 29–41.

Braithwaite, L. W., Dudzinski, M. L. & Turner, J. (1983). Studies on the arboreal marsupial fauna of eucalypt forests being harvested for woodpulp at Eden, N.S.W. II. Relationship between the fauna density, richness and diversity and measured variables of the habitat. *Australian Wildlife Research*, **10**, 231–47.

Braithwaite, L. W., Turner, J. & Kelly, J. (1984). Studies on the arboreal marsupial fauna of eucalypt forests being harvested for woodpulp at Eden, N.S.W. III. Relationship between the fauna densities, eucalypt occurrence and foliage nutrients, and soil parent materials. *Australian Wildlife Research*, **11**, 41–8.

Bridie, A., Hume, I. D. & Hill, D. M. (1994). Digestive tract function and energy requirements of the rufous hare-wallaby, *Lagorchestes hirsutus*. *Australian Journal of Zoology*, **42**, 761–74.

Brody, S. (1945). *Bioenergetics and Growth*. New York: Reinhold.

Brooker, B. & Withers, P. (1994). Kidney structure and renal indices of dasyurid marsupials. *Australian Journal of Zoology*, **42**, 463–76.

Broome, L. S. & Geiser, F. (1995). Hibernation in free-living mountain pygmy-possums, *Burramys parvus* (Marsupialia: Burramyidae). *Australian Journal of Zoology*, **43**, 373–9.

Brown, G. D. (1964). The nitrogen requirements of macropod marsupials. Ph. D. Thesis, University of Western Australia, Perth.

Brown, G. D. (1968). The nitrogen and energy requirements of the euro (*Macropus robustus*) and other species of macropod marsupials. *Proceedings of the Ecological Society of Australia*, **3**, 106–12.

Brown, G. D. (1969). Studies on marsupial nutrition. IV. The utilisation of dietary urea by the euro or hill kangaroo, *Macropus robustus* (Gould). *Australian Journal of Zoology*, **17**, 187–94.

Brown, G. D. & Main, A. R. (1967). Studies on marsupial nutrition. V. The nitrogen requirements of the euro, *Macropus robustus*. *Australian Journal of Zoology*, **15**, 7–27.

Brownfield, M. S. & Wunder, B. A. (1976). Relative medullary area: a new structural index for estimating urinary concentrating capacity of mammals. *Comparative Biochemistry and Physiology*, **SSA**, 69–75.

Brunner, H. & Coman, B. J. (1974). *The Identification of Mammalian Hair*. Melbourne: Inkata Press.

Buchmann, O. L. K. & Guiler, E. R. (1977). Behaviour and ecology of the Tasmanian devil, *Sarcophilus harrisii*. In *The Biology of Marsupials*, ed. B. Stonehouse & D. Gilmore, pp. 155–68. London: Macmillan.

Burbidge, A. A. (1995). Burrowing bettong, *Bettongia lesueur* (Quoy and Gaimard, 1824). In *The Mammals of Australia*, ed. R. Strahan, pp. 289–91. Chatswood:

Reed Books.

Burbidge, A. A., Johnson, K. A., Fuller, P. J & Southgate, R. I. (1988). Aboriginal knowledge of the mammals of the central deserts of Australia. *Australian Wildlife Research*, **15**, 9–39.

Busch, M & Kravetz, F. O. (1991). Diet composition of *Monodelphis dimidiata* (Marsupialia, Didelphidae). *Mammalia*, **55**, 619–21.

Calaby, J. H. (1958). Studies in marsupial nutrition. II. The rate of passage of food residues and digestibility of crude fibre and protein by the quokka, *Setonix brachyurus* (Quoy and Gaimard). *Australian Journal of Biological Sciences*, **11**, 571–80.

Calaby, J. H. (1960). Observations on the banded ant-eater (*Myrmecobius f. fasciatus*, Waterhouse (Marsupialia), with particular reference to its food habits. *Proceedings of the Zoological Society of London*, **135**, 183–207.

Calaby, J. H. (1966). Mammals of the upper Richmond and Clarence Rivers, New South Wales. *Technical Paper of the Division of Wildlife Survey, CSIRO Australia*, **10**, 1–55.

Calaby, J. H. (1995). Antilopine wallaroo. In *The Mammals of Australia*, ed. R. Strahan. Chatswood, NSW: Reed Books.

Calligan, J. J., Costa, M. & Furness, J. B. (1985). Gastrointestinal myoelectric activity in conscious guinea pigs. *American Journal of Physiology*, **249**, G92–9.

Carlsson, A. (1915). Zur Morphologie des *Hypsiprymnodon moschatus*. *Kungliga Svenska Vetenshapsakademiens Handlingar*, **52**, 1–48.

Carroll, E. J. & Hungate, R. E. (1954). The magnitude of the microbial fermentation in the bovine rumen. *Applied Microbiology*, **2**, 205–14.

Case, J. A. (1985). Differences in prey utilization by Pleistocene marsupial carnivores, *Thylacoleo carnifex* (Thylacoleonidae) and *Thylacinus cynocephalus* (Thylacinidae). *Australian Mammalogy*, **8**, 45–52.

Caton, J. M., Hill, D. M., Hume, I. D. & G. A. Crook. (1996). The digestive strategy of the common marmoset, *Callithrix jacchus*. *Comparative Biochemistry and Physiology*, **114A**, 1–8.

Caughley, G. J. (1964). Density and dispersion of two species of kangaroo in relation to habitat. *Australian Journal of Zoology*, **12**, 238–49.

Charles-Dominique, P. (1974). Ecology and feeding behaviour of sympatric lorisids in Gabon. In *Prosimian Biology*, ed. R. D. Martin, G. A. Doyle & A. C. Walker. Pittsburgh: University of Pittsburgh Press.

Charles-Dominique, P. (1983). Ecology and social adaptations in didelphid marsupials: comparisons with eutherians of similar ecology. In *Advances in the Study of Animal Behaviour*, ed. J. F. Eisenberg & D. G. Klieman. Philadelphia: American Society of Mammalogy.

Chatterjee, I. B. (1973). Evolution and the biosynthesis of ascorbic acid. *Science*, **182**, 1271–2.

Chen, X., Dickman, C. R. & Thompson, M. B. (1998). Diet of the mulgara *Dasycercus cristicauda* (Marsupialia: Dasyuridae), in the Simpson Desert, central Australia. *Wildlife Research*, **25**, 233–42.

Chilcott, M. J. (1984). Coprophagy in the common ringtail possum, *Pseudocheirus peregrinus* (Marsupialia: Petauridae). *Australian Mammalogy*, **7**, 107–10.

Chilcott, M. J. & Hume, I. D. (1984a). Digestion of *Eucalyptus andrewsii* foliage by the common ringtail possum, *Pseudocheirus peregrinus*. *Australian Journal of Zoology*, **32**, 605–13.

Chilcott, M. J. & Hume, I. D (1984b). Nitrogen and urea metabolism and nitrogen requirements of the common ringtail possum (*Pseudocheirus peregrinus*) fed *Eucalyptus andrewsii* foliage. *Australian Journal of Zoology*, **32**, 615–22.

Chilcott, M. J. & Hume, I. D. (1985). Coprophagy and selective retention of fluid digesta: their role in the nutrition of the common ringtail possum, *Pseudocheirus peregrinus*. *Australian Journal of Zoology*, **33**, 1–15.

Chilcott, M. J., Moore, S. A. & Hume, I. D. (1985). Effects of water restriction on nitrogen metabolism and urea recycling in the macropodid marsupials *Macropus eugenii* (tammar wallaby) and *Thylogale thetis* (red-necked pademelon). *Journal of Comparative Physiology*, **B155**, 759–67.

Chippendale, G. M. (1968). The plants grazed by red kangaroos, *Megaleia rufa* (Desmarest) in central Australia. *Proceedings of the Linnean Society of New South Wales*, **93**, 98–110.

Christensen, P. E. S. (1980). The biology of *Bettongia penicillata* (Gray 1837), and *Macropus eugenii* (Desmarest 1817) in relation to fire. *Forestry Department of Western Australia Bulletin*, No. 91 (90pp.).

Claridge, A. W. (1993). Fungal diet of the long-nosed bandicoot (*Perameles nasuta*) in south-eastern Australia. *Victorian Naturalist*, **110**, 86–91.

Claridge, A. W. & Cork, S. J. (1994). Nutritional value of hypogeal fungal sporocarps for the long-nosed potoroo (*Potorous tridactylus*), a forest-dwelling mycophagous marsupial. *Australian Journal of Zoology*, **42**, 701–10.

Claridge, A. W., Cunningham, R. B. & Tanton, M. T. (1993). Foraging patterns of the long-nosed potoroo (*Potorous tridactylus*) for hypogeal fungi in mixed-species and regrowth eucalypt forest stands in south-eastern Australia. *Forest Ecology and Management*, **61**, 75–90.

Claridge, A. W. & May, T. W. (1994). Mycophagy among Australian mammals. *Australian Journal of Ecology*, **19**, 251–275.

Claridge, A. W., McNee, A., Tanton, M. T. & Davey, S. M. (1991). Ecology of bandicoots in undisturbed forest adjacent to recently felled logging coupes: a case study from the Eden woodchip agreement area. *In Conservation of Australia's Forest Fauna*, ed. D. Lunney, pp. 331–45. Mosman: Royal Zoological Society of New South Wales.

Claridge, A. W., Tanton, M. T. & Cunningham, R. B. (1993). Hypogeal fungi in the diet of the long-nosed potoroo (*Potorous tridactylus*) in mixed-species and regrowth eucalypt forest stands in south-eastern Australia. *Wildlife Research*, **20**, 321–37.

Claridge, A. W., Tanton, M. T., Seebeck, J. H., Cork, S. J. & Cunningham, R. B. (1992). Establishment of ectomycorrhizae on the roots of two species of *Eucalyptus* from fungal spores contained in the faeces of the long-nosed potoroo (*Potorous tridactylus*). *Australian Journal of Ecology*, **17**, 207–17.

Clarke, J. L., Jones, M. E. & Jarman, P. J. (1989). A day in the life of a kangaroo: activities and movement of eastern grey kangaroos, *Macropus giganteus* at Wallaby Creek. In *Kangaroos, Wallabies and Rat-Kangaroos*, ed. G. Grigg, P. Jarman & I. Hume, pp. 611–8. Sydney: Surrey Beatty.

Clarke, P. J., Myerscough, P. J. & Skelton, N. J. (1996). Plant coexistence in coastal heaths: between- and within-habitat effects of competition, disturbance and predation in the post-fire environment. *Australian Journal of Ecology*, **21**, 55–63.

Clemens, W. A., Richardson, B. J. & Baverstock, P. R. (1989). Biogeography and phylogeny of the Metatheria. In *Fauna of Australia. Mammalia*, ed. D. W. Walton & B. J. Richardson, vol. 1B, pp. 527–48. Canberra: Australian Government Publishing Service.

Cochran, P. A. (1987). Optimal digestion in a batch-reactor gut: the analogy to partial prey consumption. *Oikos*, **50**, 268–70.

Coley, P. D., Bryant, J. P. & Chapin., F. S. III. (1985). Resource availability and plant antiherbivore defense. *Science*, **230**, 895–9.

Collins, B. G., Wooller, R. D. & Richardson, K. C. (1988). Torpor by the honey possum, *Tarsipes rostratus* (Marsupialia: Tarsipedidae), in response to food shortage and low environmental temperature. *Australian Mammalogy*, **11**, 51–7.

Comport, S. S., Ward, S. J. & Foley, W. J. (1996). Home ranges, time budgets and food-tree use in a high-density tropical population of greater gliders, *Petauroides volans minor* (Pseudoheiridae: Marsupialia). *Wildlife Research*, **23**, 401–19.

Cook, L. J., Scott, T. W., Ferguson, K. A. & McDonald, I. W. (1970). Production of poly-unsaturated ruminant body fats. *Nature*, **228**, 178.

Copley, P. B. & Robinson, A. C. (1983). Studies on the yellow-footed rock-wallaby, *Petrogale xanthopus*, Gray (Marsupialia: Macropodidae). II. Diet. *Australian Wildlife Research*, **10**, 63–76.

Cork, S. J. (1981). Digestion and metabolism in the koala (*Phascolarctos cinereus*), an arboreal folivore. Ph. D. Thesis, University of New South Wales, Sydney.

Cork, S. J. (1986). Foliage of *Eucalyptus punctata* and the maintenance nitrogen requirements of koalas, *Phascolarctos cinereus*. *Australian Journal of Zoology*, **34**, 17–23.

Cork, S. J. (1992). Polyphenols and the distribution of arboreal, folivorous marsupials in *Eucalyptus* forests of Australia. In *Plant Polyphenols*, ed. R. W. Hemingway & P. E. Laks, pp. 653–63. New York: Plenum Press.

Cork, S. J. (1996). Optimal digestive strategies for arboreal herbivorous mammals in contrasting forest types: why koalas and colobines are different. *Australian Journal of Ecology*, **21**, 10–20.

Cork, S. J. & Dove, H. (1989). Lactation in the tammar wallaby (*Macropus eugenii*). II. Intake of milk components and maternal allocation of energy. *Journal of Zoology, London*, **219**, 399–409.

Cork, S. J. & Foley, W. J. (1991). Digestive and metabolic strategies of arboreal mammalian folivores in relation to chemical defenses in temperate and tropical forests. In *Plant Defenses Against Mammalian Herbivory*, ed. R. T. Palo & C. T. Robbins, pp. 133–66. Boca Raton, Florida: CRC Press.

Cork, S. J. & Foley, W. J. (1997). Digestive and metabolic adaptations of arboreal marsupials for dealing with plant antinutrients and toxins. In *Marsupial Biology: Recent Research, New Perspectives*, ed. N. R. Saunders & L. A. Hinds), pp. 204–6. Sydney: University of New South Wales Press.

Cork, S. J. & Hume, I. D. (1983). Microbial digestion in the koala (*Phascolarctos cinereus*, Marsupialia): an arboreal folivore. *Journal of Comparative Physiology*, **B152**, 131–5.

Cork, S. J., Hume, I. D. & Dawson, T. J. (1983). Digestion and metabolism of a natural foliar diet (*Eucalyptus punctata*) by an arboreal marsupial, the koala (*Phascolarctos cinereus*). *Journal of Comparative Physiology*, **B153**, 181–90.

Cork, S. J. & Pahl, L. (1984). The possible influence of nutritional factors on diet and habitat selection by the ringtail possum (*Pseudocheirus peregrinus*). In *Possums and Gliders*, ed. A. P. Smith and I. D. Hume, pp. 269–76. Sydney: Australian Mammal Society.

Cork, S. J. & Sanson, G. D. (1990). Digestion and nutrition in the koala: a review. In *Biology of the Koala*, ed. A. K. Lee, K. A. Handasyde & G. D. Sanson, pp. 129–44. Sydney: Surrey Beatty.

Cork, S. J. & Warner, A. C. I. (1983). The passage of digesta markers through the gut of a folivorous marsupial, the koala *Phascolarctos cinereus*. *Journal of Comparative Physiology*, **B152**, 43–51.

Cowan, I. McT., O'Riordan, A. M. & Cowan, J. S. McT (1974). Energy requirements of the dasyurid marsupial mouse *Antechinus swainsonii* (Waterhouse). *Canadian Journal of Zoology*, **52**, 269–75.

Cowan, P. E. (1989). Changes in milk composition during lactation in the common brushtail possum, *Trichosurus vulpecula* (Marsupialia: Phalangeridae). *Reproduction, Fertility and Development*, **1**, 325–35.

Craig, S. A. (1985). Social organisation, reproduction and feeding behaviour of a population of yellow-bellied gliders, *Petaurus australis* (Marsupialia: Petauridae). *Australian Wildlife Research*, **12**, 1–18.

Crisp, E. (1855). On some points relating to the anatomy of the Tasmanian wolf (*Thylacinus*) and of the Cape Hunting dog (*Lycaon pictus*). *Proceedings of the Zoological Society of London, 1855*: 188–91.

Crisp, E. A., Cowan, P. E. & Messer, M. (1989). Changes in milk carbohydrates during lactation in the common brushtail possum, *Trichosurus vulpecula* (Marsupialia: Phalangeridae). *Reproduction, Fertility and Development*, **4**, 309–14.

Crisp, E. A., Messer, M. & VandeBerg, J. L. (1989). Changes in milk carbohydrates during lactation in a didelphid marsupial, *Monodelphis domestica*. *Physiological Zoology*, **62**, 1117–25.

Croft, D. B. (1987). Socio-ecology of the antilopine wallaroo, *Macropus antilopinus*, in the Northern Territory, with observations on sympatric *M. robustus woodwardii* and *M. agilis*. *Australian Wildlife Research*, **14**, 243–55.

Crossman, D. G., Johnson, C. N. & Horsup, A. B. (1994). Trends in the population of the northern hairy-nosed wombat *Lasiorhinus krefftii* in Epping Forest National Park, central Queensland. *Pacific Conservation Biology*, **1**, 141–9.

Crowe, O. & Hume, I. D. (1997). Morphology and function of the gastrointestinal tract of Australian folivorous possums. *Australian Journal of Zoology*, **45**, 357–68.

Crowley, H. M., Woodward, D. R. & Rose, R. W. (1988). Changes in milk composition during lactation in the potoroo, *Potorous tridactylus* (Marsupialia: Potoroinae). *Australian Journal of Biological Sciences*, **41**, 289–96.

Czerkawski, J. W. (1978). Reassessment of efficiency of synthesis of microbial matter in the rumen. *Journal of Dairy Science*, **61**, 1261–73.

Dash, J. A. (1988). Effect of dietary terpenes on glucuronic acid excretion and ascorbic acid turnover in the brushtail possum (*Trichosurus vulpecula*). *Comparative Biochemistry and Physiology*, **89B**, 221–6.

Dash, J. A. & Jenness, R. (1985). Ascorbate content of foliage of eucalypts and conifers utilised by some Australian and North American mammals. *Experientia*, **41**, 952–5.

Dash, J. A., Jenness, R. & Hume, I. D. (1984). Ascorbic acid turnover and excretion in two arboreal marsupials and in laboratory rabbits. *Comparative Biochemistry and Physiology*, **77B**, 391–7.

Dawson, T. J. (1973). Thermoregulatory responses of the arid zone kangaroos *Megaleia rufa* and *Macropus robustus*. *Comparative Biochemistry and Physiology*, **46A**, 153–69.

Dawson, T. J. (1989). Diets of macropodoid marsupials: general patterns and environmental influences. In *Kangaroos, Wallabies and Rat-kangaroos*, ed. G. Grigg, P. Jarman & I. Hume, pp. 129–42. Sydney: Surrey Beatty.

Dawson, T. J. (1995). *Kangaroos, Biology of the Largest Marsupials*. Sydney: University of New South Wales Press.

Dawson, T. J. & Brown, G. D. (1970). A comparison of the insulative and reflective properties of the fur of desert kangaroos. *Comparative Biochemistry and Physiology*, **37**, 23–38.

Dawson, T. J. & Dawson, W. R (1982). Metabolic scope and conductance in response to cold of some dasyurid marsupials and Australian rodents. *Comparative Biochemistry and Physiology*, **71A**, 59–64.

Dawson, T. J. & Degabriele, R. (1973). The cuscus (*Phalanger maculatus*) – a marsupial sloth? *Journal of Comparative Physiology*, **38**, 41–50.

Dawson, T. J. & Denny, M. J. S. (1969). Seasonal variation in the plasma and urine electrolyte concentration of the arid zone kangaroos *Megaleia rufa* and *Macropus robustus*. *Australian Journal of Zoology*, **17**, 777–84.

Dawson T. J., Denny, M. J. S., Russell, E. M. & Ellis, B. (1975). Water usage and diet preferences of free ranging kangaroos, sheep and feral goats in the Australian arid zone during summer. *Journal of Zoology, London*, **177**, 1–23.

Dawson, T. J. & Ellis, B. A. (1979). Comparison of the diets of yellow-footed rock-wallabies and sympatric herbivores in western New South Wales. *Australian Wildlife Research*, **6**, 245–54.

Dawson, T. J. & Ellis, B. A. (1994). Diets of mammalian herbivores in Australian arid shrublands: seasonal effects on overlap between red kangaroos, sheep and rabbits and on dietary niche breadths and electivities. *Journal of Arid Environments*, **26**, 257–71.

Dawson, T. J. & Ellis, B. A. (1996). Diets of mammalian herbivores in Australian arid, hilly shrublands: seasonal effects on overlap between euros (hill kangaroos), sheep and feral goats, and on dietary niche breadths and electivities. *Journal of Arid Environments*, **34**, 491–506.

Dawson, T. J., Finch, E., Freedman, K., Hume, I. D., Renfree, M. B. & Temple-Smith, P. D. (1989). Morphology and physiology of the Metatheria. *In Fauna of Australia*, vol. 1B (*Mammalia*), ed. D. W. Walton and B. J.

Richardson, pp. 670–715. Canberra: Australian Government Publishing Service.

Dawson, T. J., Grant, T. R. & Fanning, D. (1979). Standard metabolism of monotremes and the evolution of homeothermy. *Australian Journal of Zoology*, 27, 511–15.

Dawson, T. J. & Hulbert, A. J. (1970). Standard metabolism, body temperature, and surface areas of Australian marsupials. *American Journal of Physiology*, 218, 1233–8.

Dawson, T. J. & Olson, J. M. (1988). Thermogenic capabilities of the opossum *Monodelphis domestica* when warm and cold acclimated: similarities between American and Australian marsupials. *Comparative Biochemistry and Physiology*, 89A, 85–91.

Dawson, T. J., Tierney, P. J. & Ellis, B. A. (1992). The diet of the bridled nailtail wallaby (*Onychogalea fraenata*). II. Overlap in dietary niche breadth and plant preferences with the black-striped wallaby (*Macropus dorsalis*) and domestic cattle. *Wildlife Research*, 19, 79–87.

Dawson, T. J. & Wolfers, J. M. (1978). Metabolism, thermoregulation and torpor in shrew sized marsupials of the genus *Planigale*. *Comparative Biochemistry and Physiology*, 59A, 305–9.

Dawson, W. R. & Bennett, A. F. (1978). Energy metabolism and thermoregulation of the spectacled hare wallaby (*Lagorchestes conspicillatus*). *Physiological Zoology*, 51, 114–30.

Degabriele, R. (1981). A relative shortage of nitrogenous food in the ecology of the koala (*Phascolarctos cinereus*). *Australian Journal of Ecology*, 6, 139–41.

Degabriele, R. (1983). Nitrogen and the koala (*Plascolarctos cinereus*): some indirect evidence. *Australian Journal of Ecology*, 8, 75–6.

Degabriele, R. & Dawson, T. J. (1979). Metabolism and heat balance in an arboreal marsupial, the koala (*Phascolarctos cinereus*). *Journal of Comparative Physiology*, 134, 293–301.

Degabriele, R., Harrop, C. J. F. & Dawson, T. J. (1978). Water metabolism in the koala (*Phascolarctos cinereus*). In *The Ecology of Arboreal Folivores*, ed. G. G. Montgomery, pp. 163–72. Washington, D C: Smithsonian Institution Press.

Degen, A. A. & Kam, M. (1995). Scaling of field metabolic rate to basal metabolic rate ratio in homeotherms. *Ecoscience*, 2, 48–54.

Dehority, B. A. (1996). A new family of entodiniomorph protozoa from the marsupial forestomach, with descriptions of a new genus and five new species. *Journal of Eukaryotic Microbiology*, 43, 285–95.

Dellow, D. W. (1979). Physiology of digestion in the macropodine marsupials. Ph. D. Thesis, University of New England, Armidale.

Dellow, D. W. (1982). Studies on the nutrition of macropodine marsupials. III. The flow of digesta through the stomach and intestine of macropodines and sheep. *Australian Journal of Zoology*, 30, 751–65.

Dellow, D. W. & Hume, I. D. (1982a). Studies on the nutrition of macropodine marsupials. I. Intake and digestion of lucerne hay and fresh grass, *Phalaris aquatica*. *Australian Journal of Zoology*, 30, 391–8.

Dellow, D. W. & Hume, I. D. (1982b). Studies on the nutrition of macropodine marsupials. II. Urea and water metabolism in *Thylogale thetis* and *Macropus eugenii*, two wallabies from divergent habitats. *Australian Journal of Zoology*, **30**, 399–406.

Dellow, D. W. & Hume, I. D. (1982c). Studies on the nutrition of macropodine marsupials. IV. Digestion in the stomach and the intestine of *Macropus giganteus*, *Thylogale thetis* and *Macropus eugenii*. *Australian Journal of Zoology*, **30**, 767–77.

Dellow, D. W., Hume, I. D., Clarke, R. T. J. & Bauchop, T. (1988). Microbial activity in the forestomach of free-living macropodid marsupials: comparisons with laboratory studies. *Australian Journal of Zoology*, **36**, 383–95.

Dellow, D. W., Nolan, J. V. & Hume, I. D. (1983). Studies on the nutrition of macropodine marsupials. V. Microbial fermentation in the forestomach of *Thylogale thetis* and *Macropus eugenii*. *Australian Journal of Zoology*, **31**, 433–43.

Demment, M. W. & Van Soest, P. J. (1985). A nutritional explanation for body-size patterns of ruminant and nonruminant herbivores. *American Naturalist*, **125**, 641–72.

Dennis, A. J. & Johnson, P. M. (1995). Musky rat-kangaroo, *Hypsiprymnodon moschatus*. In *The Mammals of Australia*, ed. R. Strahan, pp. 282–4. Chatswood: Reed Books.

Denny, M. J. S. & Dawson, T. J. (1973). A field technique for studying water metabolism of large marsupials. *Journal of Wildlife Management*, **37**, 574–8.

Denny, M. J. S. & Dawson, T. J. (1975a). Comparative metabolism of tritiated water by macropodid marsupials. *American Journal of Physiology*, **228**, 1794–9.

Denny, M. J. S. & Dawson, T. J. (1975b). Effects of dehydration on body-water distribution in desert kangaroos. *American Journal of Physiology*, **229**, 251–4.

Denny, M. J. S. & Dawson, T. J. (1977). Kidney structure and function of desert kangaroos. *Journal of Applied Physiology*, 42, 636–42.

Dickman, C. R., Green, K., Carron, P. L., Happold, D. C. D. & Osbourne, W. S. M. (1983). Coexistence, convergence and competition among *Antechinus* (Marsupialia: Dasyuridae) in the Australian high country. *Proceedings of the Ecological Society of Australia*, **12**, 79–99.

Dickman, C. R. & Huang, C. (1988). The reliability of fecal analysis as a method for determining the diet of insectivorous mammals. *Journal of Mammalogy*, **69**, 108–13.

Dickman, C. R., King, D. H., Adams, M. & Baverstock, P. R. (1988). Electrophoretic identification of a new species of *Antechinus* (Marsupialia: Dasyuridae) in south-eastern Australia. *Australian Journal of Zoology*, **36**, 455–63.

Dierenfeld, E. S. (1984). Diet quality of sympatric wombats, kangaroos and rabbits during severe drought. Ph.D. thesis, Cornell University, Ithaca, NY.

Dixon, J. M. (1988). Notes on the diet of three mammals presumed to be extinct: the pig-footed bandicoot, the lesser bilby and the desert rat-kangaroo. *Victorian Naturalist*, **105**, 208–11.

Dobson, D. E. Prager, E. M. & Wilson, A. C. (1984). Stomach lysozymes of ruminants. I. Distribution and catalytic properties. *Journal of Biological Chemistry*, **259**, 11607–16.

Drake, H. L. (ed.) (1994). *Acetogenesis*. New York: Chapman and Hall.

Ducrocq, S., Buffetaut, E., Buffetaut-Tong, H., Jaeger, J., Jongkanjanasoontorn,

Y. & Suteethorn, V. (1992). First fossil marsupial from south Asia. *Journal of Vertebrate Paleontology*, **12**, 395–9.

Dudzinski, M. L., Low, W. A., Müller, W. J. & Low, B. S. (1982). Joint use by red kangaroos and shorthorn cattle in arid central Australia. *Australian Journal of Ecology*, 7, 69–74.

Ealey, E. H. M. (1967a). Ecology of the euro, *Macropus robustus* (Gould) in north-western Australia. I. The environment and changes in euro and sheep populations. *CSIRO Wildlife Research*, **12**, 9–25.

Ealey, E. H. M. (1967b). Ecology of the euro, *Macropus robustus* (Gould), in north-western Australia. II. Behaviour, movements and drinking patterns. *CSIRO Wildlife Research*, **12**, 27–51.

Ealey, E. H. M. & Main, A. R. (1967). Ecology of the euro, *Macropus robustus* (Gould), in north-western Australia. III. Seasonal changes in nutrition. *CSIRO Wildlife Research*, **12**, 53–65.

Eberhard, I. H., McNamara, J., Pearse, R. J. & Southwell, I. A. (1975). Ingestion and excretion of *Eucalyptus punctata* D. C. and its essential oils by the koala, *Phascolarctos cinereus* (Goldfuss). *Australian Journal of Zoology*, **23**, 169–79.

Edwards, G. P., Dawson, T. J. & Croft, D. B. (1995). The dietary overlap between red kangaroos (*Macropus rufus*) and sheep (*Ovis aries*) in the arid rangelands of Australia. *Australian Journal of Ecology*, **20**, 324–34.

Edwards, G. P. & Ealey, E. H. M. (1975). Aspects of the ecology of the swamp wallaby *Wallabia bicolor* (Marsupialia: Macropodidae). *Australian Mammalogy*, **1**, 307–18.

Eisenberg, J. F. (1981). *The Mammalian Radiations*. Chicago: University of Chicago Press.

Eisenberg, J. F. (1989). *Mammals of the Neotropics*. Vol. 1. *The Northern Neotropics*. Chicago: University of Chicago Press.

Elgar, M. A. & Harvey, P. H. (1987). Basal metabolic rates in mammals: allometry, phylogeny and ecology. *Functional Ecology*, **1**, 25–36.

Ellis, B. A., Russell E. M., Dawson, T. J. & Harrop, C. J. F. (1977). Seasonal changes in diet preferences of free-ranging red kangaroos, euros and sheep in western New South Wales. *Australian Wildlife Research*, **4**, 127–44.

Ellis, B. A., Tierney, P. J. & Dawson, T. J. (1992). The diet of the bridled nailtail wallaby (*Onychogalea fraenata*). I. Site and seasonal influences and dietary overlap with the black-striped wallaby (*Macropus dorsalis*) and domestic cattle. *Wildlife Research*, **19**, 65–78.

Ellis, W. A. H. & Carrick, F. N. (1992). Total body water and the estimation of fat in the koala (*Phascolarctos cinereus*). *Journal of the Australian Veterinary Association*, **69**, 229–30.

Ellis, W. A. H., Melzer, A., Green, B., Newgrain, K., Hindall, M. A. & Carrick, F. N. (1995). Seasonal variation in water flux, field metabolic rate and food consumption of free-ranging koalas (*Phascolarctos cinereus*). *Australian Journal of Zoology*, **43**, 59–68.

Engelhardt, W. v., Wolter, S., Lawrenz, H. & Hemsley, J. A. (1978). Production of methane in two non-ruminant herbivores. *Comparative Biochemistry and Physiology*, **60**, 309–11.

Evans, M. (1992a). Diet of the brushtail possum *Trichosurus vulpecula*

(Marsupialia: Phalangeridae) in central Australia. *Australian Mammalogy*, **15**, 25–30.

Evans, M. C. (1992b). The bridled nailtail wallaby: ecology and management. M. Res. Sci Thesis, University of New England, Armidale, New South Wales.

Faichney, G. J. (1968). The production and absorption of volatile fatty acids from the rumen of sheep. *Australian Journal of Agricultural Research*, **19**, 791–802.

Faichney, G. J. (1969). Production of volatile fatty acids in the sheep caecum. *Australian Journal of Agricultural Research*, **20**, 491–8.

Faichney, G. J. (1993). Digesta flow. In *Quantitative Aspects of Ruminant Digestion and Metabolism*, ed. J. M. Forbes & J. France, pp. 53–85. Wallingford: CAB International.

Faichney, G. J. & Griffiths, D. A. (1978). Behaviour of solute and particle markers in the stomach of sheep given a concentrate diet. *British Journal of Nutrition*, **40**, 71–82.

Farrell, D. J. & Wood, A. J. (1968). The nutrition of the female mink (*Mustela vison*). II. The energy requirements for maintenance. *Canadian Journal of Zoology*, **46**, 47–52.

Fenna, L. & Boag, D. A. (1974). Filling and emptying of the galliform caecum. *Canadian Journal of Zoology*, **52**, 537–40.

Finlayson, H. H. (1931). On mammals from the Dawson Valley, Queensland. Part 1. *Transactions of the Proceedings of the Royal Society of South Australia*, **50**, 67–89.

Finlayson, H. H. (1958). On Central Australian mammals (with notice of related species from adjacent tracts). Part III. The Potoroinae. *Records of the South Australian Museum*, **3**, 235–302.

Fisher, D. O. & Dickman, C. R. (1993). Body size–prey size relationships in insectivorous marsupials: tests of three hypotheses. *Ecology*, **74**, 1871–83.

Fitch, H. S. & Sandidge, L. L. (1953). Ecology of the opossum on a natural area in northeastern Kansas. *Publications of the Museum of Natural History of the University of Kansas*, **7**, 305–38.

Fitzgerald, A. E. (1984). Diet of the possum (*Trichosurus vulpecula*) in three Tasmanian forest types and its relevance to the diet of possums in New Zealand forests. In *Possums and Gliders*, ed. A. P. Smith & I. D. Hume, pp. 137–43. Sydney: Australian Mammal Society.

Flannery, T. F. (1989). Phylogeny of the Macropodoidea; a study in convergence. In *Kangaroos, Wallabies and Rat-kangaroos*, ed. G. Grigg, P. Jarman and I. Hume, pp. 1–46. Sydney: Surrey Beatty.

Flannery, T. F. (1994a). *Possums of the World*. Sydney: Geo Productions.

Flannery, T. F. (1994b). *The Future Eaters*. Chatswood: Reed Books.

Flannery, T. F. (1995). *Mammals of New Guinea*. Chatswood: Reed Books.

Flannery, T. F., Martin, R. W. & Szalay, A. (1996). *Tree Kangaroos: A Curious Natural History*. Port Melbourne: Reed Books.

Fleming, M. R. (1980). Thermoregulation and torpor in the sugar glider, *Petaurus breviceps* (Marsupialia: Petauridae). *Australian Journal of Zoology*, **28**, 521–34.

Fleming, M. R. (1985). The thermal physiology of the feathertail glider, *Acrobates pygmaeus* (Marsupialia: Burramyidae). *Australian Journal of Zoology*, **33**, 667–81.

Fleming, T. H. (1972). Aspects of the population dynamics of three species of opossums in the Panama Canal Zone. *Journal of Mammalogy*, **53**, 619–23.

Flower, W. H. (1872). Lectures on the comparative anatomy of the organs of digestion in the mammals. *Medical Times & Gazette*, **1**, 215–678.

Foley, W. J. (1987). Digestion and energy metabolism in a small arboreal marsupial, the greater glider (*Petauroides volans*), fed high-terpene *Eucalyptus* foliage. *Journal of Comparative Physiology*, **B157**, 355–62.

Foley, W. J. (1992). Nitrogen and energy retention and acid–base status in the common ringtail possum (*Pseudocheirus peregrinus*): evidence of the effects of absorbed allelochemicals. *Physiological Zoology*, **65**, 403–21.

Foley, W. J., Charles-Dominique, P. & Julien-Laferriere, D. (1999). Compensatory responses to protein-poor fruit diets in the didelphid marsupial *Caluromys philander*. *Journal of Comparative Physiology B*(in press).

Foley, W. J. & Cork, S. J. (1992). Use of fibrous diets by small herbivores: how far can the rules be 'bent'? *Trends in Ecology and Evolution*, **7**, 159–62.

Foley, W. J. & Hume, I. D. (1987a). Passage of digesta markers in two species of arboreal folivorous marsupials – the greater glider (*Petauroides volans*) and the brushtail possum (*Trichosurus vulpecula*). *Physiological Zoology*, **60**, 103–13

Foley, W. J. & Hume, I. D. (1987b). Nitrogen requirements and urea metabolism in two arboreal marsupials, the greater glider (*Petauroides volans*) and the brushtail possum (*Trichosurus vulpecula*), fed *Eucalyptus* foliage. *Physiological Zoology*, **60**, 241–50.

Foley, W. J. & Hume I. D. (1987c). Digestion and metabolism of high-tannin *Eucalyptus* foliage by the brushtail possum (*Trichosurus vulpecula*) (Marsupialia: Phalangeridae). *Journal of Comparative Physiology*, **B157**, 67–76.

Foley, W. J., Hume, I. D. and Cork, S. J. (1989). Fermentation in the hindgut of the greater glider (*Petauroides volans*) and brushtail possum (*Trichosurus vulpecula*) – two arboreal folivores. *Physiological Zoology*, **62**, 1126–43.

Foley, W. J., Hume, I. D. & Taylor, R. (1980). Protein intake and requirements of the eastern wallaroo and the eastern grey kangaroo. *Bulletin of the Australian Mammal Society*, **6**, 34–5 (Abstract).

Foley, W. J., Kehl, J. C., Nagy, K. A., Kaplan, I. R. & Borsboom, A. C. (1990). Energy and water metabolism in free–living greater gliders, *Petauroides volans*. *Australian Journal of Zoology*, **38**, 1–9.

Foley, W. J., Lassak, E. V. & Brophy, J. (1987). Digestion and absorption of *Eucalyptus* essential oils in greater glider (*Petauroides volans*) and brushtail possum (*Trichosurus vulpecula*). *Journal of Chemical Ecology*, **13**, 2115–30.

Foley, W. J., McLean, S. & Cork, S. J. (1995). The effects of plant allelochemicals on acid–base metabolism; the final common pathway. *Journal of Chemical Ecology*, **21**, 721–43.

Fonseca, S. D. & Cerqueira, R. (1991). Water and salt balance in a South American marsupial, the gray four-eyed opossum (*Philander opossum*). *Mammalia*, **55**, 421–32.

Foot, J. Z. & Romberg, B. (1965). The utilisation of roughage by sheep and the red kangaroo, *Macropus rufus* (Desmarest). *Australian Journal of Agricultural Research*, **16**, 429–35.

Forbes, D. K. & Tribe, D. E. (1969). Salivary glands of kangaroos. *Australian*

Journal of Zoology, **17**, 765–75.

Forbes, D. K. & Tribe, D. E. (1970). The utilisation of roughages by sheep and kangaroos. *Australian Journal of Zoology*, **18**, 247–56.

Ford, H. A., Paton, D. C. & Forde, N. (1979). Birds as pollinators of Australian plants. *New Zealand Journal of Botany*, **17**, 509–19.

Fox, B. J. (1982). A review of dasyurid ecology and speculation on the role of limiting similarity in community organisation. In *Carnivorous Marsupials*, ed. M. Archer, pp. 97–116, Sydney: Surrey Beatty and the Royal Zoological Society of New South Wales.

Fox, R. C. (1987). Palaeontology and the early evolution of marsupials. In *Possums and Opossums: Studies in Evolution*, ed. M. Archer, pp. 161–9. Sydney: Surrey Beatty and the Royal Zoological Society of New South Wales.

Frappell, P. B. (1984). Form and function of the gastro-intestinal tract of *Potorous tridactylus*, Gould. B. Sc. (Honours) Thesis, University of Tasmania, Hobart.

Frappell, P. B. & Rose, R. W. (1986). A radiographic study of the gastrointestinal tract of *Potorous tridactylus*, with a suggestion as to the role of the foregut and hindgut in potoroine marsupials. *Australian Journal of Zoology*, **34**, 463–71.

Fraser, E. H. & Kinnear, J. E. (1969). Urinary creatinine excretion by macropod marsupials. *Comparative Biochemistry and Physiology*, **28**, 685–92.

Freeland, W. J. & Winter, J. W. (1975). Evolutionary consequences of eating: *Trichosurus vulpecula* (Marsupialia) and the genus *Eucalyptus*. *Journal of Chemical Ecology*, **1**, 439–55.

Freitas, S. R., Astúa de Moraes, D., Santori, R. T. & Cerqueira, R. (1997). Habitat preference and food use by *Metachirus nudicaudatus* and *Didelphis aurita* (Didelphimorphia, Didelphidae) in Arestinga Forest at Rio de Janeiro. *Reviews in Brazilian Biology*, **57**, 93–8.

Freudenberger, D. O. (1992). Gut capacity, functional allocation of gut volume and size distributions of digesta particles in two macropodid marsupials (*Macropus robustus robustus* and *M. r. erubescens*) and the feral goat (*Capra hircus*). *Australian Journal of Zoology*, **40**, 551–61.

Freudenberger, D. O. & Hume, I. D. (1992). Ingestive and digestive responses to dietary fibre and nitrogen by two macropodid marsupials (*Macropus robustus erubescens* and *M. r. robustus*) and a ruminant (*Capra hircus*). *Australian Journal of Zoology*, **40**, 181–4.

Freudenberger, D. O. & Hume, I. D. (1993). Effects of water restriction on digestive function in two macropodid marsupials from divergent habitats and the feral goat. *Journal of Comparative Physiology*, **B163**, 247–57.

Freudenberger, D. O. & Nolan, J. V. (1993a). Glucose metabolism in a kangaroo (*Macropus robustus erubescens*) and a similar size eutherian herbivore, the feral goat. *Comparative Biochemistry and Physiology*, **106A**, 295–8.

Freudenberger, D. O. & Nolan, J. V. (1993b). Protein turnover in two kangaroo sub-species (*Macropus robustus robustus* and *M. r. erubescens*) from divergent habitats and the sympatric feral goat (*Capra hircus*). *Comparative Biochemistry and Physiology*, **105A**, 443–8.

Freudenberger, D. O., Wallis, I. R. & Hume, I. D. (1989). Digestive adaptations of kangaroos, wallabies and rat-kangaroos. In *Kangaroos, Wallabies and Rat-Kangaroos*, ed. G. Grigg, P. Jarman and I. Hume, pp. 179–87. Sydney:

Surrey Beatty.

Frey, H. (1991). Energetic significance of torpor and other energy-conserving mechanisms in free-living *Sminthopsis crassicaudata* (Marsupialia: Dasyuridae). *Australian Journal of Zoology*, **39**, 698–708.

Frey, H. & Fleming, M. R. (1984). Thermoregulatory behaviour of free-ranging feathertail gliders *Acrobates pygmaeus* (Marsupialia: Burramyidae) in Victoria. In *Possums and Gliders*, ed. A. P. Smith & I. D. Hume, pp. 393–401. Sydney: Australian Mammal Society.

Friend, J. A. & Whitford, D. (1993). Maintenance and breeding of the numbat (*Myrmecobius fasciatus*) in captivity. In *The Biology and Management of Australasian Carnivorous Marsupials*, ed. M. Roberts, J. Cornio, G. Crawshaw and M. Hutchins, pp. 103–24. Toronto: Metropolitan Toronto Zoo.

Frith, H. J. (1964). Mobility of the red kangaroo, *Megaleia rufa*. *CSIRO Wildlife Research*, **9**, 1–19.

Frith, H. J. & Sharman, G. B. (1964). Breeding in wild populations of the red kangaroo, *Megaleia rufa*. *CSIRO Wildlife Research*, **9**, 86–114.

Galand, G. (1989). Brush border membrane sucrase-isomaltase, maltase-glucoamylase and trehalase in mammals. Comparative development, effects of glucocorticoids, molecular mechanisms, and phylogenetic implications. *Comparative Biochemistry and Physiology*, **B158**, 413–19.

Gardner, A. L. (1993). Order Didelphimorphia. In *Mammal Species of the World: a Taxonomic and Geographic Reference*, ed. D. E. Wilson & D. M. Reeder, pp. 15–23. Washington, DC: Smithsonian Institution Press.

Garland, T., Jr, Geiser, F. & Baudinette, R. V. (1988). Comparative locomotor performance of marsupial and placental mammals. *Journal of Zoology, London*, **215**, 505–22.

Garrod, A. H. (1875). On the kangaroo called *Halmaturus luctuosus* by D'Albertis, and its affinities. *Proceedings of the Zoological Society 1875*, pp. 48–59.

Gaughwin, M. D., Judson, G. J,. Macfarlane, W. V. & Siebert, B. D. (1984). Effect of drought on the health of wild hairy-nosed wombats, *Lasiorhinus latifrons*. *Australian Wildlife Research*, **11**, 455–63.

Geiser, F. (1993). Hibernation in the eastern pygmy possum, *Cercartetus nanus* (Marsupialia: Burramyidae). *Australian Journal of Zoology*, 41, 67–75.

Geiser, F. (1994). Hibernation and daily torpor in marsupials: a review. *Australian Journal of Zoology*, **42**, 1–16.

Geiser, F. & Broome, L. S. (1991). Hibernation in the mountain pygmy possum *Burramys parvus* (Marsupialia). *Journal of Zoology* (London), **223**, 593–602.

Geiser, F. & Ruf, T. (1995). Hibernation versus daily torpor in mammals and birds: physiological variables and classification of torpor patterns. *Physiological Zoology*, **68**, 935–66.

Geiser, F., Stahl, B. & Learmonth, R. P (1992). The effect of dietary fatty acids on the pattern of torpor in a marsupial. *Physiological Zoology*, **65**, 1236–45.

Gelineo, S. (1964). Organ systems in adaptation: the temperature regulating system. In *Handbook of Physiology*, Section 4, ed. D. B. Dill, pp. 259–82. Washington, DC: American Physiological Society.

Gemmell, R. T. & Englehardt, W. v. (1977). The structure of the cells lining the stomach of the tammar wallaby (*Macropus eugenii*). *Journal of Anatomy*, **123**,

723–33.

Gibson, L. A. (1999). Nutritional ecology of the greater bilby (*Macrotis lagotis*) in south-western Queensland. Ph. D. Thesis, University of Sydney.

Gilmore, D. P. (1970). The rate of passage of food in the brush-tailed possum, *Trichosurus vulpecula. Australian Journal of Biological Sciences*, **23**, 515–18.

Gipps, J. M. (1980). Functional dental morphology in four Australian possums. B.Sc. Honours Thesis, Monash University, Melbourne.

Gipps, J. M. & Sanson, G. D. (1984). Mastication and digestion in *Pseudocheirus*. In *Possums and Gliders*, ed. A. P. Smith and I. D. Hume, pp. 237–46. Sydney: Australian Mammal Society.

Glander, K. E. (1978). Howler monkey feeding behavior and plant secondary compunds: a study of strategies. In *The Ecology of Arboreal Folivores*, ed. G. G. Montgomery, pp. 561–74. Washington, DC: Smithsonian Institution Press.

Goldingay, R. L. (1986). Feeding behaviour of the yellow-bellied glider *Petaurus australis* (Marsupialia: Petauridae) at Bombala, New South Wales. *Australian Mammalogy*, **9**, 17–25.

Goldingay, R. L. (1987). Sap feeding by the marsupial *Petaurus australis*: an enigmatic behaviour? *Oecologia*, **73**, 154–8.

Goldingay, R. L. (1989). Time budget and related aspects of the foraging behaviour of the yellow-bellied glider, *Petaurus australis. Australian Wildlife Research*, **16**, 105–12.

Goldingay, R. L. (1990). The foraging behaviour of a nectar feeding marsupial, *Petaurus australis. Oecologia*, **85**, 191–9.

Goldingay, R. L. (1994). Loud calls of the yellow-bellied glider, *Petaurus australis*: territorial behaviour by an arboreal marsupial? *Australian Journal of Zoology*, **42**, 279–93.

Goldingay, R. L., Carthew, S. M. & Whelan, R. J. (1987). Transfer of *Banksia spinulosa* pollen by mammals: implications for pollination. *Australian Journal of Zoology*, **35**, 319–25.

Goldingay, R. L. & Kavanagh, R. P. (1990). Socioecology of the yellow-bellied glider, *Petaurus australis*, at Waratah Creek, NSW. *Australian Journal of Zoology*, **38**, 327–41.

Goldingay, R. L. & Kavanagh, R. P. (1993). Home-range estimates and habitat of the yellow-bellied glider (*Petaurus australis*) at Waratah Creek, New South Wales. *Wildlife Research*, **20**, 387–404.

Goldingay, R. L. & Kavanagh, R. P. (1995). Foraging behaviour and habitat use of the feathertail glider (*Acrobates pygmaeus*) at Waratah Creek, New South Wales. *Wildlife Research*, **22**, 457–70.

Goldingay, R. & Possingham, H. (1995). Area requirements for viable populations of the Australian gliding marsupial *Petaurus australis. Biological conservation*, **73**, 161–7.

Gordon, G. (1995). Rufous spiny bandicoot. In *The Mammals of Australia*, ed. R. Strahan, pp. 191–2. Chatswood: Reed Books.

Gordon, G., Brown, A. S. & Pulsford, T. (1988). A koala (*Phascolarctos cinereus* Goldfuss) population crash during drought and heatwave conditions in south-western Queensland. *Australian Journal of Ecology*, **13**, 451–61.

Gordon, G. & Hall, L. S. (1995). Tail fat storage in arid zone bandicoots.

Australian Mammalogy, **18**, 87–90.

Gordon, G. L. R. & Phillips, M. W. (1993). Removal of anaerobic fungi from the rumen of sheep by chemical treatment and the effect of feed consumption and *in vivo* fibre digestion. *Letters in Applied Microbiology*, **17**, 220–3.

Gott, M. (1996). Ecology of the northern brown bandicoot, *Isoodon macrourus*: reproduction and resource use in a heathland population. Ph.D. Thesis, University of New South Wales, Sydney

Goudberg, N. J. (1990). The feeding ecology of three species of upland forest ringtail possums, *Hemibelideus lemuroides*, *Pseudocheirus herbertensis* and *Pseudocheirus archeri*. Ph.D. Thesis, James Cook University, Townsville.

Gowland, P. N. (1973). Aspects of the digestive physiology of the common wombat, *Vombatus ursinus* (Shaw, 1800). Unpublished Thesis, Australian National University, Canberra.

Grant, T. R. & Temple-Smith, P. D. (1987). Observations on torpor in the small marsupial *Dromiciops australis* (Marsupialia: Microbiotheriidae) from southern Chile. *In Possums and Opossums: Studies in Evolution*, ed. M. Archer. pp. 273–7. Sydney: Surrey Beatty and the Royal Zoological Society of New South Wales.

Green, B. (1989). Water and energy turnover in free-living macropodoids. In *Kangaroos, Wallabies and Rat-kangaroos*, ed. G. Grigg, P. Jarman & I. Hume, pp. 223–9. Sydney: Surrey Beatty.

Green, B. (1997). Field energetics and water fluxes in marsupials. In *Marsupial Biology: Recent Research, New Perspectives*, ed. N. R. Saunders & L. A. Hinds, pp. 143–162. Sydney: University of New South Wales Press.

Green, B. & Eberhard, I. (1979). Energy requirements and sodium and water turnovers in two captive marsupial carnivores: the Tasmanian devil *Sarcophilus harrisii*, and the native cat, *Dasyurus viverrinus*. *Australian Journal of Zoology*, **27**, 1–8.

Green, B. & Eberhard, I. (1983). Water and sodium intake, and estimated food consumption, in free-living eastern quolls, *Dasyurus viverrinus*. *Australian Journal of Zoology*, **31**, 371–80.

Green, B., King, D. & Bradley, A. (1989). Water and energy metabolism and estimated food consumption rates of free-living wambengers *Phascogale calura* (Marsupialia, Dasyuridae). *Australian Wildlife Research*, **16**, 501–7.

Green, B. & Merchant, J. C. (1988). The composition of marsupial milk. In *The Developing Marsupial*, ed. C. H. Tyndale-Biscoe & P. A. Janssens, pp. 41–56. Berlin: Springer.

Green, B., Newgrain, K., Catling, P. & Turner, G. (1991). Patterns of prey consumption and energy use in a small carnivorous marsupial, *Antechinus stuartii*. *Australian Journal of Zoology*, **39**, 539–47.

Green, B., Newgrain, K. & Merchant, J. (1980). Changes in milk composition during lactation in the tammar wallaby (*Macropus eugenii*). *Australian Journal of Biological Sciences*, **33**, 35–42.

Green, B., VandeBerg, J. L. & Newgrain, K. (1991). Milk composition in an American marsupial (*Monodelphis domestica*). *Comparative Biochemistry and Physiology*, **99B**, 663–5.

Green, K. & Crowley, H. (1989). Energetics and behaviour of active subnivean insectivores *Antechinus swainsonii* and *A. stuartii* (Marsupialia: Dasyuridae) in the Snowy Mountains. *Australian Wildlife Research*, **16**, 509–16.

Green, R. H. (1967). Notes on the devil (*Sarcophilus harrisii*) and the quoll (*Dasyurus viverrinus*) in north eastern Tasmania. *Records of the Queen Victoria Museum*, **27**, 1–13.

Green, R. H. (1980). The little pygmy-possum, *Cercartetus lepidus* in Tasmania. *Records of the Queen Victoria Museum*, **68**, 1–12.

Green, S. W. & Renfree, M. B. (1982). Changes in the milk proteins during lactation in the tammar wallaby, *Macropus eugenii*. *Australian Journal of Biological Sciences*, **35**, 145–52.

Griffiths, M. (1978). *The Biology of the Monotremes*. New York: Academic Press.

Griffiths, M. & Barker, R. (1966). The plants eaten by sheep and by kangaroos grazing together in a paddock in south-western Queensland. *CSIRO Wildlife Research*, **11**, 145–67.

Griffiths, M., Barker, R. & MacLean, L. (1974). Further observations on the plants eaten by kangaroos and sheep grazing together in a paddock in south-western Queensland. *Australian Wildlife Research*, **1**, 27–43.

Griffiths, M. & Barton, A. A. (1966). The ontogeny of the stomach in the pouch young of the red kangaroo. *CSIRO Wildlife Research*, **11**, 169–85.

Griffiths, M., Friend, J. A., Whitford, D. & Fogerty, A. C. (1988). Composition of the milk of the numbat, *Myrmecobius fasciatus* (Marsupialia: Myrmecobiidae), with particular reference to the fatty acids of the lipids. *Australian Mammalogy*, **11**, 59–62.

Griffiths, M., McIntosh, D. L. & Leckie, R. M. C. (1969). The effects of cortisone on nitrogen balance and glucose metabolism in diabetic and normal kangaroos, sheep and rabbits. *Journal of Endocrinology*, **44**, 1–12.

Guiler, E. R. (1970). Observations on the Tasmanian devil, *Sarcophilus harrisii* (Marsupialia: Dasyuridae). I. Numbers, home range, movements, and food in two populations. *Australian Journal of Zoology*, **18**, 49–62.

Guiler, E. R. (1971). Food of the potoroo (Marsupialia, Macopodidae). *Journal of Mammalogy*, **52**, 232–4.

Haines, H., Macfarlane, W. V., Setchell, C. & Howard, B. (1974). Water turnover and pulmocutaneous evaporation of Australian desert dasyurids and murids. *American Journal of Physiology*, **227**, 958–63.

Halford, D. A., Bell, D. T. & Loneragan, W. A. (1984). Diet of the western grey kangaroo (*Macropus fuliginosus*, Desin) in a mixed pasture–woodland habitat of Western Australia. *Journal of the Royal Society of Western Australia*, **66**, 119–28.

Hall, S. (1980). The diets of two coexisting species of *Antechinus* (Marsupialia: Dasyuridae). *Australian Wildlife Research*, **7**, 365–78.

Hallam, J. F., Holland, R. A. B. & Dawson, T. J. (1995). The blood of carnivorous marsupials: low hemoglobin oxygen affinity. *Physiological Zoology*, **68**, 342–54.

Hammond, K. A., Konarzewski, M., Torres, R. M. & Diamond, J. (1994). Metabolic ceilings under a combination of peak energy demands. *Physiological Zoology*, **67**, 1479–506.

Handasyde, K. A. & Martin, R. W. (1996). Field observations on the common striped possum (*Dactylopsila trivirgata*) in north Queensland. *Wildlife Research*,

23, 755–66.

Harder, J. D. & Fleck, D. W. (1997). Reproductive ecology of New World marsupials. In *Marsupial Biology: Recent Research, New Perspectives*, ed. N. R. Saunders & L. A. Hinds, pp. 175–203. Sydney: University of New South Wales Press.

Harrington, J. (1976). The diet of the swamp wallaby, *Wallabia bicolor*, at Diamond Flat, New South Wales. Unpublished Thesis, University of New England, Armidale.

Harris, P. M., Dellow, D. W. & Broadhurst, R. B. (1985). Protein and energy requirements and deposition in the growing brushtail possum and rex rabbit. *Australian Journal of Zoology*, **33**, 425–36.

Harrop, C. J. F. & Barker, S. (1972). Blood chemistry and gastro-intestinal changes in the developing red kangaroo, *Megaleia rufa* (Desmarest). *Australian Journal of Experimental Biology and Medical Science*, **50**, 245–9.

Harrop, C. J. F. & Degabriele, R. (1976). Digestion and nitrogen metabolism in the koala, *Phascolarctos cinereus*. *Australian Journal of Zoology*, **24**, 201–15.

Hartman, L., Shorland, F. B. & McDonald, I. R. (1955). The trans-unsaturated acid contents of fats of ruminants and non-ruminants. *Biochemical Journal*, **61**, 603–7.

Hayssen, V. & Lacy, R. C. (1985). Basal metabolic rates in mammals: taxonomic differences in the allometry of BMR and body mass. *Comparative Biochemistry and Physiology*, **81A**, 741–54.

Heighway, F. R. (1939). The anatomy of *Hypsiprymnodon moschatus*. Unpublished Thesis, University of Sydney, Sydney.

Heinsohn, G. E. (1966). Ecology and reproduction of the Tasmanian bandicoots (*Perameles gunnii* and and *Isoodon obesulus*). *University of California Publications in Zoology*, **80**, 1–107.

Henning, S. J. & Hird, F. J. R. (1970). Concentrations and metabolism of volatile fatty acids in the fermentative organs of two species of kangaroo and the guinea-pig. *British Journal of Nutrition*, **24**, 145–55.

Henning, S. J. & Hird, F. J. R. (1972). Diurnal variations in the concentrations of volatile fatty acids in the alimentary tracts of wild rabbits, *British Journal of Nutrition*, **27**, 57–64.

Henry, S. R. & Craig, S. A. (1984). Diet, ranging behaviour and social organisation of the yellow-bellied glider (*Petaurus australis* Shaw) in Victoria. In *Possums and Gliders*, ed. A. P. Smith and I. D. Hume, pp. 331–41. Sydney: Australian Mammal Society.

Hill, W. C. D. & Rewell, R. E. (1954). The caecum of monotremes and marsupials. *Transactions of the Zoological Society of London*, **28**, 185–240.

Hindell, M. A., Handasyde, K. A. & Lee, A. K. (1985). Tree species selection by free-ranging koala populations in Victoria. *Australian Wildlife Research*, **12**, 137–44.

Hinds, D. S. & Macmillen, R. E. (1986). Scaling of evaporative water loss in marsupials. *Physiological Zoology*, **59**, 1–9.

Hingson, D. J. & Milton, G. W. (1968). The mucosa of the stomach of the wombat (*Vombatus hirsutus*) with special reference to the cardiogastric gland. *Proceedings of the Linnean Society of New South Wales*, **93**, 69–75.

Hinks, N. T. & Bolliger, A. (1957). Glucuronuria in a herbivorous marsupial *Trichosurus vulpecula*. *Australian Journal of Experimental Biology and Medical Science*, **35**, 37–44.

Hofmann, R. R. (1973). *The Ruminant Stomach*. Nairobi: East Africa Literature Bureau.

Hollis, C. J. (1984). The ability of two macropods (*Macropus eugenii* and M. *robustus robustus*) of different body size to utilize diets of different fibre content compared with sheep. Unpublished Thesis, University of New England, Armidale.

Hollis, C. J., Robertshaw, J. D. & Harden, R. H. (1986). Ecology of the swamp wallaby (*Wallabia bicolor*) in north-eastern New South Wales. I. Diet. *Australian Wildlife Research*, **13**, 355–65.

Home, E. (1808). An account of some peculiarities in the anatomical structure of the wombat, with observations on the female organs of generation. *Philosophical Transactions of the Royal Society*, **98**, 304–12.

Home, E. (1814). *Lectures on Comparative Anatomy*, vol. I. London: Bulmer & Co.

Honigmann, H. (1941). Studies on nutrition of mammals. Part 3. X. Experiments with Australian silver-grey opossums (*Trichosurus vulpecula*). *Proceedings of the Zoological Society of London, Series A*, **111**, 1–35.

Hope, P. J., Pyle, D., Daniels, C. B., Chapman, I., Horowitz, M., Morley, J. E., Trayhurn, P., Kumaratilake, J. & Wittert, G. (1997). Identification of brown fat and mechanisms for energy balance in the marsupial, *Sminthopsis crassicaudata*. *American Journal of Physiology*, **273**, R161–7.

Hoppe, P. P. (1977). Rumen fermentation and body weight in African ruminants. In *Proceedings of the 13th International Congress of Game Biologists*, ed. T. J. Peterle, pp. 141–50. Washington, DC: Wildlife Society.

Hopper, S. D. & Burbidge, A. A. (1982). Feeding behaviour of birds and mammals on flowers of *Banksia grandis* and *Eucalyptus angulosa*. In *Pollination and Evolution*, ed. J. A. Armstrong, J. M. Powell & A. J. Richards, pp. 67–75. Sydney: Royal Botanic Gardens.

Horn, M. H. & Messer, K. S. (1992). Fish guts as chemical reactors: a model of the alimentary canals of marine herbivorous fishes. *Marine Biology*, **113**, 527–35.

Horsup, A. & Marsh, H. (1992). The diet of the allied rock-wallaby, *Petrogale assimilis*, in the wet-dry tropics. *Wildlife Research*, **19**, 17–33.

Howard, J. (1989). Diet of *Petaurus breviceps* (Marsupialia: Petauridae) in a mosaic of coastal woodland and heath. *Australian Mammalogy*, **12**, 15–21.

Hsu, M., Harder, J. D. & Lustick, S. I. (1988). Seasonal energetics of opossums (*Didelphis virginiana*) in Ohio. *Comparative Biochemistry and Physiology*, **90A**, 441–3.

Huang, C., Ward, S. & Lee, A. K. (1987). Comparison of the diets of the feathertail glider, *Acrobates pygmaeus*, and the eastern pygmy-possum, *Cercartetus nanus* (Marsupialia: Burramyidae) in sympatry. *Australian Mammalogy*, **10**, 47–50.

Hulbert, A. J. & Augee, M. L. (1982). A comparative study of thyroid function in monotreme, marsupial and eutherian mammals. *Physiological Zoology*, **55**, 220–8.

Hulbert, A. J. & Dawson T. J. (1974a). Standard metabolism and body temperature of perameloid marsupials from different environments. *Comparative Biochemistry and Physiology*, **47A**, 583–90.

Hulbert, A. J. & Dawson T. J. (1974b). Water metabolism in perameloid marsupials from different environments. *Comparative Biochemistry and Physiology*, **47A**, 617–33.

Hulbert, A. J. & Gordon, G. (1972). Water metabolism of the bandicoot *Isoodon macrourus* Gould, in the wild. *Comparative Biochemistry and Physiology*, **41A**, 27–34.

Hume, I. D. (1974). Nitrogen and sulphur retention and fibre digestion by euros, red kangaroos and sheep. *Australian Journal of Zoology*, **22**, 13–23.

Hume, I. D. (1977a). Production of volatile fatty acids in two species of wallaby and in sheep. *Comparative Biochemistry and Physiology*, **56A**, 299–304.

Hume, I. D. (1977b). Maintenance nitrogen requirements of the macropod marsupials *Thylogale thetis*, red-necked pademelon, and *Macropus eugenii*, tammar wallaby. *Australian Journal of Zoology*, **25**, 407–17.

Hume, I. D. (1978). Evolution of the Macropodidae digestive system. *Australian Mammalogy*, **2**, 37–42.

Hume I. D. (1982). *Digestive Physiology and Nutrition of Marsupials*. Cambridge: Cambridge University Press.

Hume, I. D. (1986). Nitrogen metabolism in the parma wallaby, *Macropus parma*. *Australian Journal of Zoology*, **34**, 147–55.

Hume, I. D. (1989). Optimal digestive strategies in mammalian herbivores. *Physiological Zoology*, **62**, 1145–63.

Hume, I. D. (1995). Flow dynamics of digesta and colonic fermentation. In *Physiological and Clinical Aspects of Short-Chain Fatty Acids*, ed. J. H. Cummings, J. L. Rombeau & T. Sakata, pp. 119–31. Cambridge: Cambridge University Press.

Hume, I. D., Bladon, R. V. & Soran, N. (1996). Seasonal changes in digestive performance of common ringtail possums (*Pseudocheirus peregrinus*) fed *Eucalyptus* foliage. *Australian Journal of Zoology*, **44**, 327–36.

Hume, I. D. & Carlisle, C. H. (1985). Radiographic studies on the structure and function of the gastrointestinal tract of two species of potoroine marsupials. *Australian Journal of Zoology*, **33**, 641–54.

Hume, I. D., Carlisle, C. H., Reynolds, K. & Pass, M. A. (1988). Effects of fasting and sedation on gastrointestinal tract function in two potoroine marsupials. *Australian Journal of Zoology*, **36**, 411–20.

Hume, I. D. & Dellow, D. W. (1980). Form and function of the macropod marsupial digestive tract. In *Comparative Physiology: Primitive Mammals*, ed. K. Schmidt-Nielsen, L. Bolis and C. R. Taylor, pp. 78–89. Cambridge: Cambridge University Press.

Hume, I. D. & Dunning, A. (1979). Nitrogen and electrolyte balance in the wallabies *Thylogale thetis* and *Macropus eugenii* when given saline drinking water. *Comparative Biochemistry and Physiology*, **63A**, 135–9.

Hume, I. D. & Esson, C. (1993). Nutrients, antinutrients and leaf selection by captive koalas (*Phascolarctos cinereus*). *Australian Journal of Zoology*, **41**, 379–92.

Hume, I. D., Jazwinski, E. & Flannery, T. F. (1993). Morphology and function of the digestive tract in New Guinean possums. *Australian Journal of Zoology*, **41**, 85–100.

Hume, I. D., Runcie, M. J. & Caton, J. M. (1997). Digestive physiology of the ground cuscus (*Phalanger gymnotis*), a New Guinean phalangerid marsupial. *Australian Journal of Zoology*, **45**, 561–71.

Hume, I. D. & Warner, A. C. I. (1980). Evolution of microbial digestion in mammals. In *Digestive Physiology and Metabolism in Ruminants*, ed. Y. Ruckebusch and P. Thivend, pp. 665–84. Lancaster: MTP Press.

Hungate, R. E., Phillips, G. D., McGregor, A., Hungate, D. P. & Buechner, H. K. (1959). Microbial fermentation in certain mammals. *Science*, **130**, 1192–4.

Hunsaker, D., II (1977). Ecology of New World marsupials. In *The Biology of Marsupials*, ed. D. Hunsaker II, pp. 95–196. New York: Academic Press.

Hutchinson, G. E. (1957). Concluding remarks. *Cold Spring Harbor Symposium on Quantitative Biology*, **22**, 415–27.

Hutchinson, J. C. D. & Morris S. (1936). The digestibility of dietary protein in the ruminant. I. Endogenous nitrogen excretion on a low nitrogen diet and in starvation. *Biochemical Journal*, **30**, 1682–94.

Ingleby, S. (1991). Distribution and status of the northern nailtail wallaby, *Onychogalea unguifera* (Gould 1841). *Wildlife Research*, **18**, 655–76.

Ingleby, S. and Westoby, M. (1992). Habitat requirements of the spectacled hare-wallaby (*Lagorchestes conspicillatus*) in the Northern Territory and Western Australia. *Wildlife Research*, **19**, 721 41.

Ingleby, S., Westoby, M. & Latz, P. K. (1989). Habitat requirements of the Northern nailtail wallaby, *Onychogalea unguifera* (Marsupialia: Macropodoidea) in the Northern Territory and Western Australia. In *Kangaroos, Wallabies and Rat-kangaroos*, ed. G. Grigg, P. Jarman & I. Hume, pp. 767 82. Sydney: Surrey Beatty.

Inns, R. W. (1980). Ecology of the Kangaroo Island wallaby, *Macropus eugenii* (Desmarest) in Flinders Chase National Park, Kangaroo Island. Ph.D. Thesis, University of Adelaide, Adelaide.

Janis, C. (1976). The evolutionary strategy of the Equidae and the origins of rumen and cecal digestion. *Evolution*, **30**, 757–74.

Jarman, P. J. (1994). The eating of seedheads by species of Macropodidae. *Australian Mammalogy*, **17**, 51–63.

Jarman, P. J. & Phillips, C. M. (1989). Diets in a community of macropod species. In *Kangaroos, Wallabies and Rat-kangaroos*, ed. G. Grigg, P. Jarman & I. Hume. Sydney: Surrey Beatty.

Jarman, P. J., Phillips, C. M. & Rabbidge, J. J. (1991). Diets of black-striped wallabies in New South Wales. *Wildlife Research*, **18**, 403–12.

Johnson, C. N. (1991). Utilisation of habitat by the northern hairy-nosed wombat *Lasiorhinus krefftii*. *Journal of Zoology, London*, **225**, 495–507.

Johnson, C. N. (1994a). Nutritional ecology of a mycophagous marsupial in relation to production of hypogeous fungi. *Ecology*, **75**, 2015–21.

Johnson, C. N. (1994b). Mycophagy and spore dispersal by a rat-kangaroo: consumption of ectomycorrhizal taxa in relation to their abundance. *Functional Ecology*, **8**, 464–8.

Johnson, C. N. (1994c). Distribution of feeding activity of the Tasmanian bettong

(*Bettongia gaimardi*) in relation to vegetation patterns. *Wildlife Research*, **21**, 249–55.

Johnson, C. (1994d). Fruiting of hypogeous fungi in dry sclerophyll forest in Tasmania, Australia: seasonal variation and annual production. *Mycological Research*, **98**, 1173–82.

Johnson, C. N. (1998). The evolutionary ecology of wombats. In *Wombats*, ed. R. T. Wells & P. A. Pridmore. Sydney: Surrey Beatty.

Johnson, K. (1980a). Diet of the bilby, *Macrotis lagotis* in the western desert areas of central Australia. *Bulletin of the Australian Mammal Society*, **6**, 46–7 (Abstract).

Johnson, K. A. (1977). Ecology and management of the red-necked pademelon, *Thylogale thetis*, on the Dorrigo Plateau of northern New South Wales. Ph.D. Thesis, University of New England, Armidale.

Johnson, K. A. (1980b). Spatial and temporal use of habitat by the red-necked pademelon, *Thylogale thetis* (Marsupialia: Macropodidae). *Australian Wildlife Research*, **7**, 157–66.

Johnson, K. A. (1995). The lesser bilby. In *The Mammals of Australia*, ed. R. Strahan, pp. 189–90. Chatswood: Reed Books.

Johnson, K. A. & Rose, R. (1995). Tasmanian pademelon. In *The Mammals of Australia*, ed. R. Strahan, pp. 394–6. Chatswood: Reed Books.

Johnson, P. M. & Strahan, R. (1982). A further description of the musky rat-kangaroo, *Hypsiprymnodon moschatus* Ramsay, 1876 (Marsupialia, Potoroidae), with notes on its biology. *The Australian Zoologist*, **21**, 27–46.

Johnstone, J. (1898). On the gastric glands of the Marsupialia. *Journal of the Linnean Society*, **27**, 1–14.

Jolly, S. E., Morriss, G. A., Scobie, S. & Cowan, P. E. (1996). Composition of milk of the common brushtail posssum, *Trichosurus vulpecula* (Marsupialia: Phalangeridae): concentrations of elements. *Australian Journal of Zoology*, **44**, 479–86.

Jones, C. J. & Geiser, F. (1992). Prolonged and daily torpor in the feathertail glider, *Acrobates pygmaeus* (Marsupialia: Acrobatidae). *Journal of Zoology (London)*. **227**, 101–8.

Jones, M. E. (1995). Guild structure of the large marsupial carnivores in Tasmania. Ph.D. Thesis, University of Tasmania, Hobart.

Jones, M. (1997). Character displacement in Australian dasyurid carnivores: size relationships and prey size patterns. *Ecology*, **78**, 2569–87.

Jones, M. E. & Barmuta, L. A. (1998). Diet overlap and relative abundance of sympatric dasyurid carnivores: a hypothesis of competition. *Journal of Animal Ecology*, **67**, 410–21.

Jones, M. E. & Stoddart, D. M. (1998). Reconstruction of the predatory behaviour of the extinct marsupial thylacine. *Journal of Zoology (London)*, **246**, 239–46.

Jones, W. T. & Mangan, J. L. (1977). Complexes of the condensed tannins of sainfoin (*Onobrychis vicifolia* Scop.) with fraction 1 leaf protein and with submaxillary mucoprotein and their reversal by polyethylene glycol and pH. *Journal of the Science of Food and Agriculture*, **28**, 126–36.

Kakulas, B. A. (1961). Myopathy affecting the Rottnest quokka (*Setonix brachyurus*) reversed by α-tocopherol. *Nature*, **191**, 402–3.

Kakulas, B. A. (1963a). Trace quantities of selenium ineffective in the prevention of nutritional myopathy in the Rottnest quokka (*Setonix brachyurus*). *Australian Journal of Science*, **25**, 313–14.

Kakulas, B. A. (1963b). Influence of the size of enclosure on the development of myopathy in the captive Rottnest quokka. *Nature*, **198**, 673–4.

Kakulas, B. A. (1966). Regeneration of skeletal muscle in the Rottnest quokka. *Australian Journal of Experimental Biology and Medical Science*, **44**, 673–88.

Kaldor, I. & Ezekiel, E. (1962). Iron content of mammalian breast milk; measurements in the rat and in a marsupial. *Nature*, **196**, 175.

Karasov, W. H. (1992). Daily energy expenditure and cost of activity in mammals. *American Zoologist*, **32**, 238–48.

Karasov, W. H. & Hume, I. D. (1997). Vertebrate gastrointestinal system. In *Handbook of Physiology*. Section 13: *Comparative Physiology*, col. I, ed. W. H. Dantzler, pp. 409–80. New York: Oxford University Press.

Kavanagh, R. P. (1987). Forest phenology and its effect on foraging behaviour and selection of habitat by the yellow-bellied glider, *Petaurus australis*, Shaw. *Australian Wildlife Research*, **14**, 371–84.

Kavanagh, R. P. & Lambert, M. J. (1990). Food selection by the greater glider, *Petauroides volans*: is foliar nitrogen a determinant of habitat quality? *Australian Wildlife Research*, **17**, 285–99.

Kavanagh, R. P. & Rohan Jones, W. G. (1982). Calling behaviour of the yellow-bellied glider, *Petaurus australis* Shaw (Marsupialia: Petauridae). *Australian Mammalogy*, **5**, 95–111.

Kay, R. F. & Hylander, W. L. (1978). The dental structure of mammalian folivores with special reference to Primates and Phalangeroidea (Marsupialia). In *The Ecology of Arboreal Folivores*, ed. G. G. Montgomery, pp. 173–91. Washington, DC: Smithsonian Institution Press.

Kempton, T. J. (1972). The efficiency of utilisation of a roughage diet and the rate of passage of digesta in the grey kangaroo, *Macropus giganteus* (Shaw) and the Merino sheep. Unpublished Thesis, University of New England, Armidale, Australia.

Kempton, T. J., Murray, A. M. & Leng, R. A. (1976). Rates of production of methane in the grey kangaroo and sheep. *Australian Journal of Biological Sciences*, **29**, 209–14.

Kenagy, G. J., Masman, D., Sharbaugh, S. M. & Nagy, K. A. (1990). Energy expenditures during lactation in relation to litter size in free-living golden-mantled ground squirrels. *Journal of Animal Ecology*, **59**, 73–88.

Kennedy, P. M. & Heinsohn, G. E. (1974). Water metabolism of two marsupials – the brush-tailed possum, *Trichosurus vulpecula* and the rock-wallaby, *Petrogale inornata* in the wild. *Comparative Biochemistry and Physiology*, **47A**, 829–34.

Kennedy, P. M. & Hume, I. D. (1978). Recycling of urea nitrogen to the gut of the tammar wallaby (*Macropus eugenii*). *Comparative Biochemistry and Physiology*, **61A**, 117–21.

Kennedy, P. M. & Macfarlane, W. V. (1971). Oxygen consumption and water turnover of the fat-tailed marsupials *Dasycercus cristacauda* and *Sminthopsis crassicaudata*. *Comparative Biochemistry and Physiology*, **40**, 723–32.

Kennedy, P. M. & Milligan, L. P. (1980). The degradation and utilization of endogenous urea in the gastrointestinal tract of ruminants: A review. *Canadian*

Journal of Animal Science, **60**, 205–21.

Kerle, J. A. (1984). Variation in the ecology of *Trichosurus*: its adaptive significance. In *Possums and Gliders*, ed. A. P. Smith & I. D. Hume, pp. 115–28. Sydney: Australian Mammal Society.

Kerle, J. A. & Winter, J. W. (1995). Rock ringtail possum. In *The Mammals of Australia*, ed. R. Strahan, pp. 242–3. Sydney: Reed Books.

Kerry, K. R. (1969). Intestinal disaccharidase activity in a monotreme and eight species of marsupials (with an added note on the disaccharidases of five species of sea birds). *Comparative Biochemistry and Physiology*, **52A**, 235–46.

King, D. R., Twigg, L. E. & Gardner, J. L. (1989). Tolerance to sodium monofluoroacetate in dasyurids from Western Australia. *Australian Wildlife Research*, **16**, 131–40.

Kingdon, J. (1990). Possums and conservation in Northern Queensland. Unpublished report on a project for the Winidred V. Scott Bequest and CSIRO Division of Wildlife and Ecology Tropical Forest Research Centre, Atherton, Queensland.

Kinnear, A. & Shield, J. W. (1975). Metabolism and temperature regulation in marsupials. *Comparative Biochemistry and Physiology*, **52A**, 235–45.

Kinnear, J. E. (1970). Nitrogen metabolism of macropods with special reference to the tammar (*Macropus eugenii*). Ph.D. Thesis, University of Western Australia, Perth.

Kinnear, J. E., Cockson, A., Christensen, P. & Main, A. R. (1979). The nutritional biology of the ruminants and ruminant-like mammals – a new approach. *Comparative Biochemistry and Physiology*, **64A**, 357–65.

Kinnear, J. E. & Main, A. R. (1975). The recycling of urea nitrogen by the wild tammar wallaby (*Macropus eugenii*) – a 'ruminant-like' marsupial. *Comparative Biochemistry and Physiology*, **51A**, 793–810.

Kinnear, J. E., Purohit, K. G. & Main, A. R. (1968). The ability of the tammar wallaby (*Macropus eugenii*, Marsupialia) to drink sea water. *Comparative Biochemistry and Physiology*, **25**, 761–82.

Kirkpatrick, T. H. (1965). Studies of Macropodidae in Queensland. I. Food preferences of the grey kangaroo (*Macropus major*, Shaw). *Queensland Journal of Agricultural Science*, **22**, 89–93.

Kirsch, J. A. W. & Waller, P. F. (1979). Notes on the trapping and behaviour of the Caenolestidae (Marsupialia). *Journal of Mammalogy*, **60**, 390–5.

Kitchener, D. J. (1981). Breeding, diet and habitat preference of *Phascogale calura* (Gould, 1844) (Marsupialia: Dasyuridae) in the southern wheat belt, Western Australia. *Records of the Western Australian Museum*, **9**, 173–86.

Kleiber, M. (1961). *The Fire of Life*. New York: Wiley.

Köhler, P. (1985). The strategies of energy conservation in helminths. *Molecular and Biochemical Parasitology*, **17**, 1–18.

Koteja, P. (1991). On the relation between basal and field metabolic rates in birds and mammals. *Functional Ecology*, **5**, 56–64.

Krause, W. J. (1972). The distribution of Brunner's glands in 55 marsupial species native to the Australian region. *Acta Anatomica*, **82**, 17–33.

Krause, W. J. & Leeson, C. R. (1969). Studies of Brunner's glands in the opossum. I. Adult morphology. *American Journal of Anatomy*, **126**, 255–74.

Krause, W. J. & Leeson, C. R. (1973). The stomach gland patch of the koala (*Phascolarctos cinereus*). *Anatomical Record*, **176**, 475–88.

Krause, W. J., Yamada, J. & Cutts, J. H. (1985). Quantitative distribution of enteroendocrine cells in the gastrointestinal tract of the adult opossum, *Didelphis virginiana*. *Journal of Anatomy*, **140**, 591–605.

Krefft, G. (1865). Vertebrata of the Lower Murray. *Transactions of the Philosphical Society of New South Wales, 1862–1865*: 12–14.

Krockenberger, A. K. (1993). Energetics and nutrition during lactation in the koala, *Phascolarctos cinereus*, Ph.D. Thesis, University of Sydney.

Krockenberger, A. K. (1996). Composition of the milk of the koala, *Phascolarctos cinereus*, an arboreal folivore. *Physiological Zoology*, **69**, 701–18.

Krockenberger, A. K., Hume, I. D. & Cork, S. J. (1998). Production of milk and nutrition of the dependent young of free-ranging koalas (*Phascolarctos cinereus*). *Physiological Zoology*, **71**, 45–56.

Lamont, B. B., Ralph, C. S. & Christensen, P. E. S. (1985). Mycophagous marsupials as dispersal agents for ectomycorrhizal fungi on *Eucalyptus calophylla* and *Gastrolobium bilobum*. *New Phytologist*, **101**, 651–6.

Landsberg, J. (1987). Feeding preferences of common brushtail possums, *Trichosurus vulpecula*, on seedlings of a woodland eucalypt. *Australian Wildlife Research*, **14**, 361–9.

Landwehr, G. O., Richardson, K. C. & Wooller, R. D. (1990). Sugar preferences of honey possums, *Tarsipes rostratus* (Marsupialia: Tarsipedidae) and western pygmy-possums, *Cercartetus concinnus* (Marsupialia: Burramyidae). *Australian Mammalogy*, **13**, 5–10.

Langer, P. (1979). Functional anatomy and ontogenetic development of the stomach in the macropodine species *Thylogale stigmatica* and *Thylogale thetis* (Mammalia: Marsupialia). *Zoomorphologie*, **93**, 137–51.

Langer, P. (1980). Anatomy of the stomach in three species of Potorinae (Marsupialia: Macropodidae). *Australian Journal of Zoology*, **28**, 19–31.

Langer, P. (1988). *The Mammalian Herbivore Stomach*. Stuttgart: Fischer.

Langer, P. & Chivers, D. J. (1994). Classification of foods for comparative analysis of the gastro-intestinal tract. In *The Digestive System in Mammals*, ed. D. J. Chivers and P. Langer, pp. 74–86. Cambridge: Cambridge University Press.

Langer, P., Dellow, D. W. & Hume, I. D. (1980). Stomach structure and function in three species of macropodine marsupials. *Australian Journal of Zoology*, **28**, 1–18.

Lanyon, J. M. & Sanson, G. D. (1986a). Koala (*Phascolarctos cinereus*) dentition and nutrition. I. Morphology and occlusion of cheekteeth. *Journal of Zoology, London (A)*, **209**, 155–68.

Lanyon, J. M. & Sanson, G. D. (1986b). Koala (*Phascolarctos cinereus*) dentition and nutrition. II. Implications of toothwear in nutrition. *Journal of Zoology, London (A)*, **209**, 169–81.

Lapidge, S. J. (1999). Dietary adjustment of the reintroduced yellow-footed rock-wallaby, *Petrogale xanthopus xanthopus* (Marsupialia: Macropodidae), in the northern Flinders Ranges, South Australia. *Wildlife Research* (in press).

Lawler, I. R., Foley, W. J., Eschler, B., Pass, D. M. & Handasyde, K. (1998b). Intraspecific variation in secondary metabolites determines food intake by folivorous marsupials. *Oecologia*, **116**, 160–9.

Lawler, I. R., Foley, W. J., Pass, D. M. & Eschler, B. M. (1998a). Administration of a 5HT$_3$ receptor antagonist increases the intake of diets containing *Eucalyptus* secondary metabolites by marsupials. *Journal of Comparative Physiology* B, **168**, 611–18.

Lawler, I. R., Foley, W. J., Woodrow, I. E. & Cork, S. J. (1997). The effects of elevated CO_2 atmospheres on the nutritional quality of *Eucalyptus* foliage and its interaction with soil nutrient and light availability. *Oecologia*, **109**, 59–68.

Lay, D. W. (1942). Ecology of the opossum in eastern Texas. *Journal of Mammalogy*, **23**, 147–59.

Leary, T. N. (1982). The significance of pollen in the diet of the sugar glider. Unpublished B. Bat. Res. Thesis. University of New England, Armidale.

Lee, A. K. & Cockburn, A. (1985). *Evolutionary Ecology of Marsupials*. Cambridge: Cambridge University Press.

Lehman, P. (1979). A *Myoporum – Stipa* community grazed by wombats – with particular reference to herbage production. M.Sc. Thesis, University of Adelaide, Adelaide.

Leng, R. A., Corbett, J. J & Brett, D. J. (1968). Rates of production of volatile fatty acids in the rumen of grazing sheep and their relation to ruminal concentrations. *British Journal of Nutrition*, **22**, 57–68.

Leng, R. A. & Leonard, G. J. (1965). Measurement of the rates of production of acetic, propionic and butyric acids in the rumen of sheep. *British Journal of Nutrition*, **19**, 469–84.

Levenspiel, O. (1972). *Chemical Reactor Engineering*, 2nd edn. New York: Wiley.

Lindenmayer, D. B., Boyle, S., Burgman, M. A., McDonald, D. & Tomkins, B. (1994). The sugar and nitrogen content of the gums of *Acacia* species in the mountain ash and alpine ash forests of central Victoria and its potential implications for exudivorous arboreal marsupials. *Australian Journal of Ecology*, **19**, 169–77.

Lintern, S. (1970). Aspects of nitrogen metabolism in the Kangaroo Island wallaby – *Protemnodon eugenii* (Desmarest). Ph.D. Thesis, University of Adelaide, Adelaide.

Lintern, S. M. & Barker, S. (1969). Renal retention of urea in the Kangaroo Island wallaby, *Protemnodon eugenii* (Desmarest). *Australian Journal of Experimental Biology and Medical Science*, **47**, 243–50.

Lintern-Moore, S. (1973a). Incorporation of dietary nitrogen into microbial nitrogen in the forestomach of the Kangaroo Island wallaby *Protemnodon eugenii* (Desmarest). *Comparative Biochemistry and Physiology*, **44A**, 75–82.

Lintern-Moore, S. (1973b). Utilisation of dietary urea by the Kangaroo Island wallaby *Protemnodon eugenii* (Desmarest). *Comparative Biochemistry and Physiology*, **46A**, 345–51.

London, C. J. (1981). The microflora associated with the caecum of the koala (*Phascolarctos cinereus*). M.Sc. Thesis, La Trobe University, Melbourne.

Lönnberg, E. (1902). On some remarkable digestive adaptations in diprotodont marsupials. *Proceedings of the Zoological Society of London*, **73**, 12–31.

Loudon, A., Rothwell, N. & Stock, M. (1985). Brown fat, thermogenesis and physiological birth in a marsupial. *Comparative Biochemstry and Physiology*,

81A, 815–19.

Low, B. S., Birk, E., London, C. & Low, W. A. (1973). Community utilization by cattle and kangaroos in mulga near Alice Springs, Northern Territory. *Tropical Grasslands,* **7,** 149–56.

Lundie-Jenkins, G. W. (1984). Digesta passage rates and pollen digestion in the two small nectivorous arboreal marsupials, *Tarsipes spencerae* and *Acrobates pygmaeus.* Unpublished B. Nat. Res. Thesis, University of New England, Armidale.

Lundie-Jenkins, G., Phillips, C. M. & Jarman, P. J. (1993). Ecology of the rufous hare-wallaby, *Lagorchestes hirsutus* (Gould) (Marsupialia: Macropodidae), in the Tanami Desert, Northern Territory. II. Diet and feeding strategy. *Wildlife Research,* **20,** 477–94.

Macfarlane, W. V. (1975). Ecophysiology of water and energy in desert marsupials and rodents. In *Rodents in Desert Environments,* ed. I. Prakesh & P. K. Ghosh, pp. 389–96. The Hague: Junk Publishers.

Mackenzie, W. C. (1918). *The Gastro-Intestinal Tract in Monotremes and Marsupials.* Melbourne: Critchley Parker.

Mackowski, C. M. (1988). Characteristics of eucalypts incised for sap by the yellow-bellied glider, *Petaurus australis,* Shaw (Marsupialia: Petauridae), in northeastern New South Wales. *Australian Mammalogy,* **11,** 5–13.

MacMillen, R. E. & Lee, A. K. (1969). Water metabolism of Australian hopping mice. *Comparative Biochemistry and Physiology,* **28,** 493–514.

MacMillen, R. E. & Nelson, J. E. (1969). Bioenergetics and body size in dasyurid marsupials. *American Journal of Physiology,* **217,** 1246–51.

Main, A. R. (1970). CMeasures of wellbeing in populations of herbivorous macropod marsupials. In *Dynamics of Populations,* ed. P. J. den Boer & G. R. Gradwell, pp. 159–73. Wageningen: Centre for Agricultural Publishing and Documentation.

Malajczuk, N., Trappe, J. M. & Molina, R. (1987). Interrelationships among some ectomycorrhizal trees, hypogeous fungi and small mammals: Western Australian and northwestern American parallels. *Australian Journal of Ecology,* **12,** 53–5.

Maller, O., Clark, J. M. & Kare, M. R. (1965). Short term caloric regulations in the adult opossum (*Didelphis virginiana*). *Proceedings of the Society for Experimental Biology and Medicine,* **118,** 275–7.

Mallett, K. J. & Cooke, B. D. (1986). *The Ecology of the Common Wombat in South Australia.* Adelaide: Nature Conservation Society of South Australia.

Mandels, M. & Reese, E. T. (1965). Inhibition of cellulases. *Annual Review of Phytopathology,* **3,** 85–102.

Mann, N. J., Johnson, L. G., Warrick, G. E. & Sinclair, A. J. (1995). The arachidonic acid content of the Australian diet is lower than previously estimated. *Journal of Nutrition,* **125,** 2528–35.

Manozzi-Torini, L. (1976). *Manuale di tartuficoltura: tartufi e tartuficoltura in Italia.* Bologna: Edizioni Agricole.

Mansergh, I. (1984). The mountain pygmy-possum (*Burramys parvus*), (Broom): A review. In *Possums and Gliders,* ed. A. P. Smith & I. D. Hume, pp. 413–16. Sydney: Australian Mammal Society.

Mansergh, I., Baxter, B., Scotts, D., Brady, T. & Jolley, D. (1990). Diet of the mountain pygmy-possum, *Burramys parvus* (Marsupialia: Burramyidae) and other small mammals in the alpine environment at Mt. Higginbotham, Victoria. *Australian Mammalogy*, **13**, 167–77.

Mansergh, I. M. & Broome, L. S. (1994). *The Mountain Pygmy-possum of the Australian Alps*. Sydney: University of NSW Press.

Mantell, C. L. (1949). The water soluble gums – their botany, sources and utilization. *Economic Botany*, **3**, 3–31.

Marples, T. J. (1973). Studies in the marsupial glider, *Schoinobates volans* (Kerr). IV. Feeding biology. *Australian Journal of Zoology*, **21**, 213–16.

Marquis, R. J. & Batzli, G. O. (1989). Influence of chemical factors on palatability of forage to voles. *Journal of Mammalogy*, **70**, 503–11.

Marshall, L. G. (1978). Evolution of the Borhyaenidae, extinct South American predaceous marsupials. *University of California Publications in Geological Science*, **117**, 1–89.

Martin, R. (1992). An ecological study of Bennett's tree-kangaroo (*Dendrolagus bennettianus*). Project 116, World Wide Fund for Nature, 67 pp.

Martin, R. W. (1981). Age-specific fertility in three populations of the koala, *Phascolarctos cinereus*, Goldfuss, in Victoria. *Australian Wildlife Research*, **8**, 275–83.

Martin, R. W. (1985). Overbrowsing and decline of a population of the koala, *Phascolarctos cinereus*, in Victoria. I. Food preferences and food tree defoliation. *Australian Wildlife Research*, **12**, 355–65.

Martinez del Rio, C. & Karasov, W. H. (1990). Digestion strategies in nectar- and fruit-eating birds and the composition of plant rewards. *American Naturalist*, **136**, 618–37.

Maser, C., Trappe, J. M. & Nussbaum, R. A. (1978). Fungal–small mammal interrelationships with emphasis on Oregon coniferous forests. *Ecology*, **59**, 799–809.

Maxwell, S., Burbidge, A. A. & Morris, K. D. (1996). *The 1996 Action Plan for Australian Marsupials and Monotremes*. Wildlife Australia, Canberra.

May, E. L. (1997). The roles of shivering and nonshivering thermogenesis in cold acclimation in the kowari, *Dasyuroides byrnei*. Ph.D. Thesis, University of New South Wales, Sydney.

Maynes, G. M. (1974). Occurrence and field recognition of *macropus parma*. *Australian Zoologist*, **18**, 72–87.

McAllan, B. M., Roberts, J. R. & Barboza, P. (1995). The kidney structure of the common wombat (*Vombatus ursinus*) and the hairy-nosed wombat (*Lasiorhinus latifrons*). *Australian Journal of Zoology*, **43**, 181–91.

McAllan, B. M., Roberts, J. R. & O'Shea, T. (1996). Seasonal changes in the renal morphometry of *Antechinus stuartii* (Marsupialia: Dasyuridae). *Australian Journal of Zoology*, **44**, 337–54.

McArthur, C. & Sanson, G. D. (1988). Toothwear in eastern grey kangaroos (*Macropus giganteus*) and western grey kangaroos (*Macropus fuliginosus*), and its potential influence on diet selection, digestion and population parameters. *Journal of Zoology, London*, **215**, 491–504.

McArthur, C. & Sanson, G. D. (1991). Effects of tannins on digestion in the common ringtail possum (*Pseudocheirus peregrinus*), a specialised marsupial folivore. *Journal of Zoology, London*, **225**, 233–51.

McArthur, C. & Sanson, G. D. (1993a). Nutritional effects and costs of a tannin in a grazing and a browsing macropodid marsupial herbivore. *Functional Ecology*, **7**, 690–6.

McArthur, C. & Sanson, G. D. (1993b). Nutritional effects and costs of a tannin in two marsupial arboreal folivores. *Functional Ecology*, **7**, 697–703.

McArthur, C., Sanson, G. D. & Beal, A. M. (1995). Salivary proline-rich proteins in mammals: roles in oral homeostasis and counteracting dietary tannin. *Journal of Chemical Ecology*, **21**, 663–91.

McClelland, K. (1997). Diet and digestive functions of the northern brown bandicoot, *Isoodon macrourus*. B.Sc. Honours Thesis, University of Sydney.

McDonald, I. R., Lee, A. K., Bradley, A. J. & Than, K. A. (1981). Endocrine changes in dasyurid marsupials with differing mortality patterns. *General and Comparative Endocrinology*, **44**, 292–301.

McDonald, P., Edwards, R. A., Greenhalgh, J. F. D. & Morgan, C. A. (1995). *Animal Nutrition*, 5th edn. Harlow: Longman.

McIlroy, J. C. (1973). Aspects of the ecology of the common wombat *Vombatus ursinus* (Shaw 1800). Ph.D. Thesis, Australian National University, Canberra.

McIlwee, A. P. & Johnson, C. N. (1998). Nutritional value of fungus to three marsupial herbivores in *Eucalyptus* forests, revealed by stable isotope analysis. *Functional Ecology*, **12**, 223–31.

McIntosh, D. L. (1966). The digestibility of two roughages and the rates of passage of their residues by the red kangaroo, *Megaleia rufa* (Desmerest) and the merino sheep. *CSIRO Wildlife Research*, **11**, 125–35.

McKenzie, R. A. (1978). The caecum of the koala, *Phascolarctos cinereus*. Light, scanning and transmission electron microscopic observations on its epithelium and flora. *Australian Journal of Zoology*, **26**, 249–56.

McLean, S., Foley, W. J., Davies, N. W., Brandon, S., Duo, L. & Blackman, A. J. (1993). Metabolic fate of dietary terpenes from *Eucalyptus radiata* in common ringtail possum (*Pseudocheirus peregrinus*). *Journal of Chemical Ecology*, **19**, 1625–43.

McManus, M. E., McGeachie, J. K. & Ilett, K. F. (1978). Development of the hepatic mixed function oxidase system in a marsupial, the quokka (*Setonix brachyurus*). *Toxicology and Applied Pharmacology*, **46**, 117–24.

McNab, B. K. (1978). The comparative energetics of neotropical marsupials. *Journal of Comparative Physiology*, **125**, 115–28.

McNab, B. K. (1984). Physiological convergence amongst ant-eating and termite-eating mammals. *Journal of Zoology, London*, **203**, 485–510.

McNab, B. K. (1986). Food habits, energetics and the reproduction of marsupials. *Journal of Zoology, London (A)*, **208**, 595–614.

McNab, B. K. (1988a). Energy conservation in a tree-kangaroo (*Dendrolagus matschiei* and the red panda (*Ailurus fulgens*). *Physiological Zoology*, **61**, 280–92.

McNab, B. K. (1988b). Complications inherent in scaling the basal rate of metabolism in mammals. *Quarterly Review of Biology*, **63**, 25–54.

McNab, B. K. (1992). Energy expenditure: a short history. In *Mammalian Energetics*, ed. T. E. Tomasi & T. H. Horton, pp. 1–15. Ithaca: Cornell

University Press.

Mead, R. J., Oliver, A. J. & King, D. R. (1979). Metabolism and defluorination of fluoroacetate in the brush-tailed possum (*Trichosurus vulpecula*). *Australian Journal of Biological Sciences*, **32**, 15–26.

Mead, R. J., Oliver, A. J., King, D. R. & Hubach, P. H. (1985). The co-evolutionary role of fluoroacetate in plant–animal interactions in Australia. *Oikos*, **44**, 55–60.

Menkhorst, P. W. & Collier, M. (1988). Diet of the squirrel glider, *Petaurus norfolcensis* (Marsupialia: Petauridae), in Victoria. *Australian Mammalogy*, **11**, 109–16.

Menzies, J. I. (1989). Observations on a captive forest wallaby (*Dorcopsis luctuosa*) colony. In *Kangaroos, Wallabies and Rat-kangaroos*, ed. G. Grigg, P. Jarman & I. Hume, pp. 629–31. Sydney: Surrey Beatty.

Merchant, J., Green, B., Messer, M. & Newgrain, K. (1989). Milk composition in the red-necked wallaby, *Macropus rufogriseus banksianus* (Marsupialia). *Comparative Biochemistry and Physiology*, **93A**, 483–8.

Merchant, J. C. & Libke, J. A. (1988). Milk composition in the northern brown bandicoot, *Isoodon macrourus* (Peramelidae, Marsupialia). *Australian Journal of Biological Sciences*, **41**, 495–504.

Merchant, J. C., Libke, J. A. & Smith, M. J. (1994). Lactation and energetics of growth in the brush-tailed bettong, *Bettongia penicillata* (Marsupialia: Potoroidae) in captivity. *Australian Journal of Zoology*, **42**, 267–77.

Meserve, P. L. (1981). Trophic relationships among small mammals in a Chilean semiarid thorn scrub community. *Journal of Mammalogy*, **62**, 304–14.

Messer, M., Crisp, E. A. & Czolij, R. (1989). Lactose digestion in suckling macropodids. In *Kangaroos, Wallabies and Rat-kangaroos*, ed. G. Grigg, P. Jarman and I. Hume, pp. 217–21. Sydney: Surrey Beatty.

Messer, M. & Green, B. (1979). Milk carbohydrates. II. Quantitative and qualitative changes in milk carbohydrates during lactation in the tammar wallaby (*Macropus eugenii*). *Australian Journal of Biological Sciences*, **32**, 519–31.

Miller, T. & Bradshaw, S. D. (1979). Adrenocortical function in a field population of a macropodid marsupial (*Setonix brachyurus*, Quoy & Gaimard). *Journal of Endocrinology*, **82**, 152–70.

Milton, G. W., Hingson, D. J. & George, E. P. (1968). The secretory capacity of the stomach of the wombat (*Vombatus hirsutus*) and the cardiogastric gland. *Proceedings of the Linnean Society of New South Wales*, **93**, 60–8.

Minchin, A. K. (1937). Notes on the rearing of a young koala (*Phascolarctos cinereus*). *Records of the South Australian Museum*, **6**, 1–6.

Mitchell, H. H. (1962). *Comparative Nutrition of Man and Domestic Animals*. Vol. I. New York: Academic Press.

Mitchell, P. C. (1905). On the intestinal tract of mammals. *Transactions of the Zoological Society of London*, **17**, 437–537.

Mitchell, P. C. (1916). Further observations on the intestinal tracts of mammals. *Proceedings of the Zoological Society of London*, **1916**, 183–251.

Moir, R. J. (1965). The comparative physiology of ruminant-like animals. In *Physiology of Digestion in the Ruminant*, ed. R. W. Dougherty, pp. 1–14.

London: Butterworths.

Moir, R. J. (1968). Ruminant digestion and evolution. In *Handbook of Physiology*, Section 6, *Alimentary Canal*, vol. V (*Bile; Digestion; Ruminal Physiology*), ed. C. F. Code & W. Heidel, pp. 2673–94. Washington, DC: American Physiological Society.

Moir, R. J., Somers, M. & Waring, H. (1956). Studies on marsupial nutrition. I. Ruminant-like digestion in a herbivorous marsupial *Setonix brachyurus* (Quoy and Gaimard). *Australian Journal of Biological Sciences*, **9**, 293–304.

Mole, S., Butler, L. G. & Iason, G. (1990). Defense against dietary tannin in herbivores: a survey for proline rich salivary proteins in mammals. *Biochemical Systematics and Ecology*, **18**, 287–93.

Mollison, B. C. (1960). Food regurgitation in Bennett's wallaby, *Protemnodon rufogrisea* (Desmarest) and the scrub wallaby, *Thylogale billardieri* (Desmarest). *CSIRO Wildlife Research*, **5**, 87–8.

Moore, S. J. & Sanson, G. D. (1995). A comparison of the molar efficiency of two insect-eating mammals. *Journal of Zoology, London*, **235**, 175–92.

Morrison, P. R. & McNab, B. K. (1962). Daily torpor in a Brasilian murine opossum (*Marmosa*). *Comparative Biochemistry and Physiology*, **6**, 57 68.

Morton, S. R. (1978). An ecological study of *Sminthopsis crassicaudata* (Marsupialia: Dasyuridae). II. Behaviour and social organization. *Australian Wildlife Research*, **5**, 163–82.

Morton, S. R. (1980). Field and laboratory studies of water metabolism in *Sminthopsis crassicaudata* (Marsupialia: Dasyuridae). *Australian Journal of Zoology*, **28**, 213–27.

Morton, S. R., Denny, M. J. S., & Read, D. G. (1983). Habitat preferences and diets of sympatric *Sminthopsis crassicaudata* and *S. macroura*. (Marsupialia: Dasyuridae). *Australian Mammalogy*, **6**, 29–34.

Morton, S. R. & Lee, A. K. (1978). Thermoregulation and metabolism in *Planigale maculata* (Marsupialia: Dasyuridae). *Journal of Thermal Biology*, **3**, 117–20.

Moyle, D. I. (1999). Environmental physiology of the eastern quoll in subalpine Tasmania. Ph.D. Thesis, University of Tasmania, Hobart.

Moyle, D. I., Hume, I. D. & Hill, D. M. (1995). Digestive performance and selective digesta retention in the long-nosed bandicoot, *Perameles nasuta*, a small omnivorous marsupial. *Journal of Comparative Physiology*, **B164**, 552–60.

Munks, S. A. (1990). Ecological energetics and reproduction in the common ringtail possum, *Pseudocheirus peregrinus* (Marsupialia: Phalangeroidea). Ph.D. Thesis, University of Tasmania, Hobart.

Munks, S. A., Corkrey, R. & Foley, W. J. (1996). Characteristics of arboreal marsupial habitat in the semi-arid woodlands of northern Queensland. *Wildlife Research*, **23**, 185–95.

Munks, S. A. & Green B. (1995). Energy allocation for reproduction in a marsupial arboreal folivore, the common ringtail possum (*Pseudocheirus peregrinus*). *Oecologia*, **101**, 94–104.

Munks, S. A., Green, B., Newgrain, K. & Messer, M. (1991). Milk composition in the common ringtail possum, *Pseudocheirus peregrinus* (Petauridae: Marsupialia). *Australian Journal of Zoology*, **39**, 403–16.

Murray, P. (1991). The Pleistocene megafauna of Australia. In *Vertebrate Paleontology of Australasia*, ed. P. Vickers-Rich, J. M. Monoghan, R. F. Baird & T. H. Rich, pp. 1071–164. Melbourne: Pioneer Design Studio.

Nagy, K. A. (1980). CO_2 production in animals: analysis of potential errors in the doubly labeled water method. *American Journal of Physiology*, **238**, R466–73.

Nagy, K. A. (1987). Field metabolic rate and food requirement scaling in mammals and birds. *Ecological monographs*, **57**, 111–28.

Nagy K. A. (1994). Field bioenergetics of mammals: what determines field metabolic rates. *Australian Journal of Zoology*, **42**, 43–53.

Nagy, K. A., Bradley A. J. & Morris, K. D. (1990). Field metabolic rates, water fluxes, and feeding rates of quokkas, *Setonix brachyurus*, and tammars, *Macropus eugenii*, in Western Australia. *Australian Journal of Zoology*, **37**, 553–60.

Nagy, K. A., Bradshaw, S. D. & Clay, B. T. (1991). Field metabolic rate, water flux, and food requirements of short-nosed bandicoots, *Isoodon obesulus* (Marsupialia: Peramelidae). *Australian Journal of Zoology*, **39**, 299–305.

Nagy, K. A., Lee A. K., Martin, R. W. & Fleming, M. R. (1988). Field metabolic rate and food requirement of a small dasyurid marsupial, *Sminthopsis crassicaudata*. *Australian Journal of Zoology*, **36**, 293–9.

Nagy, K. A. & Martin, R. W. (1985). Field metabolic rate, water flux, food consumption and time budget of koalas, *Phascolarctos cinereus* (Marsupialia: Phascolarctidae) in Victoria. *Australian Journal of Zoology*, **33**, 655–65.

Nagy, K. A., Meienberger, C., Bradshaw, S. D. & Wooller, R. D. (1995). Field metabolic rate of a small marsupial mammal, the honey possum (*Tarsipes rostratus*). *Journal of Mammalogy*, **76**, 862–6.

Nagy, K. A. & Peterson, C. C. (1988). Scaling of water flux rate in animals. *University of California Publications in Zoology*, **120**, 1–172.

Nagy, K. A., Sanson, G. D. & Jacobsen, N. K. (1990). Comparative field energetics of two macropod marsupials and a ruminant. *Australian Wildlife Research*, **17**, 591–9.

Nagy, K. A., Seymour, R. S., Lee, A. K. & Braithwaite, R. (1978). Energy and water budgets in free-living *Antechinus stuartii* (Marsupialia: Dasyuridae). *Journal of Mammalogy*, **59**, 60–8.

Nagy, K. A. & Suckling, G. C. (1985). Field energetics and water balance of sugar gliders, *Petaurus breviceps* (Marsupialia: Petauridae). *Australian Journal of Zoology*, **33**, 683–91.

Nakajima, Y., Shantha, T. R. & Bourne, G. H. (1969). Histochemical detection of L-gulonolactone: phenazine methosulfate oxidoreductase activity in several mammals with special reference to synthesis of vitamin C in primates. *Histochemie*, **18**, 293–301.

Naughton, J. M., O'Dea, K. & Sinclair, A. J. (1986). Animal foods in traditional Australian aboriginal diets: polyunsaturated and low in fat. *Lipids*, **21**, 684–90.

Newsome, A. E. (1964). Anoestrus in the red kangaroo, *Megaleia rufa* (Desmarest). *Australian Journal of Zoology*, **12**, 9–17.

Newsome, A. E. (1965). The distribution of red kangaroos, *Megaleia rufa* (Desmarest), about sources of persistent food and water in central Australia. *Australian Journal of Zoology*, **13**, 289–99.

Newsome, A. E. (1966). The influence of food on breeding in the red kangaroo in central Australia. *CSIRO Wildlife Research*, **11**, 187–96.

Newsome, A. E. (1975). An ecological comparison of the two arid zone kangaroos of Australia, and their anomalous prosperity since the introduction of ruminant stock to their environment. *Quarterly Review of Biology*, **50**, 389–424.

Nicol, S. C. (1976). Oxygen consumption and nitrogen metabolism in the potoroo, *Potorous tridactylus*. *Comparative Biochemistry and Physiology*, **55A**, 215–18.

Nicol, S. C. (1978). Rates of water turnover in marsupials and eutherians: a comparative review, with new data on the Tasmanian devil. *Australian Journal of Zoology*, **26**, 465–73.

Norbury, G. L. (1988). Microscopic analysis of herbivore diets – a problem and a solution. *Australian Wildlife Research*, **15**, 51–7.

Norbury, G. L. & Sanson, G. D. (1992). Problems with measuring diet selection of terrestrial, mammalian herbivores. *Australian Journal of Ecology*, **17**, 1–7.

Oates, J. F., Whitesides, G. H., Davies, A. G., Waterman, P. G., Green S. M., Dasilva, G. L. & Mole, S. (1990). Determinants of variation in tropical forest primate biomass: new evidence from west Africa. *Ecology*, **71**, 328–43.

O'Brien, T. P., Lomdahl, A. & Sanson, G. (1986). Preliminary microscopic investigations of the digesta derived from foliage of *Eucalyptus ovata* (Labill.) in the digestive tract of the common ringtail possum, *Pseudocheirus peregrinus* (Marsupialia). *Australian Journal of Zoology*, **34**, 157–76.

Obendorf, D. L. (1984a). The macropodid oesophagus. I. Gross anatomical, light microscopic, scanning and transmission electron microscopic observations of its mucosa. *Australian Journal of Zoology*, **32**, 415–35.

Obendorf, D. L. (1984b). The macropodid oesophagus. II. Morphological studies of its adherent bacteria using light and electron microscopy. *Australian Journal of Biological Sciences*, **37**, 99–116.

Obendorf, D. L. (1984c). The macropodid oesophagus. III. Observations on the nematode parasites. *Australian Journal of Zoology*, **32**, 437–45.

Obendorf, D. L. (1984d). The macropodid oesophagus. IV. Observations on the protozoan fauna of the macropodid stomach and oesophagus. *Australian Journal of Biological Sciences*, **37**, 117–22.

Oppel, A. (1896). *Lehrbuch der vergleichenden mikroskopischen Anatomie der Wirbeltiere*. Vol. I, Der Magen, pp. 286–98. Jena: Gustav Fischer.

Osawa, R. (1987). Aspects of digestive strategies in the Macropodidae. Ph.D. Thesis, University of Queensland, Brisbane.

Osawa, R. (1990). Feeding strategies of the swamp wallaby, *Wallabia bicolor*, on North Stradbroke Island, Queensland. I. Composition of diets. *Australian Wildlife Research*, **17**, 615–21.

Osawa, R. (1993). Dietary preference of koalas, *Phascolarctos cinereus* (Marsupialia: Phascolarctidae) for *Eucalyptus* spp. with a specific reference to their simple sugar contents. *Australian Mammalogy*, **16**, 87–9.

Osawa, R., Bird, P. S., Harbrow, D. J., Ogimoto, K. & Seymour, G. J. (1993). Microbiological studies of the intestinal microflora of the koala, *Phascolarctos cinereus*. I. Colonisation of the caecal wall by tannin-protein-complex-degrading enterobacteria. *Australian Journal of Zoology*, **41**, 599–609.

Osawa, R., Blanshard, W. H. & O'Callaghan, P. G. (1993). Microbiological studies of the intestinal microflora of the koala, *Phascolarctos cinereus*. II. Pap, a special maternal faeces consumed by juvenile koalas. *Australian Journal of Zoology*, **41**, 611–20.

Osawa, R., Walsh, T. P. & Cork, S. J. (1993). Metabolism of tannin–protein complex by facultatively anaerobic bacteria isolated from koala faeces. *Biodegradation*, **4**, 91–9.

Osawa, R. & Woodall, P. F. (1990). Feeding strategies of the swamp wallaby *Wallabia biocolor*, on North Stradbroke Island, Queensland. II. Effects of seasonal changes in diet quality on intestinal morphology. *Australian Wildlife Research*, **17**, 623–32.

Osawa, R. & Woodall, P. F. (1992). A comparative study of macroscopic and microscopic dimensions of the intestine in five macropods (Marsupialia: Macropodidae). II. Relationship with feeding habits and fibre content of the diet. *Australian Journal of Zoology*, **40**, 99–113.

Osgood, W. H. (1921). A monographic study of the American marsupial *Caenolestes*. *Field Museum Natural History, Zoology Series*, **14**, 1–162.

Owen, R. (1834) Notes on the anatomy of a new species of kangaroo (*Macropus parryi*, Bern.). *Proceedings of the Zoological Society of London*, Part II, 152.

Owen, R. (1868). *On the Anatomy of Vertebrates*. Vol. III. *Mammals*, pp. 411–20. London: Longmans, Green & Co.

Owen, W. H. & Thomson, J. A. (1965). Notes on the comparative ecology of the common brush-tail and mountain possums in eastern Victoria. *Victorian Naturalist*, **82**, 216–17.

Pahl, L. (1984). Diet preference, diet composition and population density of the ringtail possum (*Pseudocheirus peregrinus cooki*) in several plant communities in southern Victoria. In *Possums and Gliders*, ed. A. P. Smith and I. D. Hume, pp. 252–60. Sydney: Australian Mammal Society.

Pahl, L. I. & Hume, I. D. (1990). Preferences for *Eucalyptus* species of the New England Tablelands and initial development of an artificial diet for koalas. In *Biology of the Koala*, ed. A. K. Lee, K. A. Handasyde & G. D. Sanson, pp. 123–8. Sydney: Surrey Beatty.

Palamara, J., Phakey, P. P., Rachinger, W. A., Sanson, G. D. & Oranus, H. J. (1984). On the nature of the opaque and translucent enamel regions of some Macropodinae (*Macropus giganteus*, *Wallabia bicolor* and *Peradorcus concinna*). *Cell and Tissue Research*, **238**, 329–37.

Parker, D. S. & McMillan, R. T. (1976). The determination of volatile fatty acids in the caecum of the conscious rabbit. *British Journal of Nutrition*, **35**, 365–71.

Parsons, F. G. (1903). On the anatomy of the pig-footed bandicoot (*Chaeropus castanotis*). *Journal of the Linnean Society of London, Zoology*, **29**, 64–80.

Pass, D. M., Foley, W. J. & Bowden, B. (1998). Vertebrate herbivory on *Eucalyptus* – identification of specific feeding deterrents for common ringtail possums (*Pseudocheirus peregrinus*) by bioassay-guided fractionation of *Eucalyptus ovata* foliage. *Journal of Chemical Ecology*, **24**, 1513–27.

Paton, B. C. & Janssens, P. A. (1981). Metabolic changes associated with the switch from a milk to a vegetable diet in the tammar wallaby *Macropus eugenii* (Desmarest). *Comparative Biochemistry and Physiology*, **70B**, 105–13.

Pearson, D. J. (1989). The diet of the rufous hare-wallaby (Marsupialia: Macropodidae) in the Tanami Desert. *Australian Wildlife Research*, **16**, 527–35.

Pehrson, A. (1983). Digestibility and retention of food components in caged mountain hares *Lepus timidus* during the winter. *Holarctic Ecology*, **6**, 395–403.

Penry, D. L. (1993). Digestive constraints on diet selection. In *Diet Selection. An Interdisciplinary Approach to Foraging Behaviour*, ed. R. N. Hughes, pp. 32–55. Oxford: Blackwell Scientific Publications.

Penry, D. L. & Jumars, P. A. (1987). Modeling animal guts as chemical reactors. *American Naturalist*, **129**, 69–96.

Pernetta, J. C. (1976). Diets of the shrews *Sorex araneus* L. and *Sorex minutus* L. in Wytham grassland. *Journal of Animal Ecology*, **45**, 899–912.

Peterson, C. C., Nagy, K. A. & Diamond, J. (1990). Sustained metabolic scope. *Proceedings of the National Academy of Science, USA*, **87**, 2324–8.

Petter, J. J., Schilling, A. & Pariente, G. (1971). Observations écoéthologiques sur deux lémuriens malgaches nocturnes: *Phaner furcifer* et *Microcebus coquereli. La Terre et la Vie*, **25**, 287–327.

Plakke, R. K. & Pfeiffer, E. W. (1965). Influence of plasma urea on urine concentration in the opossum (*Didelphis marsupialis virginiana*), *Nature*, **207**, 866–7.

Plakke, R. K. & Pfeiffer, E. W. (1970). Urea, electrolyte and total solute excretion following water deprivation in the opossum (*Didelphis marsupialis virginiana*). *Comparative Biochemistry and Physiology*, **34**, 325–32.

Pond, C. M. (1978). Morphological aspects and the ecological and mechanical consequences of fat deposition in wild animals. *Annual Review of Ecology and Systematics*, **9**, 519–70.

Poole, W. E., Sharman, G. B., Scott, K. J. & Thompson, S. Y. (1982). Composition of milk from red and grey kangaroos with particular reference to vitamins. *Australian Journal of Biological Sciences*, **35**, 607–15.

Priddel, D. (1986). The diurnal and seasonal patterns of grazing of the red kangaroo, *Macropus rufus*, and the western grey kangaroo, *M. fuliginosus*. *Australian Wildlife Research*, **13**, 113–20.

Prince, R. I. T. (1976). Comparative studies of aspects of nutritional and related physiology in macropod marsupials. Ph.D. Thesis, University of Western Australia, Perth.

Prince, R. I. T. (1995). Banded hare-wallaby. In *The Mammals of Australia*, ed. R. Strahan, pp. 406–8. Chatswood, New South Wales: Reed Books.

Proctor-Gray, E. (1984). Dietary ecology of the coppery brushtail possum, green ringtail possum and Lumholtz's tree-kangaroo in north Queensland. In *Possums and Gliders*, ed. A. P. Smith and I. D. Hume, pp. 129–35. Sydney: Australian Mammal Society.

Provenza, F. D. (1995). Postingestive feedback as an elementary determinant of food preference and intake in ruminants. *Journal of Range Management*, **48**, 2–17.

Provenza, F. D. (1996). Acquired aversions as the basis for varied diets of ruminants foraging on rangelands. *Journal of Animal Science*, **74**, 2010–20.

Provenza, F. D., Pfister, J. A. & Cheney, C. D. (1992). Mechanisms of learning in diet selection with reference to phytotoxicosis in herbivores. *Journal of Range Management*, **45**, 36–45.

Purohit, K. G. (1971). Absolute duration of survival of tammar wallaby (*Macropus eugenii*, Marsupialia) on sea water and dry food. *Comparative Biochemistry and Physiology*, **39A**, 473–81.

Quin, D. G. (1988). Observations on the diet of the southern brown bandicoot, *Isoodon obesulus* (Marsupialia: Peramelidae), in southern Tasmania. *Australian Mammalogy*, **11**, 15–25.

Quin, D. G. (1995). Population ecology of the squirrel glider (*Petaurus norfolcensis*) and the sugar glider (*P. breviceps*) (Marsupialia: Petauridae) at Limeburners Creek, on the central north coast of New South Wales. *Wildlife Research*, **22**, 471–505.

Quin, D., Goldingay, R., Churchill, S. & Engel, D. (1996). Feeding behaviour and food availability of the yellow-bellied glider in north Queensland. *Wildlife Research*, **23**, 637–46.

Ramsay, B. A. (1966). Field nutrition in the Rottnest quokka. M. Sc. Thesis, University of Western Australia, Perth.

Rand, A. L. (1937). Results of the Archbold Expeditions. No. 17. Some original observations on the habits of *Dactylopsila trivirgata*, Gray. *American Museum Novitates*, **957**: 1–7.

Read, D. G. (1987a). Diets of sympatric *Planigale gilesi* and *P. tenuirostris* (Marsupialia: Dasyuridae): relationships of season and body size. *Australian Mammalogy*, **10**, 11–21.

Read, D. G. (1987b). Rate of food passage in *Planigale* spp. (Marsupialia: Dasyuridae). *Australian Mammalogy*, **10**, 27–8.

Read, D. G. & Fox, B. J. (1991). Assessing the habitat of the parma wallaby, *Macropus parma* (Marsupialia: Macropodidae). *Wildlife Research*, **18**, 469–78.

Redenbach, C. J. (1982). A dietary comparison of three sympatric forest-dwelling macropodids. B.Nat.Res. Honours. Thesis, University of New England, Armidale, New South Wales.

Redford, K. H. & Dorea, J. G. (1984). The nutritional value of invertebrates with emphasis on ants and termites as food for mammals. *Journal of Zoology, London*, **203**, 385–95.

Redford, K. H. & Eisenberg, J. F. (1992). Mammals of the Neotropics. Vol II. The Southern Cone. Chicago: University of Chicago Press.

Redgrave, T. G. & Vickery, D. M. (1973). The polyunsaturated nature of horse and kangaroo fats. *Medical Journal of Australia*, **2**, 1116–18.

Reeds, P. J., Fuller, M. F. & Nicholson, B. A. (1985). Metabolic basis of energy expenditure with particular reference to protein. In *Substrate and Energy Metabolism*, ed. J. S. Garrow & D. Halliday, pp. 46–57. London: John Libbey.

Reichardt, P. B., Bryant, J. P., Clausen, T. P. & Weiland, G. D. (1984). Defence of winter-dormant Alaska paper birch against snowshoe hares. *Oecologia*, **65**, 58–69.

Reid, I. A. (1977). Some aspects of renal physiology in the brush-tailed possum,

Trichosurus vulpecula. In *The Biology of Marsupials*, ed. B. Stonehouse & D. Gilmore, pp. 393–410. London: Macmillan.

Reimer, A. B. & Hindell, M. A. (1996). Variation in body condition and diet of the eastern barred bandicoot (*Perameles gunnii*) during the breeding season. *Australian Mammalogy*, **19**, 47–52.

Rensberger, J. M. (1973). An occlusion model for mastication and dental wear in herbivorous mammals. *Journal of Paleontology*, **47**, 51–68.

Rich, T. H. (1991). Monotremes, placentals and marsupials: their record in Australia and its biases. In *Vertebrate Paleontology of Australasia*, ed. P. Vickers-Rich, J. M. Monaghan, R. F. Baird & T. H. Rich, pp. 893–1070. Melbourne: Pioneer Design Studio.

Richardson, K. C. (1980). The structure and radiographic anatomy of the alimentary tract of the tammar wallaby, *Macropus eugenii* (Marsupialia). I. The stomach. *Australian Journal of Zoology*, **28**, 367–79.

Richardson, K. C. (1989). Radiographic studies on the form and function of the gastrointestinal tract of the woylie (*Bettongia penicillata*). In *Kangaroos, Wallabies and Rat-kangaroos* ed. G. Grigg, P. Jarman and I. Hume, pp. 205–15. Sydney: Surrey Beatty.

Richardson, K. C., Bowden, T. A. J. & Myers, P. (1987). The cardiogastric gland and alimentary tract of caenolestid marsupials. *Acta Zoologica (Stockholm)*, **68**, 65–70.

Richardson, K. C., Wooller, R. D. & Collins, B. G. (1986). Adaptations to a diet of nectar and pollen in the marsupial *Tarsipes rostratus* (Marsupialia: Tarsipedidae). *Journal of Zoology, London (A)*, **208**, 285–97.

Richardson, K. C. & Wyburn, R. S. (1980). The structure and radiographic analysis of the alimentary tract of the tammar wallaby, *Macropus eugenii* (Marsupialia). II. The intestines. *Australian Journal of Zoology*, **28**, 367–79.

Richardson, K. C. & Wyburn, R. S. (1983). Electromyographic events in the stomach and small intestine of a small kangaroo, the tammar wallaby (*Macropus eugenii*). *Journal of Physiology*, **342**, 453–63.

Richardson, K. C. & Wyburn, R. S. (1988). Electromyography of the stomach and small intestine of the tammar wallaby, *Macropus eugenii*, and the quokka, *Setonix brachyurus*. *Australian Journal of Zoology*, **36**, 363–71.

Richmond, C. R., Langham, W. H. & Trujillo, T. T. (1962). Comparative metabolism of tritiated water by mammals. *Journal of Cellular and Comparative Physiology*, **59**, 45–53.

Ride, W. D. L. & Tyndale-Biscoe, C. H. (1962). Mammals. In *The Results of an Expedition to Bernier and Dorre Islands, Shark Bay, Western Australia*, ed. A. J. Fraser, pp. 54–97. Fauna Bulletin No. 2, Fisheries Department of Western Australia.

Rishworth, C., McIlroy, J. C. & Tanton, M. T. (1995). Diet of the common wombat, *Vombatus ursinus*, in plantations of *Pinus radiata*. *Wildlife Research*, **22**, 333–9.

Robbins, C. T. (1983). *Wildlife Feeding and Nutrition*. New York: Academic Press.

Robinson, J. G. & Redford, K. H. (1986). Body size, diet and population density of neotropical forest mammals. *American Naturalist*, **128**, 665–80.

Rogers, Q. R., Morris, J. G. & Freedland, R. A. (1977). Lack of hepatic enzymatic adaptation to low and high levels of dietary protein in the adult cat. *Enzyme*, **22**, 348–56.

Rose, R. W. (1986). The habitat, distribution and conservation status of the Tasmanian bettong, *Bettongia gaimardi* (Desmarest). *Australian Wildlife Research*, **13**, 1–6.

Rose, R. W. (1987). Reproductive energetics of two Tasmanian rat-kangaroos (Potoroinae, Marsupialia). *Symposium of the Zoological Society of London*, **57**, 149–65.

Rosenmann, M. & Ampuero, R. (1981). Hibernacion en *Dromiciops australis*. *Archivos de Biologia y Medicina Experimentales (Chile)*, **14**, R294.

Rübsamen, K., Hume, I. D., Foley, W. J. & Rübsamen, U. (1983). Regional differences in electrolyte, short chain fatty acid and water absorption in the hindgut of two species of arboreal marsupials. *Pflugers Archiv*, **399**, 68–72.

Rübsamen, K., Hume, I. D., Foley, W. J. & Rübsamen, U. (1984). Implications of the large surface area to body mass ratio on the heat balance of the greater glider (*Petauroides volans*: Marsupialia). *Journal of Comparative Physiology*, **B154**, 105–11.

Rübsamen, K., Nolda, V. & Engelhardt, W. v. (1979). Difference in the specific activity of tritium labelled water in blood, urine and evaporative water in rabbits. *Comparative Biochemistry and Physiology*, **62A**, 279–82.

Rübsamen, U., Hume I. D. & Rübsamen, K. (1983). Effect of ambient temperature on autonomic thermoregulation and activity patterns in the rufous rat-kangaroo (*Aepyprymmus rufescens*). *Journal of Comparative Physiology*, **B153**, 175–9.

Russell, E. M. (1974). The biology of kangaroos (Marsupialia – Macropodidae). *Mammal Reviews*, **4**, 1–59.

Sabat, P. & Bozinovic, F. (1994). Seasonal changes in digestive enzymatic activity in the small Chilean marsupial *Thylamys elegans*: intestinal disaccharidases. *Revista Chilena de Historia Natural*, **67**, 221–8.

Sabat, P., Bozinovic, F. & Zambrano, F. (1995). Role of dietary substrates on intestinal disaccharidases, digestibility, and energetics in the insectivorous mouse – opposum (*Thylamys elegans*). *Journal of Mammalogy*, **76**, 603–11.

Sakaguchi, E. & Hume, I. D. (1990). Digesta retention and fibre digestion in brushtail possums, ringtail possums and rabbits. *Comparative Biochemistry and Physiology*, **96A**, 351–4.

Sampson, J. C. (1971). The biology of *Bettongia penicillata*, Gray 1837. Ph.D. Thesis, University of Western Australia, Perth.

Sanson, G. D. (1978). The evolution and significance of mastication in the Macropodidae. *Australian Mammalogy*, **2**, 23–8.

Sanson, G. D. (1980). The morphology and occlusion of the molariform cheek teeth in some Macropodinae (Marsupialia: Macropodidae). *Australian Journal of Zoology*, **28**, 341–65.

Sanson, G. D. (1985). Functional dental morphology and diet selection in dasyurids (Marsupialia: Dasyuridae). *Australian Mammalogy*, **8**, 239–47.

Sanson, G. D. (1989). Morphological adaptations of teeth to diets and feeding in the Macropodoidea. In *Kangaroos, Wallabies and Rat-kangaroos*, ed. G. Grigg, P. Jarman and I. Hume, pp. 151–68. Sydney: Surrey Beatty.

Sanson, G. D. (1991). Predicting the diet of fossil mammals. In *Vertebrate Paleontology of Australasia*, ed. P. Vickers-Rich, J. M. Monaghan, R. F. Baird & T. H. Rich, pp. 201–28. Melbourne: Pioneer Design Studio.

Sanson, G. D., Nelson J. E. & Fell, P. (1985). Ecology of *Peradorcus concinna* in Arnhemland in a wet and a dry season. *Proceedings of the Ecological Society of Australia*, **13**, 65–72.

Santori, R. T., Astúa de Moraes, D. & Cerqueira, R. (1995). Diet composition of *Metachirus nudicaudatus* and *Didelphis aurita* (Marsupialia, Didelphoidea) in southeastern Brazil. *Mammalia*, **59**, 511–16.

Santori, R. T., Astúa de Moraes, D., Grelle, C. E. V. & Cerqueira, R. (1996). Natural diet at a Restinga forest and laboratory food preferences of the opossum *Philander frenata* in Brazil. *Studies in Neotropical Fauna and Environments*, **32**, 1–5.

Santori, R. T., Cerqueira, R. & da Cruz Kleske, C. (1995). Digestive anatomy and efficiency of *Didelphis aurita* and *Philander opossum* (Didelphimorphia, Didelphidae) in relation to food habits. *Reviews in Brazilian Biology*, **55**, 323–9.

Saz, H. J. (1981). Energy metabolism of parasitic helminths: adaptations to parasitism. *Annual Reviews of Physiology*, **43**, 323–41.

Schäfer, E. A. & Williams, D. J. (1876). On the structure of the mucous membrane of the stomach in the kangaroos. *Proceedings of the Zoological Society of London*, **12**, 165–77.

Schlager, F. E. (1981). The distribution and status of the rufous rat-kangaroo, *Aepyprymnus rufescens* and the long-nosed potoroo, *Potorous tridactylus* in northern New South Wales. Report to the New South Wales National Parks and Wildlife Foundation. Department of Ecosystem Management, University of New England, Armidale. Report No. 18.

Schmidt-Nielsen, B. (1958). Urea excretion in mammals. *Physiological Reviews*, **38**, 139–68.

Schmidt-Nielsen, B. & O'Dell, R. (1959). Effect of diet on distribution of urea and electrolytes in kidneys of sheep. *American Journal of Physiology*, **197**, 856–60.

Schmidt-Nielsen, B. & Schmidt-Nielsen, K. (1950). Do kangaroo rats thrive when drinking sea water? *American Journal of Physiology*, **160**, 291–4.

Schmidt-Nielsen, K. (1964). *Desert Animals*. Oxford: Clarendon Press.

Schmidt-Nielsen, K., Crawford, E. C., Newsome, A. E., Rawson, K. S. & Hammel, H. T. (1967). Metabolic rate of camels: effects of body temperature and dehydration. *American Journal of Physiology*, **212**, 341–6.

Schmidt-Nielsen, K. & Newsome, A. E. (1962). Water balance in the mulgara (*Dasycercus cristicauda*), a carnivorous desert marsupial. *Australian Journal of Biological Sciences*, **15**, 683–9.

Schreiber, G. & Richardson, S. J. (1997). The evolution of gene expression, structure and function of transthyretin. *Comparative Biochemistry and Physiology B*, **116**, 137–60.

Schultz, W. (1976). Magen-Darm-Kanal der Monotremen und Marsupialier. *Handbuch der Zoologie*, **8**, 1–177.

Scott, I. M., Yousef, M. K. & Johnson, H. D. (1976). Plasma thyroxine levels of mammals: desert and mountain. *Life Sciences*, **19**, 807–12.

Seebeck, J. H., Bennett, A. F. & Scotts, D. J. (1989). Ecology of the Potoroidae: A Review. In *Kangaroos, Wallabies and Rat-kangaroos*, ed. G. Grigg, P. Jarman and I. Hume, pp. 67–88. Sydney: Surrey Beatty.

Seebeck, J. H., Warneke, R. M. & Baxter, B. J. (1984). Diet of the bobuck, *Trichosurus caninus* (Ogilby) (Marsupialia: Phalangeridae) in a mountain forest in Victoria. In *Possums and Gliders*, ed. A. P. Smith and I. D. Hume, pp. 145–54. Sydney: Australian Mammal Society.

Sharpe, D. J. & Goldingay, R. L. (1998). Feeding behaviour of the squirrel glider at Bungawalbin Nature Reserve, north-eastern New South Wales. *Wildlife Research*, **25**, 243–54.

Shield, J. (1965). A breeding season difference in two populations of the Australian macropod marsupial (*Setonix brachyurus*). *Journal of Mammalogy*, **45**, 616–25.

Shield, J. W. (1959). Rottnest field studies concerned with the quokka. *Journal of the Royal Society of Western Australia*, **42**, 76–82.

Short, J. (1989). The diet of the brush-tailed rock-wallaby in New South Wales. *Australian Wildlife Research*, **16**, 11–18.

Short, J. & Turner, B. (1991). Distribution and abundance of spectacled hare-wallabies and euros on Barrow Island, Western Australia. *Wildlife Research*, **18**, 421–9.

Short, J. & Turner, B. (1993). The distribution and abundance of the burrowing bettong (Marsupialia: Macropodoidea). *Wildlife Research*, **20**, 525–34.

Sibbald, I. R., Sinclair, D. G., Evans, E. V. & Smith, D. L. T. (1962). The rate of passage of feed through the digestive tract of the mink. *Canadian Journal of Biochemistry and Physiology*, **40**, 1391–4.

Sibly, R. M. (1981). Strategies of digestion and defaecation. In *Physiological Ecology: An Evolutionary Approach*, ed. C. R. Townsend and P. Callow, pp. 109–39. Sunderland, Mass: Sinauer.

Slaven, M. R. & Richardson K. C. (1988). Aspects of the form and function of the kidney of the honey possum, *Tarsipes rostratus*. *Australian Journal of Zoology*, **36**, 465–71.

Smith, A. (1982a). Is the striped possum (*Dactylopsila trivirgata*; Marsupialia Petauridae) an arboreal anteater? *Australian Mammalogy*, **5**, 229–34.

Smith, A. P. (1982b). Diet and feeding strategies of the marsupial sugar glider in temperate Australia. *Journal of Animal Ecology*, **51**, 149–66.

Smith, A. (1984). Diet of Leadbeater's possum, *Gymnobelideus leadbeateri* (Marsupialia). *Australian Wildlife Research*, **11**, 265–73.

Smith, A. & Russell, R. (1982). Diet of the yellow-bellied glider *Petaurus australis* (Marsupialia: Petauridae) in north Queensland. *Australian Mammalogy*, **5**, 41–5.

Smith, A. P. (1986). Stomach contents of the long-tailed pygmy-possum, *Cercartetus caudatus* (Marsupialia: Burramyidae). *Australian Mammalogy*, **9**, 135–7.

Smith, A. P. (1995). Leadbeater's Possum. In *The Mammals of Australia*, ed. R. Strahan, pp. 224–6. Chatswood: Reed Books.

Smith, A. P. & Broome, L. (1992). The effects of season, sex and habitat on the diet of the mountain pygmy-possum (*Burramys parvus*). *Wildlife Research*, **19**, 755–68.

Smith, A. P. & Green, S. W. (1987). Nitrogen requirements of the sugar glider (*Petaurus breviceps*), an omnivorous marsupial, on a honey-pollen diet. *Physiological Zoology*, **60**, 82–92.

Smith, A. P., Nagy, K. A., Fleming, M. R. & Green, B. (1982). Energy requirements and water turnover in free-living Leadbeater's possums, *Gymnobelideus leadbeateri* (Marsupialia: Petauridae). *Australian Journal of Zoology*, **30**, 737–49.

Smuts, D. B. (1935). The relation between the basal metabolism and the endogenous nitrogen metabolism with particular reference to the estimation of the maintenance requirement of protein. *Journal of Nutrition*, **9**, 403–33.

Smuts, D. B. & Marais, J. S. C. (1938). The endogenous nitrogen excretion of sheep with special reference to the maintenance requirement of protein. *Onderstepoort Journal of Veterinary Science*, **11**, 131–9.

Snipes, R. L., Snipes, H. & Carrick, F. N. (1993). Surface enlargement in the large intestine of the koala (*Phascolarctos cinereus*): morphometric parameters. *Australian Journal of Zoology*, **41**, 393–7.

Soderquist, T. R. & Serena, M. (1994). Dietary niche of the western quoll, *Dasyurus geoffroii*, in the jarrah forest of Western Australia. *Australian Mammalogy*, **17**, 133–6.

Sonntag, C. F. (1921). Contributions to the visceral anatomy and myology of the Marsupialia. *Proceedings of the Zoological Society of London*, **2**, 851–82.

Southgate, R. I. (1990). Habitats and diet of the greater bilby, *Macrotis lagotis*, Reid (Marsupialia: Peramelidae). In *Bandicoots and Bilbies*, ed. J. H. Seebeck, P. R. Brown, R. L. Wallis and C. M. Kemper, pp. 303–9. Sydney: Surrey Beatty.

Southwell, I. A. (1978). Essential oil content of koala food trees. In *The Koala. Proceedings of the Taronga Symposium*, ed. T. J. Bergin, pp. 62–74. Sydney: Zoological Parks Board of New South Wales.

Southwell, I. A., Flynn, T. M. & Degabriele, R. (1980). Metabolism of α and β-pinene, ρ-cymene and 1,8-cineole in the brushtail possum, *Trichosurus vulpecula*. *Xenobiotica*, **10**, 17–23.

Sperber, I. (1944). Studies on the mammalian kidney. *Zoologiska Bidrag fran Uppsala*, **22**, 249–431.

Statham, H. L. (1982). *Antechinus stuartii* (Dasyuridae, Marsupialia) diet and food availability at Petroi, north-eastern New South Wales. In *Carnivorous Marsupials*, ed. M. Archer, pp. 151–63. Sydney: Royal Zoological Society of New South Wales.

Statham, H. L. (1984). The diet of *Trichosurus vulpecula* (Kerr) in four Tasmanian forest locations. In *Possums and Gliders*, ed. A. P. Smith and I. D. Hume, pp. 213–19. Sydney: Australian Mammal Society.

Stehli, F. G. & Webb, S. D. (eds.) (1985). *The Great American Biotic Interchange*. New York: Plenum Press.

Stevens, C. E. & Hume I. D. (1995). *Comparative Physiology of the Vertebrate Digestive System*, 2nd edn. Cambridge: Cambridge University Press.

Stewart, C. M., Melvin, J. F., Ditchurne, N., Than, S. M. & Zerdoner, E. (1973). The effect of season of growth on the chemical composition of cambial saps of *Eucalyptus regnans* trees. *Oecologia*, **12**, 349–72.

Stirling, E. C. (1891). Description of a new genus and species of Marsupialia,

Notoryctes typhlops. Transactions of the Royal Society of South Australia, **14**, 154–87.

Storr, G. M. (1964a). Studies on marsupial nutrition. 4. Diet of the quokka, *Setonix brachyurus* (Quoy and Gaimard) on Rottnest Island, Western Australia. *Australian Journal of Biological Sciences*, **17**, 469–81.

Storr, G. M. (1964b). The environment of the quokka (*Setonix brachyurus*) in the Darling Range, Western Australia. *Journal of the Royal Society of Western Australia*, **47**, 1–2.

Strahan, R. (ed.) (1995). *The Mammals of Australia*. Chatswood: Reed Books.

Streilein, K. E. (1982). Behaviour, ecology, and distribution of South American marsupials. In *Mammalian Biology in South America*, ed. M. A. Mares and H. H. Genoways, pp. 231–50. Pittsburgh: University of Pittsburgh.

Suckling, G. C. (1984). Population ecology of the sugar glider, *Petaurus breviceps*, in a system of fragmented habitats. *Australian Wildlife Research*, **11**, 49–75.

Sussman, R. W. & Raven, P. H. (1978). Pollination by lemurs and marsupials: an archaic co-evolutionary system. *Science*, **200**, 731–4.

Szalay, F. S. (1982). A new appraisal of marsupial phylogeny and classification. In *Carnivorous Marsupials*, ed. M. Archer, pp. 621–40. Sydney: Royal Zoological Society of New South Wales.

Szalay, F. S. & Trofimov, B. A. (1996). The Mongolian late Cretaceous *Asiatherium* and the early phylogeny and paleobiology of Metatheria. *Journal of Vertebrate Paleontology*, **16**, 474–509.

Taylor, R. J. (1983). The diet of the eastern grey kangaroo and wallaroo in areas of improved and native pasture in the New England Tablelands. *Australian Wildlife Research*, **10**, 203–11.

Taylor, R. J. (1985). Effects of pasture improvement on the nutrition of eastern grey kangaroos and wallaroos. *Journal of Applied Ecology*, **22**, 717–25.

Taylor, R. J. (1991). Plants, fungi and bettongs: a fire dependent co-evolutionary relationship. *Australian Journal of Ecology*, **6**, 409–11.

Taylor, R. J. (1992). Seasonal changes in the diet of the Tasmanian bettong (*Bettongia gaimardi*), a mycophagous marsupial. *Journal of Mammalogy*, **73**, 408–14.

Taylor, R. J. (1993). Habitat requirements of the Tasmanian bettong (*Bettongia gaimardi*), a mycophagous marsupial. *Wildlife Research*, **20**, 699–710.

Tedford, R. H. (1966). A review of the macropodid genus *Sthenurus*. *University of California Publications, Department of Geology*, **57**, 1–72.

Tedman, R. A. (1990). Some observations on the visceral anatomy of the bandicoot *Isoodon macrourus* (Marsupialia: Peramelidae). In *Bandicoots and Bilbies*, ed. J. H. Seebeck, P. R. Brown, R. L. Wallis and C. M. Kemper, pp. 107–16. Sydney: Surrey Beatty.

Thomas, O. (1887). On the wallaby commonly known as *Lagorchestes fasciatus*. *Proceedings of the Zoological Society of London*, 1887, pp. 544–7.

Thompson, S. D. (1988). Thermoregulation in the water opossum (*Chironectes minimus*): an exception that 'proves' a rule. *Physiological Zoology*, **61**, 450–60.

Thomson, J. A. & Owen, W. H. (1964). A field study of the Australian ringtail possum *Pseudocheirus peregrinus* (Marsupialia: Phalangeridae). *Ecological Monographs*, **34**, 27–52.

Townsend, C. R. & Hughes, R. H. (1981). Maximising net energy returns from foraging. In *Physiological Ecology: An Evolutionary Approach*, ed. C. R. Townsend and P. Callow, pp. 86–108. Sunderland, Mass: Sinauer.

Traill, B. J. & Coates, T. D. (1993). Field observations on the brush-tailed phascogale *Phascogale tapoatafa* (Marsupialia: Dasyuridae). *Australian Mammalogy*, **16**, 61–5.

Tribe, D. E. & Peel, L. (1963). Body composition of the kangaroo (*Macropus* sp.). *Australian Journal of Zoology*, **11**, 273–89.

Troughton, E. (1965). *Furred Animals of Australia*, 8th edn. Sydney: Augus & Robertson.

Turner, V. (1982). Marsupials as pollinators in Australia. In *Pollination and Evolution*, ed. J. A. Armstrong, J. M. Powell & A. J. Richards, pp. 55–66. Sydney: Royal Botanic Gardens.

Turner, V. (1984a). *Eucalyptus* pollen in the diet of the feathertail glider, *Acrobates pygmaeus* (Marsupialia: Burramyidae). *Australian Wildlife Research*, **11**, 77–81.

Turner, V. (1984b). Banksia pollen as a source of protein in the diet of two Australian marsupials, *Cercartetus nanus* and *Tarsipes rostratus*. *Oikos*, **43**, 53–61.

Twigg, L. E. & King, D. R. (1991). The impact of fluoroacetate-bearing vegetation on native Australian fauna: a review. *Oikos*, **61**, 412–30.

Twigg, L. E., King, D. R. & Mead, R. J. (1990). Tolerance to fluoroacetate by populations of *Isoodon* spp. and *Macrotis lagotis* and its implications for fauna management. In *Bandicoots and Bilbies*, ed. J. H. Seebeck, P. R. Brown, R. L. Wallis & C. M. Kemper, pp. 185–92. Sydney: Surrey Beatty.

Tyndale-Biscoe, C. H. & Calaby, J. H. (1975). Eucalypt forests as refuge for wildlife. *Australian Forestry*, **38**, 117–33.

Tyndale-Biscoe, H. (1973). *Life of Marsupials*. London: Edward Arnold.

Ullrey, D. E., Robinson, R. T. & Whetter, P. A. (1981a). Composition of preferred and rejected *Eucalyptus* browse offered to captive koalas, *Phascolarctos cinereus* (Marsupialia). *Australian Journal of Zoology*, **29**, 839–46.

Ullrey, D. E., Robinson, P. T. & Whetter P. A. (1981b). *Eucalyptus* digestibility and digestible energy requirements of adult male koalas, *Phascolarctos cinereus* (Marsupialia). *Australian Journal of Zoology*, **29**, 847–52.

Underwood, E. J. (1977). *Trace Elements in Human and Animal Nutrition*, 4th edn. New York: Academic Press.

van Deusen, H. M. & Keith, K. (1966). Range and habitat of the bandicoot, *Echymipera clara*, in New Guinea. *Journal of Mammalogy*, **47**, 721–3.

Van Dyck, S. M. (1995). Mahogany glider. In *The Mammals of Australia*, ed. R. Strahan, pp. 232–3. Chatswood, New South Wales: Reed Books.

Van Soest, P. J. (1965). Symposium on factors influencing the voluntary intake of herbage by ruminants: voluntary intake in relation to chemical composition and digestibility. *Journal of Animal Science*, **24**, 834–43.

van Tets, I. G. (1996). Pollen in the diet of Australian mammals. Ph.D. Thesis, University of Wollongong.

van Tets, I. & Hulbert, A. (1999). A comparison of the nitrogen requirements of the eastern pygmy-possum, *Cercartetus nanus*, on a pollen and on a mealworm diet. *Physiological and Biochemical Zoology*, **72** (in press).

van Tets, I. G. & Whelan, R. J. (1997). *Banksia* pollen in the diet of Australian mammals. *Ecography*, **20**, 499–505.

Vieira, E. M. & Palma, A. R. T. (1996). Natural history of *Thylamys velutinus* (Marsupialia, Didelphidae) in central Brazil. *Mammalia*, **60**, 481–4.

Vogtsberger, L. M. & Barrett, G. W. (1973). Bioenergetics of captive red foxes. *Journal of Wildlife Management*, **37**, 495–500.

Wake, J. (1980). The field nutrition of the Rottnest Island quokka. Ph.D. Thesis, University of Western Australia, Perth.

Wakefield, N. A. (1971). The brush-tailed rock-wallaby (*Petrogale penicillata*) in Western Victoria. *Victorian Naturalist*, **88**, 92–102.

Wallis, I. R. (1990). The nutrition, digestive physiology and metabolism of potoroine marsupials. Ph.D. Thesis, University of New England, Armidale.

Wallis, I. R. (1994). The rate of passage of digesta through the gastrointestinal tract of potoroine marsupials: more evidence about the role of the potoroine foregut. *Physiological Zoology*, **67**, 771–95.

Wallis, I. R. & Farrell, D. J. (1992). Energy metabolism in potoroine marsupials. *Journal of Comparative Physiology*, **B162**, 478–87.

Wallis, I. R. & Green, B. (1992). Seasonal field energetics of the rufous rat-kangaroo (*Aepyprymnus rufescens*). *Australian Journal of Zoology*, **40**, 279–90.

Wallis, I. R., Green, B. & Newgrain, K. (1997). Seasonal field energetics and water fluxes of the long-nosed potoroo (*Potorous tridactylus*) in southern Victoria. *Australian Journal of Zoology*, **45**, 1–11.

Wallis, I. R. & Hume, I. D. (1992). The maintenance nitrogen requirements of potoroine marsupials. *Physiological Zoology*, **65**, 1246–70.

Wallis, R. L. (1976). Torpor in the dasyurid marsupial *Antechinus stuartii*. *Comparative Biochemistry and Physiology*, **53A** 318–22.

Waring, H., Moir, R. J. & Tyndale-Biscoe, C. H. (1966). Comparative physiology of marsupials. *Advances in Comparative Physiology and Biochemistry*, **2**, 237–376.

Warner, A. C. I. (1981a). Rate of passage of digesta through the gut of mammals and birds. *Nutrition Abstracts and Reviews, Series B*, **51**, 789–820.

Warner, A. C. I. (1981b). The mean retention times of digesta markers in the gut of the tammar, *Macropus eugenii*. *Australian Journal of Zoology*, **29**, 759–71.

Waterlow, J. C. (1984). Protein turnover with special reference to man. *Quarterly Journal of Experimental Physiology*, **69**, 409–38.

Watts, C. H. S. (1969). Distribution and habits of the rabbit bandicoot. *Transactions of the Royal Society of South Australia*, **93**, 135–41.

Wellard, G. A. & Hume, I. D. (1981a). Digestion and digesta passage in the brushtail possum, *Trichosurus vulpecula* (Kerr). *Australian Journal of Zoology*, **29**, 157–66.

Wellard, G. A. & Hume, I. D. (1981b). Nitrogen metabolism and nitrogen requirements of the brushtail possum *Trichosurus vulpecula* (Kerr). *Australian Journal of Zoology*, **29**, 147–56.

Wells, R. T. (1968). Some aspects of the environmental physiology of the hairy – nosed wombat *Lasiorhinus latifrons*. B.Sc. (Honours) Thesis, University of

Adelaide, Adelaide.

Wells, R. T. (1973). Physiological and behavioural adaptations of the hairy-nosed wombat (*Lasiorhinus latifrons* Owen) to its arid environment. Ph.D. Thesis, University of Adelaide, Adelaide.

Wells, R. T. (1978a). Thermoregulation and activity rhythms in the hairy-nosed wombat, *Lasiorhinus latifrons* (Owen), (Vombatidae). *Australian Journal of Zoology*, **26**, 639–51.

Wells, R. T. (1978b). Field observations of the hairy-nosed wombat, *Lasiorhinus latifrons* (Owen). *Australian Wildlife Research*, **5**, 299–303.

Wells, R. T., Horton, D. R. & Rogers, P. (1982). *Thylacoleo carnifex* Owen (Thylacoleonidae, Marsupialia): marsupial carnivore? In *Carnivorous Marsupials*, ed. M. Archer, pp. 573–85. Sydney: Surrey Beatty and the Royal Zoological Society of New South Wales.

Werdelin, L. (1986). Comparison of skull shape in marsupial and placental carnivores. *Australian Journal of Zoology*, **34**, 109–17.

Werdelin, L. (1987). Jaw geometry and molar morphology in marsupial carnivores: analysis of a constraint and its macroevolutionary consequences. *Paleobiology*, **13**, 342–50.

West, G. B., Brown, J. H. & Enquist, B. J. (1997). A general model for the origin of allometric scaling laws in biology. *Science*, **276**, 122–60.

White, R. G., Hume, I. D. & Nolan, J. V. (1988). Energy expenditure and protein turnover in three species of wallabies (Marsupialia: Macropodidae). *Journal of Comparative Physiology*, **B158**, 237–46.

Whitelaw, F. G., Hyldgaard-Jensen, J., Reid, R. S. & Kay, M. G. (1970). Volatile fatty acid production in the rumen of cattle given an all-concentrate diet. *British Journal of Nutrition*, **24**, 179–95.

Wiens, D., Renfree, M. B. & Wooller, R. D. (1979). Pollen loads of honey possums (*Tarsipes spenserae*) and non-flying mammal pollination in southwestern Australia. *Annals of the Missouri Botanical Garden*, **66**, 830–8.

Wilckens, M. (1872). *Untersuchungen über den Magen der wiederkauenden Hausthiere*. Berlin: von Wiegardt & Hempel.

Wilkinson, P. (1979). Urinary creatinine excretion by the macropod marsupials, red-necked pademelon (*Thylogale thetis*) and tammar wallaby (*Macropus eugenii*). Unpublished Thesis, University of New England, Armidale.

Wilson, A. D. (1991). The influence of kangaroos and forage supply on sheep productivity in the semi-arid woodlands. *Rangeland Journal*, **13**, 69–80.

Winkel, K. & Humphery-Smith, I. (1988). Diet of the marsupial mole, *Notoryctes typhlops* (Stirling 1889) (Marsupialia: Notoryctidae). *Australian Mammalogy*, **11**, 159–61.

Winter, J. W. & Johnson, P. M. (1995). Northern bettong, *Bettongia tropica*. In *The Mammals of Australia*, ed. R. Strahan, pp. 294–5. Chatswood: Reed Books.

Withers, P. C. (1992a). *Comparative Animal Physiology*. Fort Worth: Saunders College Publishing.

Withers, P. C. (1992b). Metabolism, water balance and temperature regulation in the golden bandicoot (*Isoodon auratus*). *Australian Journal of Zoology*, **40**, 523–31.

Withers, P. C., Richardson K. C. & Wooller, R. D. (1990). Metabolic physiology of euthermic and torpid honey possums, *Tarsipes rostratus*. *Australian Journal of Zoology*, **37**, 685–93.

Wolin, M. J. & Miller, T. L. (1994). Acetogenesis from carbon dioxide in human gastrointestinal ecosystems. In *Acetogenesis*, ed. H. L. Drake. New York: Chapman and Hall.

Wood, D. H. (1970). An ecological study of *Antechinus stuartii* (Marsupialia) in a south-east Queensland rain forest. *Australian Journal of Zoology*, **18**, 185–207.

Wood, J. E. (1954). Food habits of furbearers in the upland Post Oak region in Texas. *Journal of Mammalogy*, **35**, 406–15.

Woodall, P. F. & Skinner, J. D. (1993). Dimensions of the intestine, diet and faecal water loss in some African antelope. *Journal of Zoology (London)*, **229**, 457–71.

Woodburne, M. O. & Case, J. A. (1996). Dispersal, vicariance, and the late Cretaceous to early Tertiary land mammal biogeography from South America to Australia. *Journal of Mammalian Evolution*, **3**, 121–61.

Wood Jones, F. (1924). *The Mammals of South Australia*. Part II. *The Bandicoots and the Herbivorous Marsupials*. Adelaide: R. E. E. Rogers, Government Printer.

Woods, J. T. (1956). The skull of *Thylacoleo carnifex*. *Memoirs of the Queensland Museum*, **13**, 125–40.

Woollard, P. (1971). Differential mortality of *Antechinus stuartii* (Macleay): nitrogen balance and somatic changes. *Australian Journal of Zoology*, **19**, 347–53.

Wooller, R. D., Renfree, M. B., Russell, E. M., Dunning, A., Green, S. W. & Duncan, P. (1981). Seasonal changes in a population of the nectar-feeding marsupial *Tarsipes spencerae* (Marsupialia: Tarsipedidae). *Journal of Zoology, London*, **195**, 267–79.

Wooller, R. D., Richardson, K. C. & Collins, B. G. (1993). The relationship between nectar supply and the rate of capture of a nectar-dependent small marsupial *Tarsipes nostratus*. *Journal of Zoology, London*, **229**, 651–658.

Wooller, R. D., Russell, E. M. & Renfree, M. B. (1984). Honey possums and their food plants. In *Possums and Gliders*, ed. A. P. Smith and I. D. Hume, pp. 439–43. Sydney: Australian Mammal Society.

Woolley, P. A. (1991). Reproduction in *Dasykaluta rosamondae* (Marsupialia: Dasyuridae): field and laboratory observations. *Australian Journal of Zoology*, **39**, 549–68.

Woolley, P. A. & Allison, A. (1982). Observations on the feeding and reproductive status of captive feather-tailed possums, *Distoechurus pennatus* (Marsupialia: Burramyidae). *Australian Mammalogy*, **5**, 285–7.

Woolnough, A. P. & Carthew, S. M. (1996). Selection of prey by size in *Ningaui yvonneae*. *Australian Journal of Zoology*, **44**, 319–26.

Wright, W., Sanson, G. D. & McArthur, C. (1991). The diet of the extinct bandicoot *Chaeropus ecaudatus*. In *Vertebrate Paleontology of Australasia*, ed. P. Vickers-Rich, J. M. Monoghan, R. F. Baird, R. F. & T. H. Rich, pp. 229–45. Melbourne: Pioneer Design Studio.

Wyburn, R. S. & Richardson, K. C. (1989). Motility patterns of the stomach and intestine of small macropodids. In *Kangaroos, Wallabies and Rat-kangaroos*, ed. G. Grigg, P. Jarman and I. Hume, pp. 197–203. Sydney: Surrey Beatty.

Yadav, M. (1979). The kidney types of some Western Australian macropod marsupials. *Mammalia*, **43**, 225–33.

Yadav, M., Stanley, N. F. & Waring, H. (1972). The microbial flora of the gut of the pouch-young and the pouch of a marsupial, *Setonix brachyurus*. *Journal of General Microbiology*, **70**, 437–42.

Yamada, J., Krause, W. J., Kitamura, N. & Yamashita, T. (1987). Immunocyto-chemical demonstration of gastric endocrine cells in the stomach gland patch of the koala, *Phascolarctos cinereus*. *Anatomie Anz., Jena*, **163**, 311–18.

Yamada, J., Richardson, K. C. & Wooller, R. D. (1989). An immunohistochemical study of gastrointestinal endocrine cells in a nectarivorous marsupial, the honey possum (*Tarsipes rostratus*). *Journal of Anatomy*, **162**, 157–68.

Yamakoshi, Y., Murata, M., Shimizu, A. & Homma, S. (1992). Isolation and characterization of macrocarpal-B, macrocarpal-G antibacterial compounds from *Eucalyptus macrocarpa*. *Bioscience, Biotechnology and Biochemistry*, **56**, 1570–6.

Yokota, S. D., Benyajati, S. & Dantzler, W. H. (1985). Comparative aspects of glomerular filtration rate in vertebrates. *Renal Physiology*, **8**, 193–221.

Yousef, M. K., Dill, D. B. & Mayes, M. G. (1970). Shifts in body fluids during dehydration in the burro, *Equus asinus*. *Journal of Applied Physiology*, **29**, 345–9.

Zoidis, A. M. & Markowitz, H. (1992). Findings from a feeding study of the koala (*Phascolarctos cinereus adustus*) at the San Francisco Zoo. *Zoo Biology*, **11**, 417–31.

Index

f = figure, t = table